Complete Knowledge on Secrets of the Universe and Life

Dr. Bernhard Helpful
(Researcher on Fundamental Laws of Nature)

DEDICATION & COPYRIGHT

To my father Samuel, my wife Jocelyn and my children Jonah, Joelle, Jethro, Jonty and Jelena. To the loving memory of my mother Grace who passed away on February 13, 2016.

Dr. Bernhard Helpful (Author's pen name) corresponding email address: Dr.Bernhard.Helpful@gmail.com

Cover design: Fractal image of Julia set. Julia ("laces") set and Fatou ("dusts") set are two complementary sets defined from a function. The behavior of function on Fatou set is "regular" while that on Julia set is "chaotic".

Publisher: Self Publishing

ISBN: 9781706580843

ACKNOWLEDGEMENTS

Many thanks to Jeanette "Jenny" O'Hagan (writer-speaker with medical degree), Rodney "Rod" Williams (civil engineer with mathematical degree) and Anthony "Tony" O'Hagan (software engineer with mathematical degree) for occasional invaluable advices, editing and proofreading of this book.

Jenny can be found at Linkedin https://www.linkedin.com/in/jeanetteohaganwrites/. Rod can be found at Linkedin https://www.linkedin.com/in/rod-williams-1b88a346/. Tony from Digital Ministry https://www.crossover.org.au/digital-missionaries/ with his personal homepages Internet Missions http://internetmissions.org and Crusade for Christ https://give.cru.org/0877152 also works part time at Power to Change https://www.powertochange.org.au.

For a good cause, please donate generously at Give to a Missionary https://www.powertochange.org.au/give/give-to-a-missionary by selecting O'Hagan, Tony & Jenny.

PREFACE This book caters for the general public. It introduces a new branch of mathematics coined "Mathematics for Incompletely Predictable Problems (MIPP)". The two landmark research papers synonymous with this branch of mathematics are reproduced in Appendices 1 and 2. MIPP makes all mathematical arguments valid and complete in the first paper (based on first key step of converting Riemann zeta function into its continuous format version) and the second paper (based on second key step of applying Information-Complexity conservation to Sieve of Eratosthenes). MIPP will result in solving Incompletely Predictable problems of Riemann hypothesis (& explaining manifested properties of two types of Gram points) and solving Polignac's & Twin prime conjectures. Open problems in Number theory of Riemann hypothesis, Polignac's and Twin prime conjectures have been unsolved for over 150 years. They are finally solved when sine qua non treated as Incompletely Predictable problems in 2019. Riemann hypothesis belongs to one of seven Millennium Prize Problems in mathematics stated by Clay Mathematics Institute on May 24, 2000. The author serendipitously cross path with the Institute's website Riemann Hypothesis on Thursday March 10, 2016 in permanently altering its information content from previous incorrect [sic] "…the first 10,000,000,000 solutions" to current correct [sic] "…the first 10,000,000,000,000 solutions". This book contains materials relating to Alphabet and Language of Science with emphasis on its important role in innovative 'Spherical Model of Science' and 'Spherical Model of Numbers'; explaining the Fundamental Laws; speculating on the role of Umbral ("Shadow"), Mathieu and Monstrous Moonshine in String theory potentially uniting Einstein General Relativity and Quantum gravity; and refuting the common misconception that solving Riemann hypothesis will lead to E-Commerce apocalypse.

CONTENTS

PROLOGUE

Re: http://www.claymath.org/millennium-problems/riemann-hypothesis

Miss Angel (pseudonym) Person <admin@claymath.org>

Thursday March 10, 2016 7:56 PM

To: Bernhard Helpful <Dr.Bernhard.Helpful@gmail.com>

Dear Dr. Helpful,

Thank you very much for your email and for drawing our attention to the error on our website, which I have now corrected. I think the text was probably picked up from an old site. Please let us know when even more zeroes need to be added!

With best regards,

Angel

Miss Angel (pseudonym) Person
Administrative Manager
Office of the President, Clay Mathematics Institute
Andrew Wiles Building
Radcliffe Observatory Quarter
Woodstock Road
Oxford OX2 6GG U.K.
Email: admin@claymath.org
Tel: +44 (0)1865 615155
www.claymath.org

On Mar 10, 2016 at 07:58, Bernhard Helpful <Dr.Bernhard.Helpful@gmail.com> wrote:
Thursday 10 March 2016
Dr. Bernhard Helpful
123456 Fictitious Street

Pretty Place, Queensland 123456
Australia
Email: Dr.Bernhard.Helpful@gmail.com

admin@claymath.org
Clay Mathematics Institute
70 Main St
Suite 300
Peterborough, NH 03458
USA

Dear Sir / Madam,

I refer to your website on Riemann Hypothesis "http://www.claymath.org/millennium-problems/riemann-hypothesis" last updated on Tuesday 12:34 pm 1 March 2016 – see attached document. I believe that its last paragraph "This has been checked for the first 10,000,000,000 solutions. A proof that it is true for every interesting solution would shed light on many of the mysteries surrounding the distribution of prime numbers." is incorrect in that "...the first 10,000,000,000 solutions" should be "...the first 10,000,000,000,000 solutions" instead.

Please let me know whether the above observation by me is correct or not.

Many thanks.

Kind regards,

Bernhard Helpful

<Riemann Hypothesis Clay Mathematics Institute Last update 01May2016 1234pm.pdf>

2

Relevant parts of this book contain memoirs outlining the personal experiences of author in 2012 of having his premature daughter Jelena and in 2019 of having his medical career laced with medical conspiracy whereby earning sufficient income to financially support his family is in tatters. It is often said that Laws of Nature are written using Language of Mathematics. **Then holors are universal "Alphabet of Science", and holor algebra & holor calculus are universal "Language of Science".** Doing mathematics not only relax the author's mind in time of stressful life events but also provide him with a second career pathway option. Under the classification system of 'Elementary-Emergent Fundamental Laws', these Laws will consist of **Elementary** Fundamental Laws dealing with **Nonliving Things** and **Emergent** Fundamental Laws dealing with **Living Things**. *By simple logical thinking, the broad fields of Medicine, Physiology and Religion will predominantly obey Emergent Fundamental Laws, and the broad fields of Mathematics and Science will predominantly obey Elementary Fundamental Laws. Then 'Contents' of this book is succinctly summarized as being constituted by firstly, the fields of Medicine, Physiology and Religion that mainly obey Emergent Fundamental Laws; and secondly, by the fields of Mathematics and Science that mainly obey Elementary Fundamental Laws.* Thus the common denominator 'Fundamental Laws' will unify the three fields of Medicine, Physiology and Religion with the seemingly different two fields of Mathematics and Science. In fact, all other 'broad fields' could logically be united to each other in a similar fashion under the banner 'Fundamental Laws'.

Here is a brief synopsis of this book employing a commonly used book classification system. Title: *Complete Knowledge on Secrets of the Universe and*

Life. Category: *Literary Non-fiction.* Genre: *Popular Science and Medicine.* Conjuring analogous terms such as 'Beautiful Mathematics' (for Completely Predictable problems) versus 'Sexy Mathematics' (for Incompletely Predictable problems) will fulfil designated role of "rewriting science" using friendly mathematical language for the general public.

This book provides Simplexity and Simplicity plus Complexity explanations to the successful proofs for Incompletely Predictable problems of Riemann hypothesis, Polignac's and Twin prime conjectures in 2019. Readers who perceive themselves as having poor mathematical grounding are reassure that they are able to comprehend core materials without having to fully understand confronting mathematical equations in this book. Whether the reader believe in God or not, it should provide interesting reading for him or her. For believers, God is omnipotent, omniscient and omnipresent. Chapters 1 to 10 as beginning part outline materials on Alphabet & Language of Science emphasizing its crucial "Universal Language" role in the innovative 'Spherical Model of Science' and 'Spherical Model of Numbers'. Chapters 11 to 15 introduce materials for the three open problems with Chapter 11 explaining Fundamental Laws. Chapters 16 to 21 describe in layman's terms how rigorous proofs for the three open problems are derived as two complete research papers first published in viXra (reproduced in Appendix 1 and 2). The exotic A228186 Hybrid integer is outlined in Chapter 16. Chapters 22 to 25 as end part contain materials that speculate on important role of Umbral ("Shadow"), Mathieu & Monstrous Moonshine in String theory potentially uniting Einstein General Relativity & Quantum gravity, and refute a common misconception that solving Riemann hypothesis will lead to E-Commerce apocalypse.

XXXXXXXXXXXXXXXXXXXXXXXXXXXXXXXXXXXX

12:34:56pm Saturday July 4, 2020

Message-of-Knowledge Moonshine Mathematics

{NOTE: The following is a fictitious message.}

Warmest greetings from all of us Eurekians to all of you Earthlings.

We hail from planet Eureka, third planet of the Utopia Star System. We arrive in peace to visit your wonderful planet using our Time Travel Machine.

On this momentous occasion, we will simultaneously sent to all you people of planet Earth, third planet of the Solar System, a brief Message-of-Knowledge Moonshine Mathematics. For basic housekeeping rules applicable to both Terrestrial and Extraterrestrial beings alike in relation to intellectual property and copyright materials, we Eurekians expressively stated here that we have acquired the super advanced Wacko-Mental-Fantasy Technology to accomplish this illusory deed of multiple simultaneous Message-of-Knowledge transmissions as if these messages are now, have been, or will always be, apparently broadcast out on the particular nominated Common Planetary Time of 12:34:56 pm Saturday July 4, 2020.

Our mother planet Eureka is spatially located exactly 123456 light-years away from your planet Earth. Your local time in Washington, District of

Columbia, capital of United States of America, is now exactly 12(Earth Hour):34(Earth Minute):56(Earth Second) pm on Saturday July 4, 2020 (Earth date). We notice that this traditional date is your 'Independence Day' celebration in America but the equivalent Saturday July 4, 2020 (Eureka date) has previously been declared and voted by a majority of Eurekians to be the inaugural 'Complete Knowledge Day' celebration. The particular 12:34:56 pm Earth time also instantaneously and transiently coincide exactly with our 12:34:56 pm Eureka time on this special day of great significance for many inhabitants on both planet Earth and planet Eureka.

Photo of Cotopaxi volcano in Ecuador taken on 08:01:30 am September 11, 2015 by César Muñoz/Andes. Image originally posted to Flickr by Agencia Andes at https://flickr.com/photos/75116651@N03/21142451178. Used here under Creative Commons Attribution-Share Alike 2.0 Generic license.

Prior to this Coinciding Point, the absolute numerical values of our

Eureka time and date have always been larger than those same parameters from Earth. After this Coinciding Point, our Eureka time and date in question when put in the identical context will logically always be smaller. Using comparative time clock, we have observed that our Eureka Civilization is temporally and briefly exactly ten times older than your Earth Civilization on this Common Planetary Time of 12:34:56 pm Saturday July 4, 2020 because our planet Eureka takes (i) 10 Earth years to rotate once around our Utopia Star and (ii) 240 Earth hours to rotate once around its axis. Therefore 1 Eureka second, hour, day, week, month, or year is the exact equivalence of 10 Earth seconds, hours, days, weeks, months, or years. It will be in Earth Year 20200 – which is another mind-blowing 18180 (20200 minus 2020) Earth years in future and equating to our Eureka Year 3838 (2020 plus 1818) – before your Earth Civilization reaches the equivalent pinnacle status of our Eureka Civilization as dictated on this undoubtedly proudest day in Eureka history when celebrating our very first Complete Knowledge Day. This special day is coolly signified by our "Picturesque Snowcapped Volcano Mountain Model of Mathematics and Science" Logo, which symbolizes the Extraterrestrial-Terrestrial Elementary-Emergent Fundamental Laws.

We Eurekians have duly taken into consideration your Earth's primitive calendar system requiring timely leap year and leap cycle insertions to correct the 'drift' in the above calculations. The leap year (years evenly divisible by 4) is a year containing one additional day (February 29) added to keep the calendar year synchronized with the astronomical or seasonal year. Because seasons and astronomical events do not repeat in a whole number, calendars that have the same number of days in each year drift over time with respect to the event that the year is supposed to track. By inserting an additional day into the year, the drift can be corrected. Over a period of four centuries, the accumulated error of adding a leap day every four years amounts to about three extra days. Your Gregorian calendar therefore removes three leap days every 400 years, which is the length of its leap cycle. This is done by removing February 29 in the three century years (multiples of 100) that cannot be exactly divided by 400. For instance, the years 2000 and 2400 are leap years, while 1800, 1900, 2100, 2200, 2300 and 2500 are common years. By this rule, the average number of days per year is $365 + \frac{1}{4} - \frac{1}{100} + \frac{1}{400} = 365.2425$ and this figure was incorporated in our calculations. Therefore, 1 Eureka year contains exactly 3652.425 Earth days.

We Eurekians now have a giant confession to make. In your Earth year of 2019, our incredible See-And-Hear-All Technology has enable us to transparently glean the two landmark research papers from planet Earth for rigorous proof on the-initially-proposed-in-1859 (Earth date) Riemann hypothesis including providing explanations on its closely related Gram points [and proofs for Polignac's and Twin prime conjectures] – all elegantly achieved by the sole author through treating these problems as *sine qua non* Incompletely Predictable problems. These papers were published in viXra. For seemingly time immemorial, many a wise humanoid from our planet Eureka have contemplated suicide while frustratingly trying to solve that same elusive Eureka-equivalent Millennium Prize Problem of Riemann hypothesis. We fully admit to cheating in how we finally obtained the solution to this our very last frontier Problem. In this regard we command our utmost respect for, and offer our eternal gratitude to, the no-mean-feat success by the clever author of these two papers. Only after this stupendous knowledge

8

acquisition have we Eurekians finally achieved bragging rights to be dubbed a perfect advanced civilization having conquered all available knowledge of the Universe.

Based on various compelling scientific evidences and proofs, we Eurekians have now settled the absolute true answer outcomes of the great Creationism-versus-Evolutionism debate. This had occurred just recently in the twilight years striving to achieve our perfect civilization. The Religious or Creationist belief is that God create all Terrestrial [and Extraterrestrial] living creatures, plants, and organisms in our Universe; with all believers given the miracle of peace, salvation, and guidance by God. The Atheist or Evolutionist belief is that all those living things evolve as per your proposed-in-1859 (Earth year) Darwin Theory of Evolution [and Natural Selection]. We regrettably cannot prematurely divulge the truths stemming from this great debate to you Earthlings at your current relatively early stage of Earth Civilization $[\frac{2020}{20200}$ X 100% = 10% Age of Eureka Civilization] as this action may result in unforeseeable or unpredictable consequences. To make you Earthlings feel better, we now crack the following Eurekian-style surreal joke that perhaps only a selected few Earthlings can fully comprehend.

In this Message-of-Knowledge Moonshine Mathematics, we have purposefully employed one of the principles from Moonshine Mathematics which is akin to 'Presence of All-The-Stars-Lining-Up Coincidental Numbers' such as repeatedly using the number 123456. Unwittingly, you Earthlings had serendipitously utilized the 1859 (Earth year) number twice – in Riemann hypothesis and Darwin Theory of Evolution [and Natural Selection]. Ha-Ha-Ha! Ho-Ho-Ho! What a marvelous joke for us Eurekians!

In the decision on finite-supply fossil fuel accelerating usage on your planet Earth causing Global Warming stemming from rising Carbon dioxide (CO_2) emission induced Greenhouse effect, we Eurekians give you the following friendly advises based on previous painful lessons learned from various Global Warming related climatic changes occurring on our planet Eureka. This Global Warming event does exist and its harmful effects, not least causing uncomfortable 'warm butt' [colloquial Earth language for 'warm buttock'] sensation, must be curtailed by leaders of major countries on planet Earth via timely United Nations initiated global cooperation in devising the Save-Planet-Earth Global Economy containing judicious incentives directed at manufacturing various (renewable) Solar-energy, Wind-energy, Hydro-energy, and (clean) fusion & fission Nuclear-energy contraptions that will ultimately allow permanent sourcing of infinite, or near-infinite, supply of energy. This is akin to devising an eventual global 'Giant Perpetual Energy Machine' by transitioning from gradual phasing out of (dirty) exhaustible fossil fuel usage to clean and renewable energy usage. Current fossil fuel usage must be targeted to mainly be utilized specifically for purpose of promoting the supply of clean and renewable energy in order that this entity will one day become fully sustainable and self-sufficient.

That is all for this message. We Eurekians now bid all Earthlings a fond farewell.

XXXXXXXXXXXXXXXXXXXXXXXXXXXXXXXXXXX

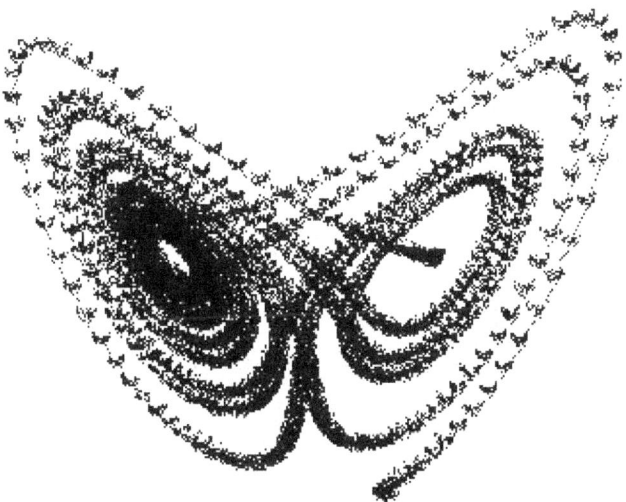

Based on computer simulation, one can easily communicate the picture above as typical representation of iconic 'Butterfly effect' from Chaos theory.

For purpose of our discussion, the human language of communication can be conveniently divided into (i) English Language [although this could be German, French, Chinese or Mandarin Language] and (ii) Science Language; and each further subdivided into their respective spoken and written language counterparts. Spoken language is communication by voice in the distinctively human manner, using arbitrary auditory symbols in conventional ways with conventional meanings. However, many other animals can generate sounds as a non-linguistic means of communication, for example, the language of birds. On Earth, human is the only intelligent animal species capable of written language utilizing any set or system of arbitrary visual symbols as used in a more or less uniform fashion by a number of people, who are thus enabled to communicate intelligibly with one another.

As this book was originally composed in (written) English language, I will adopt (without being discriminatory) the basic language for communication to be the English Language. We will note that there are numerous parallels and analogies; for example the CROP effect [outlined in Chapter 10 of this book], between English Language (EL) and Science Language (SL). However, the two are also intricately and unavoidably intertwined – we need EL as a tool to understand SL and vice versa (to a lesser extent). For instance, many people commonly quote Einstein's famous equation $E = mc^2$ ("E equals to m c square") using SL. Without EL to explain that (in SI unit) E stands for Energy (kJ), m for mass (kg), and c for the speed of light (2.99792458×10^8 m/s); this SL equation would be meaningless. Significantly, in a deeper level of analysis (once again needing EL), this equation also implies that matter and energy are equivalent and that a small amount of matter (mass) can be converted into a gigantic amount of energy. This is illustrated by the heat energy liberated by an exploding atomic bomb or the electrical energy generated by a nuclear power plant – example, for 1 kg of mass:

$E = 1000$ g X 2.9979×10^8 m/s X 2.9979 m/s X 10^8 m/s

$= 8.98755 \times 10^{19}$ gm2/s2 (Nm or J)

$= 8.98755 \times 10^{16}$ kJ

1.1 English Language

English Language (EL) is an integral part of human life. It is a vehicle

12

for growth. It involves exploring beings, objects, events, ideas and experiences. It is making sense of the world using the meaning and context of self, family and cultural group. It is personal, social and functional. It involves attention, reception, perception, association, interpretation, organization, information storage and reflection. It affects our perception, degree of understanding, our acquisition of and degree of knowledge, thinking and problem-solving abilities and social skills. It is an active process learned through its use. It includes risk-taking and error as one experiments with various terms, interpretations of terms, and language forms and conventions. It involves symbols and systems of organizing the symbols (patterns and rules) within a context of usage. Language events (viewing, listening, speaking, reading, writing, thinking, interacting) are interactive, interrelated and interdependent.

In the narrower sense, when people talk about EL, they talk about reading, writing, listening, speaking, viewing, communicating, social skills and conventions, and thinking skills and strategies. In other words, language is:

- what we hear/see, how we hear/see it, how we organize it, how we integrate it, how we interpret it, how we remember it.
- what we say, how we say it, when we say it, where we say it, why we say it, to whom we say it.
- what we do, how we do it, when we do it, where we do it, why we do it, to whom we do it.
- what we think, how we think, why we think in certain ways, when we think in certain ways, where we think in certain ways.

For those of us who have naturally and automatically acquired language, without having to consciously think about it, it is difficult to recognize and understand the intricate and extremely complicated processes involved in the acquisition of language. Language learning involves engagement, attention, modeling, demonstrating, discussing, reflecting, strategy planning and application, self-awareness, self-monitoring, and self-evaluation. For instance, there are many reasons why a person might have difficulty in developing an area of language learning called reading skills. One of the most common reasons is that the person has what is known as a learning disability. There are many types of learning disabilities that can cause problems with learning to read or learning in general – dyslexia and hyperlexia are examples of such learning disabilities. Not all troubles with reading are caused by learning disabilities. In children and adolescents who perform poorly in school for various reasons, it is important to exclude (and treat) other causes of reading skills difficulties such as a primary visual, hearing, or motor disability, emotional disturbance, social or cultural conditions, or mental retardation. Having done so, approximately 10 to 20% have a neurologically based disorder of the type called a learning disability that I alluded to above. These children (and some adults whose disabilities were missed when they themselves were children) are of at least average intelligence (many are far above average), and their academic problems are not caused by the above-mentioned "other causes of reading skills difficulties such as…". Instead, the reason for their learning problems seems to be that their brains are "wired" in a way slightly different from the average person's. About 20% of children with learning disabilities also have a related problem, attention deficit disorder (ADD) or attention deficit hyperactivity disorder (ADHD). Its symptoms include hyperactivity, distractibility, and impulsiveness. ADD or ADHD must be evaluated and treated separately from the learning disability.

14

1.2 Dyslexia and Hyperlexia

Dyslexia is a familial disorder with autosomal dominant inheritance that occurs more frequently in males. Example of a useful reference website is The British Dyslexia Association [http://www.bdadyslexia.org.uk/]. The word 'dyslexia' comes from the Greek and means 'difficulty with words'. It is an inability to read, spell, and write words, despite the ability to see and recognize letters due to a difference in the brain area that deals with language where it affects the underlying skills that are needed for learning to read, write and spell. Brain imaging techniques show that dyslexic people process information differently.

Around 4% of the population are severely dyslexic. A further 6% have mild to moderate problems. Dyslexia occurs in people from all backgrounds and of all abilities, from people who cannot read to those with university degrees. Dyslexic people, of all ages, can learn effectively but often need a different approach. Dyslexia is a puzzling mix of both difficulties and strengths. It varies in degree and from person to person. Dyslexic people often have distinctive talents as well as typical clusters of difficulties.

Hyperlexia is a syndrome observed in children who have the following characteristics:

- A precocious ability to read words, far above what would be expected at their chronological age or an intense fascination with letters or numbers.
- Significant difficulty in understanding verbal language.
- Abnormal social skills, difficulty in socializing and interacting

appropriately with people.

A useful reference website on Hyperlexia is the American Hyperlexia Association [https://autismawarenesscentre.com/resources/american-hyperlexia-association/]. In addition, some children who are hyperlexic may exhibit the following characteristics:

- Learn expressive language in a peculiar way, echo or memorize the sentence structure without understanding the meaning (echolalia), reverse pronouns.
- Rarely initiates conversations.
- An intense need to keep routines, difficulty with transitions, ritualistic behavior.
- Auditory, olfactory and/or tactile sensitivity.
- Self-stimulatory behavior.
- Specific, unusual fears.
- Normal development until 18 - 24 months, then regression.
- Strong auditory and visual memory.
- Difficulty answering "Wh—" questions, such as "what," "where," "who," and "why".
- Think in concrete and literal terms, difficulty with abstract concepts.
- Listen selectively, appear to be deaf.

Hyperlexia has characteristics similar to autism, behavior disorder, language disorder, emotional disorder, Attention Deficit Disorder, hearing impairment, giftedness or, paradoxically, mental retardation. To develop effective teaching strategies and more typical childhood development, it is important to differentiate hyperlexia from other disorders.

16

A thorough speech and language pathologist who is familiar with the syndrome of hyperlexia is a crucial first step. Psychological tests which emphasize visual processes rather than verbal skills aid in identifying hyperlexia. Hearing, neurological, psychiatric, blood chemistry and genetic evaluations can be performed to rule out other disorders but are not needed to identify hyperlexia.

Although children with hyperlexia demonstrate numerous common characteristics, it is important to remember that each child is unique & exhibit different types and degrees of language difficulties and therefore, would require an individualized program to address individual needs. Generally (without going into specific details) these children learn language via:
   1) gestalt processing
   2) visual means
   3) patterns
   4) echolalic or rote memorization techniques
   5) routine.

A more comprehensive in depth analysis on the subjects of dyslexia and hyperlexia, especially diagnosis and treatment aspects, is beyond the scope of this book. Interested readers are urged to look up respective websites for more information where the core materials in this section is based on.

1.3 Model of Learning Disabilities

A publication from the National Information Center for Children and Youth with Disabilities (NICHCY) is "Reading and Learning Disabilities", Briefing Paper FS17, 4th Edition, February 2004. The following information, used with express written permission from NICHCY, is based on this publication containing **adaptation of** Dr. Larry Silver's article "A Look at Learning Disabilities in Children and Youth" which appeared in the November 1991 newsletter of Learning Disability Association of Montgomery County, Inc.

The present model of learning disabilities was established by the late 1960s. This model distinguishes four stages of information processing used in learning: input, integration, memory, and output. Input is the process of recording in the brain information that comes from the senses. Integration is the process of interpreting this information. Memory is its storage for later retrieval. Output of information is achieved through language or motor (muscular) activity. Learning disabilities can be classified by their effects at one or more of these stages. Each child has individual strengths and weaknesses at each stage.

(1) Input The first major type of problem at the input stage is a visual perception disability. Some students have difficulty in recognizing the position and shape of what they see. Letters may be reversed or rotated; for example, the letters d, b, p, q, and g might be confused. The child might also have difficulty distinguishing a significant form from its background. People with this disability often have reading problems. They may jump over words, read the same line twice, or skip lines. Other students have poor depth perception or poor distance judgement. They might bump into things, fall over chairs, or knock over drinks.

The other major input disability is in auditory perception. Students may have difficulty understanding because they do not distinguish subtle

18

differences in sounds. They confuse words and phrases that sound alike –
for example, "blue" with "blow" or "ball" with "bell". Some children find it
hard to pick out an auditory figure from its background; they may not
respond to the sound of a parent's or teacher's voice, and it may seem that
they are not listening or paying attention. Others process sound slowly and
therefore cannot keep up with the flow of conversation, inside or outside the
classroom. Suppose a parent says, "It's getting late. Go upstairs, wash your
face, and get into your pyjamas. Then come back down for a snack." A child
with this disability might hear only the first part and stay upstairs.

(2) Integration

Integration disabilities take several forms, corresponding to the three
stages of sequencing, abstraction, and organization.

A student with a sequencing disability might recount a story by starting in
the middle, going to the beginning, and then proceeding to the end. The child
might also reverse the order of letters in words, seeing "dog" and reading
"god". Such children are often unable to use single units of a memorized
sequence correctly. If asked what comes after Wednesday, they have to start
counting from Sunday to get the answer. In using a dictionary, they must
start with "A" each time. The second type of integration disability involves
abstraction. Students with his problem have difficulty in inferring meaning.
They may read a story but not be able to generalize from it. They may confuse
different meanings of the same word used in different ways. They find it
difficult to understand jokes, puns, or idioms.

Once recorded, sequenced, and understood, information must be organized – integrated into a constant flow and related to what has previously been learned. Students with an organization disability find it difficult to make bits of information cohere into concepts. They may learn a series of facts without being able to answer general questions that require the use of these facts. Their lives in and outside of the classroom reflect this disorganization.

(3) Memory

Disabilities also develop at the third stage of information processing, memory. Short-term memory retains information briefly while we attend to it or concentrate upon it. For example, most of us can retain the 10 digits of a long distance telephone number long enough to dial, but we forget it if we are interrupted. When information is repeated often enough, it enters long-term memory, where it is stored and can be retrieved later. Most memory disabilities affect short-term memory only; students with these disabilities need many more repetitions than usual to retain information.

(4) Output

At the fourth stage, output, there are both language and motor disabilities. Language disabilities almost always involve what is called "demand language" rather than spontaneous language. Spontaneous language occurs when we initiate speaking – select the subject, organize our thoughts, and find the correct words before opening our mouths. Demand language occurs when someone else creates the circumstances in which communication is required. A question is asked, and we must simultaneously organize our thoughts, find the right words, and answer. A child with a language disability may speak normally when initiating conversation but respond hesitantly in demand situations – pause, ask for the question to be repeated, give a confused answer, or fail to find the right words.

Motor disabilities are of two types: poor coordination of large muscle groups, which is called gross motor disability; and poor coordination of small muscle groups, which is called fine motor disability. Gross motor disabilities make children clumsy. They stumble, fall, and bump into things; they may have difficulty in running, climbing, riding a bicycle, buttoning shirts, or tying shoelaces. The most common type of fine motor disability is difficulty in coordinating the muscles needed for writing. Children with this problem write slowly, and their handwriting is often unreadable. They may also make spelling, grammar, and punctuation errors.

1.4 Cybernetica View on Language

Here is another perspective on language based on work done by Professor Valentin F. Turchin (February 14, 1931 – April 7, 2010). Written permission to do so was obtained from his son Professor Peter Turchin. This work is titled Language http://pespmc1.vub.ac.be/LANG.html (created on September 1991, modified on October 6, 1997) in Principia Cybernetica Web. A language is a system which, if properly controlled, can produce objects called messages. A language is a convention according to which certain material objects, to be referred to as linguistic objects, define certain actions, which are referred to as their meanings. There are two basic types of linguistic objects: commands and statements. Commands are used in the context of control, where the meaning of a command issued by the controlling system is the resulting action of the controlled system. The meaning of a statement is the piece of knowledge (true or false), that is a hierarchical generator of predictions.

Languages is classified as below by two parameters

(1) Concrete language vs Abstract language

(2) Unformalized language vs Formalized language, leading to the following four types of language-related activities as depicted below:

	Concrete language	Abstract language
Unformalized language	Art	Philosophy
Formalized Language	Descriptive sciences	Theoretical sciences, mathematics

Concrete language versus Abstract language:

Human language is a multilevel system. On the lower levels, which are close to our sensual perception, our notions are almost in one-to-one correspondence with some conspicuous elements of perception. In our theories we construct higher levels of language. The concepts of the higher levels do not replace those of the lower levels, as they should if the elements of the language reflected things "as they really are", but constitute a new linguistic reality, a superstructure over the lower levels. Predictions produced by the higher levels are formulated in terms of the lower levels. It is a hierarchical system, where the top cannot exist without the bottom. We loosely call the lower-level concepts of the linguistic pyramid concrete, and the higher-level abstract. This is a very imprecise terminology because abstraction alone is not sufficient to create high level concepts. Pure abstraction from specific qualities and properties of things leads ultimately to the loss of contents, to such concepts as "something". Abstractness of a concept in the language is actually its "constructness", the height of its position in the hierarchy, the degree to which it needs intermediate linguistic objects to have meaning and be used. Thus in algebra, when we say that x is a variable, we abstract ourselves from its value, but the possible values themselves are numbers, which are not "physical" objects but linguistic objects formed by abstraction present in the process of counting. This intermediate linguistic level of numbers must become reality before we use abstraction on the next level. Without it, i.e. by a direct abstraction from countable things, the concept of a variable could not come into being. In the next metasystem transition we deal with abstract algebra, like group theory, where abstraction is done over various operations. As before, it could not appear without preceding metasystem level, which is now the school algebra.

Unformalized language versus Formal language:

The second parameter to describe concepts of a language is the degree to which the language embedding the concept or concepts is formalized. A language is formal, or formalized, if the rules of manipulation of linguistic objects depend only on the "form" of the objects, and not on their "human meanings". The "form" here is simply the material carrier of the concept, i.e. a linguistic object. The "human meaning" is the sum of associations it evokes in the human brain. While "forms" are all open for examination and manipulation, i.e. are objective, "human meanings" are subjective, and are communicated indirectly. Operations in formal languages can be delegated to mechanical devices, machines. A machine of that kind becomes an objective model of reality, independent from the human brain which created it. This makes it possible to construct hierarchies of formal languages, in which each level deals with a well-defined, objective reality of the previous levels. Exact sciences operate using such hierarchies, and mathematics makes them its object of study.

Four types of language-related activities

Art is characterized by unformalized and concrete language. Words and language elements of other types are important only as symbols which evoke definite complexes of mental images and emotions. Philosophy is characterized by abstract informal thinking. The combination of high-level abstract constructs used in philosophy with a low degree of formalization requires great effort by the intuition and makes philosophical language the most difficult type of the four. Philosophy borders with art when it uses artistic images to stimulate the intuition. It borders with theoretical science when it develops conceptual frameworks to be used in construction of formal scientific theories. The language of descriptive science must be concrete and precise; formalization of syntax by itself does not play a large part, but rather acts as a criterion of the precision of semantics.

1.5 Science Language

The word "holor" is a term coined by Parry Hiram Moon and Domina Eberle Spencer in their book Theory of Holors: A generalization of tensors (Cambridge University Press, New York 1986). We quote from Paragraph 6 in the Preface section of this book: "The authors look forward to the golden day when university curricula will not separate vector and tensor analysis into different subjects but will cover all holors in a uniform treatment. The present book constitutes a step in this direction. …"

Holors are made up of all types of hypernumbers plus the special case of the scalar (which is not a hypernumber). Therefore, a holor is a mathematical entity that is made up of one or more independent elements (merates), and includes complex numbers, vectors, matrices, tensors, quaternions, and other hypernumbers; and scalars. The merates themselves may be real or complex numbers or they may be more complicated quantities such as vectors (& subvectors), matrices (& submatrices) or other hypernumbers (& subhypernumbers). This simplified classification will embody all the invented number systems; for instance, quaternions (as holors with four elements) and other hypercomplex numbers are part of the wider hypernumber system. [Loosely speaking (although not very fruitful); instead of merates, we could use the traditional but more ambiguous word "components" to designate elements regarded as vectors, and, similarly, "coordinates" for elements regarded as scalars.]

It is my personal opinion that scientific community should fully adopt and standardize the holor notation using the so-called index notation, as outlined by Moon and Spencer in their book. Such a single notation, using both subscripts and superscripts, would apply to all values of valence (N) and all values of plethos (n). A nilvalent holor ($N = 0$) corresponds to a scalar, univalent holor ($N = 1$) a vector, bivalent holor ($N = 2$) an ordinary matrix, trivalent holor ($N = 3$) a three-dimensional matrix, and so forth – that is, valence may be extended to any integer value up to infinity. Similarly, for each and every holor, except holors with valence $N = 0$ (scalars), its plethos may have any positive integer value starting from 1 up to infinity – for instance, for the case of holors with valence $N = 1$; the pletho $n = 1$ may represent a vector in 1-space, $n = 2$ a vector in 2-space or an ordinary complex number, $n = 3$ a vector in 3-space, $n = 4$ a quaternion or the four-dimensional space-time as encountered in (for example) Einstein's Theory of

25

General Relativity, and so forth. Note that the infinite, or close to infinite, dimensional cases occur in the study of many systems in both the quantum and classical world, and also in the high dimensionality of complex adaptive (and nonadaptive) systems and chaotic dynamical systems.

The nomenclature of all holors, both tensor and nontensor, is simplified by using plethos and valence; and when all possible types of holors are considered, holor algebra and holor calculus in the most general sense can be developed to help explain their theoretical aspects and practical applications. The classification of holors includes the tensors and the nontensors, with the nontensors consisting of akinetors, oudors, geometric objects and nongeometric objects. It is a useful mental exercise to think of holors as roughly equivalent to the universal "Alphabet of Science" and holor algebra & holor calculus as the universal "Language of Science" (Science Language).

1.6 Analogies between English Language (EL) and Science Language (SL)

The terms phylum, class, order, family, genus, and species constitute a hierarchy in zoology. In a similar fashion, the terms alphabet, words, sentences, paragraphs, sections, chapters, treatise (books), and literature constitute a hierarchy in EL. Finally, so does the terms holors, equations, theories and meta-theories (theories of theories) constitute a hierarchy in SL. We emphasize that the following analogies and parallels are derived from personal ideas and there may be other differing views on these matters.

Just as the historical development of EL (for example, etymology – the study of the origin and developments of words) is a rich subject in its own right, so is the historical development of SL. A book on the history of SL may begin, for instance, with: "The birth of the holor concept may be set at 1673 when John Wallis suggested the geometric representation of complex numbers by points in a plane. ..."

EL uses the 26-letter alphabet called Roman (commonly called the ABC's). For pronunciation purposes, the vowel sounds (open sounds with free breath) are represented by the letters a, e, i, o, u and sometimes w and y. The closed sounds, called consonants, are made with the breath wholly or partly checked. Stopped consonants require complete stoppage of the breath – they are b, d, g, k, p and t. Other consonants require only partial stoppage of breath – they are l, m, n, r, w and y. The spirants are open consonants that require friction in the oral passages – they are f, s, v and z. H is an aspirant, or breathed, consonant. The analogous equivalence of vowels and consonants in SL is all holors of different plethos and valence.

Grammar is the system by which a language functions. Description of the way in which EL functions includes traditional grammar, structural grammar, and transformational (generative) grammar. Traditional grammar defines parts of speech by their meaning and function. Structural grammar defines them primarily by their order in a sentence. Transformational grammar shifts the emphasis from analysis of parts of speech to the way people produce all of the possible sentences of the language.

Semantics refer to the scientific study of the meaning of words and sentences, and is closely associated with the disciplines of linguistics, logic and philosophy. One aspect of word meaning involves the ways words can be semantically related to other words. Another aspect of word meaning is polysemy, the property of having many meanings. The meanings of sentences result from word meaning and syntax – the way words are put together. Formal semantics, which comes from philosophy, is concerned with truth conditions – the view that to know the meaning of a sentence is to know all situations and conditions under which it is true. Entailments and presuppositions are another part of sentence meaning. Entailments are relations that connect statements. Presuppositions are statements that are assumed to be true.

Grammar and semantics may be thought of as the rough equivalence of studying various aspects of SL, for instance, as outlined in the Contents of the book "Theory of Holors: A Generalization of Tensors" by Parry Hiram Moon and Domina Eberle Spencer, Cambridge University Press, 1986):

I   Holors
– Index notation, Holor algebra, Gamma products
II   Transformations
– Tensors, Akinetors, Geometric spaces
III   Holor calculus
– The linear connection, The Riemann-Christoffel tensors
IV   Space structure
– Non-Riemannian spaces, Riemannian space, Euclidean space

Words consist of the sequences of one or more sounds or morphemes constituting the basic units of meaningful speech used in forming a sentence. Spelling is the (correct) way we combine letters to write words. Vocabulary, which is always changing and growing, can be defined as the total number of words in a language. The analogous equivalence of words in SL is holors, and of vocabulary in SL is classification of holors depicted below [based on Figure 5.02, page 127 of the book by Moon et al and used with permission from Cambridge University Press]:

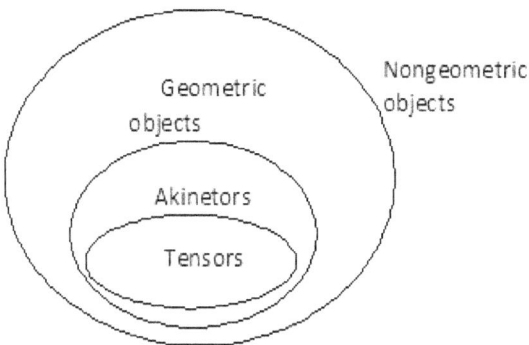

Sentence is a series of words arranged into a meaningful unit. It begins with a capital letter and ends with a punctuation mark – a period (full stop), a question mark, or an exclamation point. A grammatically complete sentence consists of an independent group of words that has a subject and a predicate. The predicate must include a finite verb. The subject of a sentence consists of a noun or another word used as a noun, plus its modifiers. Sentences must be main clauses. (A clause is a group of words with a subject and a predicate.) Sentence can be analyzed through sentence classifications, sentence patterns and diagraming.

Sentence classifications:

(1) Purpose - declarative sentence, interrogative sentence, imperative sentence, exclamatory sentence

(2) Grammatical form - simple sentence, compound sentence, complex

sentence

The variety of English sentences may be shown by examining how several basic sentence patterns (kernals) can be changed. Basic sentence patterns can be added to, reduced, combined, and rearranged in many ways. Changing the parts of a sentence to vary a basic pattern is called transformation. The following are the 5 sentence patterns:

Pattern I	Subject	Intransitive verb	(Optional adverb)	
Pattern II	Subject	Transitive verb	Direct object	(Optional objective complement)
Pattern III	Subject	Transitive verb	Indirect object	Direct object
Pattern IV	Subject	Linking verb	Predicate noun	
Pattern V	Subject	Linking verb	Predicate adjective	

Diagraming is a form of sentence analysis that shows how each part of a sentence is related to another. The forms of diagraming vary from the traditional model to extremely complex diagrams used by modern linguists. One rough analogous equivalence of sentences in SL is the classification of noninvariant and invariant (nontensor and tensor) equations depicted below [based on Figure 4.03, page 111 of the book by Moon et al and used with permission from Cambridge University Press]:

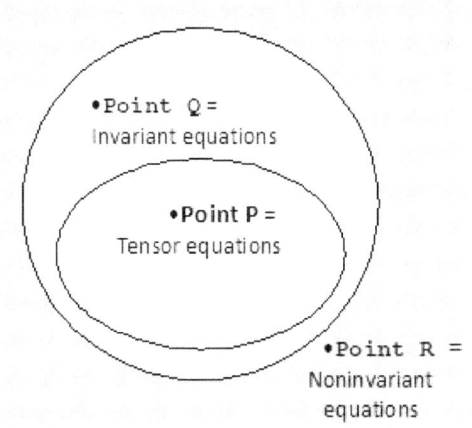

Point P = tensor equation = invariant equation

Point Q = nontensor equation = invariant equation

Point R = nontensor equation = noninvariant equation

Most nontensor equations will be in the (shaded) outer region as represented by Point R.

The comma, semicolon, colon, spacing between words, and punctuation marks in a sentence can be thought of as somewhat akin to the rough equivalence of mathematical operators such as the 4 basic arithmetic operators (addition, subtraction, multiplication and division), differentiation (partial derivatives or total/complete derivatives, ordinary single/first-order derivatives or higher-order derivatives), and integration (definite or indefinite integrals, and single or multiple integrals).

Another way of looking at the type of equations is the differential equations. It is an equation involving derivatives or differentials of an unknown function. When partial derivatives occur, the equation is a partial differential one, otherwise an ordinary one. The order of the highest

31

derivative occurring is the order of the equation and the highest power of the function or its derivative is the degree. Equations of the first degree are called linear, others are nonlinear. To solve a given differential equation, we first try to determine whether the equation is:

(a) a single equation – an ordinary differential equation,

(b) a single equation – a partial differential equation, and

(c) a system of simultaneous differential equations.

Then further classifications can be developed in each case. Because many integrals cannot be evaluated in terms of unknown functions and it thus follows that not all differential equations can be solved. However, numerical, graphical, or mechanical methods with the help of high-speed electronic computers can be resorted to if classical devices fail.

In EL, a question is a sentence in an interrogative form (ending with a question mark) addressed to someone in order to elicit information. Derivatives in SL may be thought as the rough equivalent of asking questions in EL, with finding the answers to the questions as the rough equivalent of finding the integral solutions to the derivatives.

Paragraph is a section of a written work that consists of one or more sentences constructed and arranged to function as a unit. The subject or topic of a paragraph is often stated in the first sentence, called a topic sentence. An effective paragraph must be unified, ordered, and complete. Chapter is a main division, usually numbered, of a book, treatise, or the like.

Newtonian mechanics is based on Newton's 3 laws of motion:

First Law: A body continues in a state of rest or uniform motion in a straight line unless it is acted upon by external forces. This is sometimes called Galileo's law of inertia & often regarded as contained in Second Law.

Second Law: The rate of change of momentum of a moving body (with mass m) is proportional to and in the same direction as the force (F) acting on it, that is, $F = d(mv)/dt = ma$ where a is the acceleration.

Third Law: If one body exerts a force on another, there is an equal and opposite force, called a reaction on the first body by the second. This is also called the law of action and reaction.

Relativistic mechanics (Special relativity) is required for calculations for moving objects with velocities approaching the speed of light. In the limit as this velocity approach the lower classical speed limit, Newtonian mechanics closely approximate relativistic mechanics. Therefore, Newtonian mechanics may be regarded as the rough analogy of (just) a paragraph contained in a section (with a few paragraphs) entitled relativistic mechanics which itself may be considered as part of a chapter on electromagnetic force. The 4 chapters of a book are regarded roughly equivalent to the 4 forces of nature.

Treatise is a book or writing dealing with some particular subject. The writing may be done in "paperless" books when saved on a computer system. The traditional book consists of written or printed sheets of paper or some other material fastened together along one edge so it can be opened at any point. Unification of 4 forces culminating in ultimate Theory of Everything

(TOE) maybe thought of as a (the most) complete treatise or book.

Literature in its broadest sense, is everything that has ever been written. It has two main divisions: fiction and nonfiction. Fiction is writing that an author creates from the imagination – however, authors may include facts about real persons or events, but they combine these facts with imaginary situations. Most fiction is narrative writing, such as novels and short stories. Fiction also includes science fiction, drama and poetry. Nonfiction is factual writing about real-life situations. The chief forms of nonfiction include the essay, history, biography, autobiography, and diary. (In a broader sense, nonfiction may also be taken to include specialized technical subjects such as written Physics and Mathematics materials.)

Elements of every literary work includes (1) characters (2) plots (3) theme or statement and (4) style. A good writer tries to balance these elements to create a unified work of art. In this section, with respect to Science Language, we have been concentrating on theories and meta-theories in relation to the four forces of nature and their unification into the ultimate Theory of Everything. This is only a small part of science and is somewhat akin to analyzing a small part of nonfiction aspect of literature in English Language.

In conclusion:

There are close analogies and parallels between English Language and Science Language. Science Language, written using the Alphabet of Science, can be applied to **EVERY CONCEIVABLE TYPE AND BRANCH OF SCIENCE** including the ultimate Theory of Everything. This is with the proviso that English Language is intertwined with, and required to explain, Science Language. Significantly, a very different branch of science to a particular scientist from a different field will not look so strange or be so difficult to understand once he or she master the "common" Language of Science which is based on the "common" Alphabet of Science.

3 Rigorous Proof on Riemann Hypothesis for Dummies

A brief synopsis on the historical timeline with respect to the author solving the proposed-in-1859 Riemann hypothesis is given below:

(1) On Thursday October 9, 2014 my then 14 year-old eldest son Jonah emailed me along the following lines: Hey dad, I have a Maths challenge for you. Do this exercise for the infinite sum 1+2+3+4+5+6+7+8+.... Keep going and you will easily get the infinity value answer. Now watch this YouTube video "X" [the source of which is purposefully not revealed here] which will tell you the astonishingly answer: $1 + 2 + 3 + 4 + 5 + ... = -\frac{1}{12}$.

Postscript: 2019 represents Jonah's second year of his undergraduate Engineering degree in 2019 in a renowned Australian university.

For the uninitiated person, the mathematical 'trickery' [which invalidates this answer] employed in that unidentified video would easily fool that person into believing this answer to be correct. However, by putting all things in proper prospective, we know that this infinite sum endowed with $-\frac{1}{12}$ value actually relates to assigning -1 value to Riemann zeta (ζ) function. With analytic continuation of this function, one can show that $\zeta(-1) = -\frac{1}{12}$ gives a way to "assign" a finite result to the divergent series $1 + 2 + 3 + 4 + 5 + ...$ [= $+\infty$], which is useful in certain contexts such as String theory.

(2) Approximately two years later by mid-2016, that 2014 email had stimulated me to publish the initial "drafts" [as two published papers in Journal of Mathematics Research] for rigorous proof for Riemann hypothesis and explaining its closely related Gram points. The "combined" viXra manuscript of these two papers with full corrections were written by me in 2019 (with a second viXra manuscript written with rigorous proofs for Polignac's and Twin prime conjectures).

Thus Riemann hypothesis is an age-old problem in mathematics which was proposed in 1859 and was only solved 160 years later in 2019. To explain all these in a nutshell, we will have to lay down some groundwork using simple English and Mathematical languages.

First: complex numbers. "What is the square root of minus one $(\sqrt{-1})$?" We call it 'i'; with i multiplied by i equals to -1. If the real number line ...-4, -3, -2, -1, 0, 1, 2, 3, 4... is represented as a horizontal axis, then the numbers ...-4i, -3i, -2i, -i, 0, i, 2i, 3i, 4i... can be thought of as the vertical axis on this diagram. The whole plane taken together is then called the complex plane. This is a two-dimensional set of numbers.

Every complex number can be represented in the form a+bi. For real numbers, we simply take b=0 and the resultant real numbers 'a' only lie on the horizontal axis. Complex numbers can be "simple" lying only on the vertical axis when a=0 and 'b' have non-zero values, or "complicated" lying on the whole plane when 'a' and 'b' both have non-zero values.

Second: functions. In mathematics, a function can be thought of as a black box which, when we put a number into it, spits out a different number. A function is represented by a letter - usually "f". If we put a number x into the function we call f, then what f spits out is written as "f(x)". A convenient example to express f(x) in terms of x is the $f(x) = x^2$ simple function. Whatever x we put in, we will get x^2 out; viz.

f(1) = 1 X 1 = 1,

f(-1) = -1 X -1 = 1,

f(2) = 2 X 2 = 4,

f(-2) = -2 X -2 = 4,

f(3) = 3 X 3 = 9,

f(-3) = -3 X -3 = 9, and so on.

We are most familiar with real functions, or functions where we put a real number in and always get a real number out. However, we can also put weird complex numbers into a function. For example, if $f(x) = x^2$ and we let x = i, which is the square root of minus one, then we will get

f(i) = 1 X 1 X $\sqrt{-1}$ X $\sqrt{-1}$ = -1,

f(-i) = -1 X -1 X $\sqrt{-1}$ X $\sqrt{-1}$ = 1 X -1 = -1,

f(2i) = 2 X 2 X √-1 X √-1 = 4 X -1 = -4,

f(-2i) = -2 X -2 X √-1 X √-1 = 4 X -1 = -4,

f(3i) = 3 X 3 X √-1 X √-1 = 9 X -1 = -9,

f(-3i) = -3 X -3 X √-1 X √-1 = 9 X -1 = -9, and so on.

This is more generally known as complex functions whereby we can put any complex number a+bi in and potentially get another complex number c+di out.

The Riemann zeta (ζ) function is just such a complex function except that it needed the [additional] proxy or surrogate Dirichlet eta (η) function representation when its real number sigma (σ) variable have values [in the interval] between 0 and 1. Both ζ and η functions are best thought of as "super-complicated" complex functions with complex number input a+bi [traditionally denoted by s (= σ + it) instead] and complex number output ζ(s) or η(s) giving rise to another complex number c+di.

Third: zero of a function. This is a point a+bi where f(a+bi) = 0. If f(x) = x^2 then the only zero is obviously at 0, where f(0) = 0. For zeta and eta functions, this is more complicated. They basically have two types of zeros both of infinite magnitude: the trivial zeroes that are easily predictable and occur at all negative even integers, that is, -2, -4, -6, -8, ...; and the nontrivial zeroes, which are all the other ones that are not easily predictable and

39

[allegedly] occur only for 'σ' [or 'a'] = 0.5 (or $\frac{1}{2}$) value at various 't' [or 'b']
values with the first six values being 14.134725, 21.022040, 25.010858,
30.424876, 32.935062, and 37.586178 [when rounded off to six decimal
places].

Actually, it is a bit more complicated than this again! The nontrivial zeros
of zeta or eta functions are best visualized as the zeros occurring at the
'Origin' of the relevant graph. There is also the Gram [y=0] points (also
known as the usual / traditional Gram points), referring to zeros occurring
on the horizontal x-axis; and our newly created Gram [x=0] points, referring
to zeros occurring on the vertical y-axis.

Fourth: The 1859 Riemann hypothesis. This hypothesis essentially refer
to "mathematically needing to prove in a rigorous manner that all nontrivial
zeros only occur at σ = $\frac{1}{2}$ value [which is the 'Origin']". A very difficult
problem indeed as, unlike other general problems which can be solved in
more than one ways, this "super-special" problem can be solved in only
ONE WAY and it is extremely difficult to work out.

Fifth: The 2019 rigorous proof for Riemann hypothesis. The one and only
one way for this rigorous proof to materialize is to analytically utilize only
equations derived from zeta or eta functions that contain [but do not allow
calculation for the actual locations of] all nontrivial zeros. To accomplish this
proof, it will still require subsequent mind blowing "trickery-like"
mathematical techniques such as Ratio Study and Dimensional Analysis to
be employed in ONE VERY SPECIFIC WAY.

Sixthly and Lastly: The beneficiary by-products arising from this rigorous proof for Riemann hypothesis. The usual primary ("direct") by-products are often quoted as "With this one solution, we have proven five hundred theorems or more at once". This apply to the many important theorems in number theory (mostly about prime numbers) that rely on properties of zeta or eta functions such as where trivial and nontrivial zeros are, and are not, located. A classic example of this primary by-product is the resulting absolute and full delineation of the prime number theorem, which relates to prime counting function. This function is denoted by $\pi(x)$ and is defined as the number of prime numbers less than or equal to x. The rigorous proof for Riemann hypothesis is instrumental in proving the efficacy of techniques that estimate $\pi(x)$ efficiently and reasonably well.

4 Nonliving Things and Living Things

In this chapter, we outline our innovative secondary ("indirect") by-product derived out of solving our three open problems in Number theory of Riemann hypothesis, Polignac's and Twin prime conjectures viz. Classification System on Fundamental Laws of Nature aka Classification System of Nonliving and Living Things [which is more broadly known as Extraterrestrial-Terrestrial Elementary-Emergent Fundamental Laws]. This is one of the 'star of our show' (so-to-speak) conveniently summarized below as a list of four entities [numbered in hierarchical order from top to bottom]:

(I) Complex Living Things (Incompletely predictable with "dynamic spatial and temporal irregularity" and Supramaximal Complexity).

(II) Simple Living Things (Completely predictable with "dynamic spatial and temporal regularity" and Supraminimal Complexity).

(III) Complex Nonliving Things (Incompletely predictable with "static spatial and temporal irregularity" and Supraminimal Simplicity).

(IV) Simple Nonliving Things (Completely predictable with "static spatial and temporal regularity" with Supramaximal Simplicity).

There is an apparent increase in orderliness (corresponding to a decrease in entropy) in the direction progressing from Hierarchy Level IV to III to II to I. There could be possible roles played by Evolution &/or Creation processes declaring Nonliving Things in Level III [spontaneously arising through evolution or created by God's hands] "giving birth" to Living Things in Level II.

IV Complex Emergent Living Things have Supramaximal Complexity: Incompletely predictable with dynamic spatial and temporal irregularity
III Simple Emergent Living Things have Supraminimal Complexity: Completely predictable with dynamic spatial and temporal regularity

↑
Evolution and/or Creation Processes
↑

II Complex Elementary Nonliving Things have Supraminimal Simplicity: Incompletely predictable with static spatial and temporal irregularity
I Simple Elementary Nonliving Things have Supramaximal Simplicity: Completely predictable with static spatial and temporal regularity

What is the true meaning or significance of the word 'Thing'? It is often used vaguely in the context of 'Physical Things' as opposed to 'Nonphysical Things'. Physical Things can be inanimate material objects (Nonliving Things) as distinct from living sentient beings and primitive life-forms (both deemed to be 'Living Things'). We can best perceive Nonphysical Things as 'Abstract Things' existing in thought or as an idea but not having a physical or concrete existence.

Nonliving Things are ubiquitous throughout our Universe. Currently we know that both simple and complex living organisms constituting all Living Things are ubiquitous only (terrestrially) on planet Earth, with 'simple' and 'complex' terms here roughly equated with 'non-intelligent' and 'intelligent' organisms. Whether living organisms of any kind exist (extraterrestrially) outside of planet Earth is a matter of conjecture at this moment in time.

Typhoon Lekima satellite image: Tropical Storm Lekima captured on 14:30 UTC August 7, 2019. From: Japanese Meterological Agency (used in accordance with guidelines as stipulated in its Legal Notice).

NASA Solar System Exploration image of planet Pluto's magestic mountains, frozen plains and thin nitrogen atmosphere on September 17, 2015 (used in accordance with NASA Media Usage Guidelines).

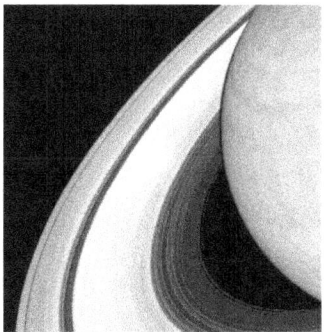

Image of rings of planet Saturn taken by Cassini spacecraft on June 26, 2016 (used in accordance with NASA Media Usage Guidelines)

Hubble Space Telescope view of Jupiter taken on June 27, 2019 revealing the giant planet's trademark Great Red Spot (used in accordance with NASA Media Usage Guidelines).

One must associate Living Things as 'dynamically' active (and ALIVE) in usual sense that they all physically [spatially] mature, change, grow, and (eventually) die over time [in a temporal manner], often at vastly different rates. However, there are numerous examples of many [but not all] Nonliving Things being 'dynamically' active (but NOT ALIVE) in the sense that they physically [spatially] change over time [in a temporal manner]. Some iconic examples with [vastly] varying dynamical activity range are the eternally evolving (i) global weather systems of planet Earth, and (ii) surface landscapes of planet Jupiter as seen by NASA's Juno spacecraft in 2016, of (dwarf) planet Pluto as seen by NASA's New Horizons spacecraft in 2015, and of planet Saturn's magnificent rings as seen by NASA's Cassini spacecraft in 2004.

It has traditionally been stated that the totally unpredictable weather forecast of planet Earth on medium to long term timescale is due to Chaos theory. In particular, this refer to the "butterfly effect" which is essentially 'sensitivity to initial conditions' – something as small as the flutter of a butterfly's wing can ultimately cause a typhoon [also known as tropical cyclone or hurricane] halfway around the world. An example of a Fractal object is the beautiful rings of planet Saturn seen in images transmitted from Casini. They manifest Cantor-like states with fractal dimensions consistently calculated to be about 1.6 to 1.7 using box-counting method by Jun Li and Martin Ostoja-Starzewski – this phenomenon was submitted in a letter to the editor titled "Saturn's Rings are Fractal" on June 30, 2012 [https://arxiv.org/ftp/arxiv/papers/1207/1207.0155.pdf].

Having said all the caveats in the preceding paragraphs, our nominated stance in this book is that it is far more meaningful to preferentially associate Nonliving Things with "static" spatial and temporal regularity [for simple cases] and irregularity [for complex cases], and Living Things with "dynamic" spatial and temporal regularity [for simple cases] and irregularity [for complex cases]. Thus the terms "static" and "dynamic" are arbitrarily used here to respectively connotate 'not alive' and 'alive'.

In the crudest sense, "Simple systems" (artificially deemed to mainly manifest 'linearity') may usefully be perceived [but not with total accuracy] as lying on opposite end to "Complex systems" (artificially deemed to mainly manifest 'non-linearity'). Much of following views are partially based upon vast resources posted on Internet (World Wide Web) by many professional people hailing from all scientific backgrounds. We make an important note here that many of those Internet resources are very dynamic in the sense that they may (of course) be altered and up-dated or even removed over time. Thus the phrase "If I have seen a little further it is by standing on the shoulders of giants" used by Isaac Newton in a 1676-dated letter to his rival Robert Hooke is eminently and equally applicable to all modern researchers, scientists, and authors alike from the current 21st Century.

A useful and simplified definition for "Complex systems" is that they are structures, entities, agents or processes that involve 'non-linearity'. This includes both sets of many interacting objects (structures), entities and agents, as well as non-linear dynamical systems (processes). Thus the broad field of Complexity theory would include a wide variety of topics that have become prominent research area in recent years, such as artificial life, cellular automata, chaos, evolutionary computation, fractals, genetic algorithms, neural networks (natural and artificial), parallel computing, percolation, renormalization theory, and spin glasses.

There is no agreed-upon definition of complexity, simply because it manifests itself in so many different ways. Accordingly, operational descriptions of complexity are helpful. There are certain basic characteristics that must be considered. These are: (a) purpose and function, (b) size and configuration, (c) structure, including composition and make-up, and (d) the type of dynamics. The first two can be called "static complexity", the third "embedded complexity", and the last "dynamic complexity".

For total complexity these factors should coexist; however, this need not be. The dynamics of a complex system may vary, and depending on the circumstances, dynamic stability may be steady, transient, or chaotic. Structural complexity in itself is an important characteristic of a system, and it follows that this complexity must be considered in the light of the surroundings in which the system finds itself.

The following could be model statements:

(1) Complexity can occur in natural or man-made (artificial) organic and inorganic systems, as well as in social-economic structures.

(2) Complex dynamical systems may be very large or very small; indeed, in some complex systems, large and small components live cooperatively.

(3) The physical shape may be regular or irregular.

(4) As a rule the larger the number of the parts of the system, the more likely it is for complexity to occur.

(5) Complexity can occur in energy-conserving systems, as well as in energy-dissipating systems.

(6) The system is neither completely deterministic nor completely random (stochastic), and exhibits both characteristics.

(7) The causes and effects of the events that the system experiences are not proportional.

(8) The different parts of complex systems are linked and affect one another in a synergistic manner.

(9) There is positive or negative feedback.

(10) The level of complexity depends on the character of the system, its environment, and the nature of the interactions between them.

(11) Complex systems are open in the sense that they can exchange material, energy, momentum, and information with their surroundings.

(12) Complex systems tend to undergo irreversible processes.

(13) Complex systems are dynamic and not in equilibrium; they are like a journey, not a destination, and they may pursue a moving target.

(14) Many complex systems are not well behaved and frequently undergo sudden changes that suggest that the functional relations that represent them

are not differentiable.

(15) Paradoxes exist, such as fast and slow events, regular as well as irregular forms, and organic and inorganic bodies in cohabitation.

Synergistics, an interdisciplinary field of research, is notably concerned with the cooperation of individual parts of a system that produces macroscopic spatial, temporal or functional structures. It deals with deterministic as well as stochastic processes.

One can pursue complexity along one of three paths: the spiritual, the philosophical, and the scientific. Our emphasis will be on the scientific, but this too can be pursued along different paths, and the Chaos and Fractals approach is the most powerful and general paradigm for the study of complex system. This may involve issues such as an integrated interpretation of nonlinear dynamics, nonequilibrium thermodynamics, information theory, and fractal geometry. Not all complex systems are self-organizing, but all self-organizing systems are complex. While we do not completely understand the evolution of biological systems, we realise that there is a difference between organic and inorganic systems. Jacques Monod [Monod, J. (1971). Chance and Necessity. New York: Alfred A. Knopf, Inc. Reissued as a paperback in 1972 by Vintage Books, New York], the 1965 Nobel laureate in physiology and medicine, calls the characteristic of having a purpose, such as the preservation and multiplication of the species, "teleonomic". The following describe some other scientific approaches to complexity.

In 1972, French mathematician Rene Thom suggested a geometric or topological approach for the study of complex systems undergoing discontinuous changes. This approach is called "catastrophe theory" because according to Rene, a system can be described by one of seven elementary catastrophes to which he ascribed romantic names like the "swallowtail catastrophe", the "butterfly catastrophe", the "cusp catastrophe" and others.

A revolutionary concept – that of cellular automata – was conceived in 1950 by John von Neumann based on ideas of Stanislaw Ulam. Looked at simplistically, a cellular automaton is a set of cells arranged in computer space. These automata change according to predetermined rules. One particular advantage of cellular automata is that they constitute discrete dynamical systems amenable to treatments with computers, whereas complex systems modelled with nonlinear differential equations are difficult to deal with. The current prominence of cellular automata is due to the contributions of Edward Fredkin, Tommaso Toffoli, and Stephen Wolfram. For example, the dynamic behaviour of animal population could be modelled using cellular automaton.

Complex systems also entail incomplete knowledge and imprecision. One approach to this is through fuzzy logic, developed by Professor Lotfi Zadeh of the University of California (Berkeley) in 1965. A recent trend in studying complex systems is the so-called "self-organized criticality".

Meta-quantification of complexity, or "partial measurements" of complexity is a useful way to provide a scale whereby one could measure complexity. Complexity of a system is not an intrinsic property of the system, but is a composite of a variety of factors. In addition, complexity is a "contingent" property of the system, is a very apt statement. While not all complex systems are chaotic, many are, so that the metrics of Chaos such as fractal dimension, Lyapunov exponential coefficients, Kolmogorov-Sinai entropy, and others can help us gauge the degree of complexity. Various indicators of Chaos are other sources for interesting discussion as well.

The following are some (three) of the paths that have been followed for quantifying complexity:

Hierarchical approach. Physicist B. West and virologist J. Salk [West, B. J. and Salk, J. (1987). Complexity, Organization, and Uncertainty. European Journal of Operational Research, 30, 117 - 128] proposed two hierarchies [depicted in order from Left to Right] (a) & (b) below for complexity, and (c) below for three stages in ascending order for sciences:

(a) Complexity in nature: Particles --> Atoms --> Molecules --> Replicating molecules --> Cells --> Organizations --> Populations --> Human mind.

(b) Complexity in knowledge: Physics --> Chemistry --> Molecular biology --> Biology --> Physiology --> Sociology --> Psychobiology.

(c) Three stages: Physical sciences --> Life sciences --> Human sciences.

J. G. Miller [Miller, J. G. (1987). Living systems. McGraw Hill, New York] in 1978 proposed his generalised living system along the following sequence:

'Increasing level of complexity in living system' roughly correlating with 'Increasing approximate number of years since period of origin' and 'Increasing approximate median diameter'. He also outlined a generalised living system interacting and intercommunicating with its environment as follows:

Subsystems which process both matter-energy and information – Reproducer, Boundary.

Matter-energy processing subsystems – Ingestor, Distributor, Decomposer, Producer, Matter-energy storage, Extruder, Motor, Supporter.

Information processing subsystems – Input transducer, Internal transducer, Channel and net, Decoder, Associator, Memory, Decider, Encoder, Output transducer.

Molecular biologists [e.g. Gerard, R. W. (1957). Units and concepts of biology.Science, NY, 125, 429.] would also include lower nonliving to higher living entities in the biological hierarchy [in order from Left to Right]: atom --> molecule --> macromolecule --> subcellular organelle --> living cell --> multi-cellular functioning organ --> whole living organism --> a population of organisms --> ecosystem.

Increasing populations and technological innovations leading to evolution of inter-organizational complexity; and technological advancement with industrial phase space can all follow a hierarchical restructuring as well.

The geometric approach to complexity is expounded next: Complexity can manifest itself both in form and in function, and frequently in both, for example, in building a paper jet plane, both its shape and function have changed from the original sheet of paper (Origami paper jet airplane). The geometry of this plane has a non-integer (fractal) dimension.

Finally, algorithmic complexity approach is the mathematical way of measuring complexity. An algorithm is a precisely defined, step-by-step computational procedure that is not ambiguous. The game plan is to model a complex system that has associated with it a large number of data points by establishing the minimum length of the algorithm that describes the arrangement of the number of the data set. The mathematical methodology constitutes the SKC theory, which stands for the initials of Solomonoff, Kolmogorov, and Chaitin, who worked on it independently in the mid-1960s. Their algorithmic complexity theorem provides a means of measuring randomness that, in turn, gives us information about the extent of the complexity. The SKC theory may be applied to characterise complex systems that are between complete randomness and complete determinism. Important applications of algorithmic complexity are in the coding of genetic DNA molecules, information theory, and Artificial Intelligence. It is a powerful metric for characterising complexity.

Complexity characterises the behavior of a system or model whose components interact in multiple ways and follow local rules, meaning there is no reasonable higher instruction to define the various possible interactions. The term is generally used to characterize something with many parts where those parts interact with each other in multiple ways, culminating in a higher order of emergence greater than the sum of its parts. The study of these complex linkages at various scales is the main goal of complex systems theory. Simplicity is the state or quality of being simple. Simplexity is an emerging theory that proposes a possible complementary relationship between complexity and simplicity. Simplicity emerging from complexity concept: **In complexity theory, the central theme is that simplicity can emerge from complexity.** Nature' patterns are "emergent phenomena" and this could be analyzed with topics such as the shape of water drops, the dynamic behaviour of animal populations, and the strange patterns in plant-petal numerology. Not surprisingly, they will manifest Chaos and Fractals such as fractional dimension and strange attractor with many specific spatial and temporal properties. Note: The [unrelated] term 'complicity' is the participation in a completed criminal act of an accomplice, a partner in crime who aids or encourages (abets) other perpetrators of that crime, and who shared with them an intent to act to complete the crime.

"Autistic savant" is one of the most fascinating cognitive phenomena in psychology. It refers to individuals with autism who have extraordinary skills not exhibited by most persons. There are many forms of savant abilities, with the most common forms involving mathematical calculations, memory feats, artistic abilities, and musical abilities. Example, a mathematical ability which many autistic individuals display is incredible calendar memory. Another example, there are known autistic individuals capable of 'photographic' memory in being able to artistically create paintings of surrounding scenery (from stored visual memory) with astonishing accuracy.

In mathematics, Monstrous Moonshine or Moonshine Theory is the unexpected connection between the monster group M and modular functions (in particular, the j-function). The term "Monstrous Moonshine" was coined by John Conway and Simon Norton in 1979. Dual British and

Canadian mathematician John McKay in the late 1970s told British mathematician John Conway about an odd coincidence ['Presence of All-The-Stars-Lining-Up Coincidental Number'] in that the first coefficient of q (namely 196884) from j-function was precisely one more than the degree of the smallest faithful complex representation of the monster group (namely 196883). Conway replied that this was "moonshine" (in the sense of being a crazy or far-fetched or foolish idea). Furthermore, there are many other coefficients soon discovered to be the exact sums of various chosen special dimensions of the monster group. Therefore the term "moonshine" not only just refers to the monster group M; it also refer to the perceived craziness of the intricate relationship between M and the theory of modular functions.

The term "Extraterrestrial-Terrestrial Elementary-Emergent Fundamental Laws (ETEEF Laws)" is specifically coined by us here in 2019 to denote Elementary and Emergent Fundamental Laws that are applicable not just on planet Earth (Terrestrial), but equally applicable outside of planet Earth (Extraterrestrial). Propositional-wise with its main designated role as sound "Classification System on Fundamental Laws of Nature"; both Laws can be further subdivided (respectively) into Simple and Complex Elementary Fundamental Laws, and Simple and Complex Emergent Fundamental Laws with the word "Simple" connotating "Completely Predictable", "Complex" connotating "Incompletely Predictable", "Elementary" connotating "Nonliving Things", and "Emergence" connotating "Living Things".

We coin the all-encompassing term "Moonshine Mathematics" to be the nickname for, and be largely synonymous with, this Complex Elementary Fundamental Laws. We assign Moonshine Mathematics to include not just Monstrous Moonshine but also other closely related subject areas such as Umbral ("Shadow") Moonshine and Mathieu Moonshine – all these currently undergoing active research with possible relevance in String theory potentially uniting Einstein General Relativity and Quantum gravity. More broadly, our Hybrid integer sequence A228186 and the 'Special-Class-of-Mathematical-Problems with Solitary-Proof-Solution'; both endowed with odd or paradoxical behaviors are additionally incorporated as belonging to Moonshine Mathematics with relevant extra consolidatory and reconciliatory explanations given below. For comparison, we now touch on an example of 'Mathematical Problems with Multiple-Proof-Solutions' below.

Well over 2000 years ago (c. 300 BC) Euclid proved that there were infinitely many primes predominantly by *reductio ad absurdum*} (proof by contradiction) method. Since then dozens of proofs have been devised - three of which are chronologically presented with the strangest candidate likely to be Furstenberg's Topological Proof. These are (i) Goldbach's Proof using Fermat numbers (written in a letter to Euler, July 1730), (ii) Furstenberg's Topological Proof, and (iii) Filip Saidak's Proof.

In essence, our Extraterrestrial-Terrestrial Elementary-Emergent Fundamental Laws (ETEEF Laws) of Nature could intuitively be perceived to achieve ultimate goal of being the simplest yet broadest all-encompassing general Classification System on Laws of Nature. With most, if not all, proposed Classification Systems, there are bound to be gray areas (referring to ill-defined situations or areas of activity not readily conforming to a category or set of rules); and our ETEEF Laws will not be immune to this intrinsic and unavoidable phenomenon. Therefore we start off this section with carefully chosen examples explained with concocted jargons for each defined component of ETEEF Laws that will help shed light and understanding on Moonshine Mathematics.

Simple Elementary Fundamental Laws usually involve "Completely Predictable" Laws on Nonliving Things (e.g.) governing the appearance of all infinite even numbers or infinite odd numbers, viz. even numbers 0, 2, 4, 6, 8, 10, 12,... always end with a digit of 0, 2, 4, 6 or 8; and odd numbers 1, 3, 5, 7, 9, 11, 13,... always end with a digit of 1, 3, 5, 7, or 9. We can always generate and completely predict all consecutive even numbers and odd numbers with simple formulae and calculations as they occur with "complete regularity". A further example to illustrate this property can be provided by the "Completely Predictable" single origin intercept and infinite x-axis intercepts of $y = \sin(x)$ equation.

Complex Elementary Fundamental Laws usually involve "Incompletely Predictable" Laws on Nonliving Things (e.g.) governing the appearance of all infinite prime numbers or infinite non-trivial zeros of $\zeta(s)$, viz. prime numbers 2, 3, 5, 7, 11, 13, 17,... are always evenly divisible only by 1 or itself & they must be whole numbers greater than 1; and all nontrivial zeros are always origin intercepts occurring only when $\sigma = \dfrac{1}{2}$. We can always generate and completely predict all consecutive prime numbers and non-trivial zeros but only with complex formulae and calculations as they occur with "complete irregularity". Perhaps controversial, we include our A228186

Hybrid integer sequence as further example under Complex Elementary Fundamental Law for the reason that it has odd or paradoxical behavior. As alluded to earlier, this Complex Elementary Fundamental Laws has been proposed by us to be synonymous with Moonshine Mathematics.

Simple Emergent Fundamental Laws usually involve "Completely Predictable" Laws on Living Things (e.g.) governing the behaviors of simple body organs or "ideal" dynamical population growth of living organisms, and the structural patterns of various organisms, viz. Cardiac Output of the human heart is physiologically determined as Stroke Volume multiplied by Heart Rate; Fibonacci's Rabbits breeding, Plant leaf structure, and Nautilus shell spiral arrangements could be [completely] modelled by Fibonacci Numbers and Golden section in Nature; with all above intuitively manifesting "complete spatial and temporal regularity".

Complex Emergent Fundamental Laws usually involve "Incompletely Predictable" Laws on Living Things (e.g.) governing the behavior or structure of the most complex body system in the Universe, namely the Nervous System, viz. intelligence, consciousness; and structural neuronal arrangement arising out of our human Brain could be [incompletely] modelled by Artificial Intelligence; and Neural Network Connections, with all intuitively manifesting "complete spatial and temporal irregularity".

The five common scenarios on how our world could end are expanded below.

(1) Nuclear holocaust. Nuclear weapons – missiles, bombs, shells, or land mines that use fission or fusion of nuclear material yielding enormous quantities of heat, light, blast, and radiation. The first atomic bomb (or fission bomb), dropped on Hiroshima in 1945, used uranium-235. Later models used plutonium-239 with more explosive power. The hydrogen bomb (fusion bomb or thermonuclear bomb) consists of an atom bomb surrounded by a layer of hydrogenous material, such as lithium deuteride. The atom bomb creates the necessary temperature (about 10^8 °C) needed to ignite the fusion reaction. Hydrogen bombs have an even greater explosive power measured in tens of megatons (millions of tons) of TNT. The neutron bomb (or enhanced radiation bomb) is a nuclear weapon designed to maximize neutron radiation. It is lethal to all forms of life, but, having reduced blast, leaves buildings, etc., relatively undamaged.

The sudden collapse of the Soviet Union in 1991 as a superpower communist country has changed the world. It is a vast potential supermarket of nuclear, chemical, and biological weapons and may made the world's biggest stockpile of nuclear weapons available to terrorists. On Tuesday 11 September, 2001 we have "America Under Attack", where in the worst terrorist attack on U.S. soil (with resulting massive number of casualties and the total death tolls numbering over three thousand people), four U.S. passenger planes were apparently hijacked and crashed, including two jets that flew into the twin towers of the World Trade Center in New York City, a plane that struck the Pentagon in Washington and a final plane that crashed southeast of Pittsburgh. When it reopen on Monday 17 September 2001, the New York Stock Exchange suffered its biggest one day fall in history with the Dow Jones index plunging 684.80 points (7.13% - the 14th largest percentage fall) to close at 8920.7. The tech-heavy Nasdaq index fell 115.82 points to 1579.55, while the broader S&P 500 index lost 53.77 points to 1038.77. The Dow gave up 1369.70 (14.26% drop) to close 8235.81 for that particular week, its biggest ever one-week points loss. In total, the week's losses wiped a massive $US1.2 trillion from the market value of all NYSE, Nasdaq and American Stock Exchange issues. Needless to say, all this plunged the world into a period of temporary global economic recession. Just as President George Bush Senior has to deal with the Persian Gulf War (sometimes called Operation Desert Storm) in 1991, now his son President George Bush Junior has to deal with the war declared against global terrorists (War against Terrorism, dubbed Operation Enduring Freedom).

Perhaps the real Armageddon is the (nuclear) World War III causing mankind to vanish from the face of the earth. If you believe in (Darwin) evolution theory and the Big Bang theory as the start of our universe, life may evolve again a very long time after this! Perhaps not, if you believe in the creation story as told by Genesis in the Bible.

Could weapons of mass disorientation end our world? Information-bombs might just fit this criterion. This is terrorism in the information age, where the I-bombs are electromagnetic pulse (EMP) generator or high-powered microwave (HPM) device designed to fry all nearby electronics, including computers and storage media. Instead of targeting physical infrastructure and human lives, these weapons aim to knock out computing and communications infrastructure that controls so much of modern society. In advanced countries like US and Australia, concerted attacks on digital nervous system could cause massive economic loss and social upheaval, and endanger lives entrusted to the smooth operation of technology.

Thank God inhumane practices, such as mass scale ethnic cleansing of Albanians by Serbs in the war-torn province of Kosovo in Yugloslavia, in early months of 1999, is never meant to be repeated again. NATO airforce and, eventually, ground troops intervention was required to restore peace. Towards the end of March 1999, NATO commander reported that for the first time in history, war was waged on the Internet, as Yugloslav computer hackers bombarded NATO's communication network with hundreds of unwanted e-mails, thus almost crippling communication system.

(2) Doomsday virus. The Ebola virus, of the Genus Filovirus, causes the fatal Hemorrhagic Fever disease in humans. It occurs as outbreaks in the African countries of Zaire and Sudan. The vector is unknown but it can be transmitted from man to man. Germ warfare agents like Clostridium botulinum toxin, anthrax, and viruses such as the Ebola virus are therefore potential doomsday machines.

Botulism occurs in three forms: foodborne, wound, and infant botulism. It is caused by the sporulating, anaerobic gram-positive bacillus Clostridium botulinum; which elaborates 7 types of antigenically distinct toxins. The toxins interfere with release of acetylcholine at peripheral nerve endings. Major complications include respiratory failure due to diaphragmatic paralysis & pulmonary infections. This is the greatest threat to life.

Anthrax (malignant pustule; woolsorter's disease) is a highly infectious disease of animals, especially ruminants, transmitted to man by contact with the animals or their products. The causative organism, Bacillus anthracis, is a large, gram-positive, facultatively anaerobic, encapsulated rod. Inhaling spores under adverse conditions (e.g., the presence of an acute respiratory infection) may result in pulmonary anthrax (woolsorter's disease), which is often fatal.

Shoko Asahara, leader of the lunatic religious cult, Aum Shinrikyo, ordered the nerve gas Sarin attack on the Tokyo subway on March 20, 1995 which killed 12 people and injured more than 5000. The Doomsday cult had stockpiled vast quantities of germ warfare agents, including anthrax, and had even sent a team to Zaire to collect the deadly Ebola virus. In late November 1997, the United Nations (UN) international weapon inspection team reported that the fanatical leader of Iraq, Saddam Hussein, has synthesized sufficient quantities of the deadly chemical weapon, VX agent, to kill every man, woman and child on the face of our planet.

Nuclear weapons, chemical weapons and biological weapons mentioned above are designed to cause massive casualties and damages. They are usually considered as weapons of mass destruction. In the hands of fanatical governments or terrorists, they are especially frightening and dangerous.

(3) Killer Asteroid. About sixty-five million years ago an asteroid from space struck the earth. The asteroid is thought to have a diameter of 10 km and weigh one quadrillion metric tons; and the velocity at the time of impact is reckoned to have been several hundreds of thousands of miles per hour. The crater resulting from such a collision would be some 100 km or more in diameter, probably hidden today on the ocean floor. An impact explosion of this kind, perhaps exploding with the force of about 300 million hydrogen bombs, would have ejected an enormous volume of terrestrial and asteroid material into the atmosphere, producing a cloud of dusts and solid particles that would have encircled the earth for many months, possibly years. The loss of sunlight would have eliminated photosynthesis, and resulted in death of plants and the subsequent extinction of herbivores and their predators and scavengers. This catastrophic event, causing the extinction of the dinosaurs, is thought to take place around the end of Cretaceous period (144 to 66.4 million years ago). If a similar event were to happen today, it could plummet the earth into similar conditions, thus causing human extinction.

(4) Threat of Global warming. The greenhouse effect would eventually cause significant global warming affecting global climatic conditions, leading to potential catastrophe like great flooding; the mechanism of which is postulated to be possibly due to the melting of large expanse of icefield in the Antarctic polar region. Carbon dioxide in the atmosphere, acting like a blanket trapping heat, is essential for life. This effect of carbon dioxide is known as the greenhouse effect, which is the gas influence on earth's temperature. The conversion of the world's fossil carbon storage into atmospheric gases is popularly known as the greenhouse gas emission. It is

producing a global ecosystem of a kind unknown in Earth's history. Although a degree of greenhouse effect is desirable, steadily rising concentration of carbon dioxide in the troposphere is causing grave concerns for scientists. Scientific evidence indicates this will lead to significant climatic changes with global warming, changes in rainfall patterns and increased climate variability being part of the package.

Even though many of the ideas on Chaos and Fractals have not been explored at this point, it is worthwhile diverging for a moment to talk on the Lorenz attractor, named after Edward Lorenz, an American meteorological scientist at MIT, who in 1961 was studying weather by means of computer simulations (Lorenz, E. N. (1963). "Deterministic Nonperiodic Flow", Journal of Atmospheric Sciences, 357, 130-141). He had used computer graphing to study convection – the way heat moves through air. Although convection equations continued to demonstrate sensitive dependence on initial states, his graph displayed an astounding predictability – an "attractor".

Any single moment in time is represented by one point on the attractor. But differential equations cannot merge, so Lorenz realized he had discovered a mathematical object which was in fact an infinitely complex set of surfaces, never intersecting. The Lorenz's graph is three-dimensional. The line on the graph described what might be seen as a twisted figure eight, and it traced it over and over again. The pattern never repeated itself precisely. The line never even crossed itself. The line in the Lorenz attractor as we usually find it pictured seems to cross itself only because we are seeing it in two dimensions – where it seems to cross, it is passing in front of itself. The Lorenz attractor (a strange attractor) in the zx-plane (a Poincare section) is often described as "owl face", "mask" or "butterfly". Analogue computer can model differential equations, and the resulting phase portraits monitored on an oscilloscope screen or voltmeter. In fact, Lorenz's original work is based on this method. Today, digital computer can easily model the Lorenz attractor. We will have more to say on the Lorenz attractor and its (three) differential equations down below.

The trajectories of the weather variables follow the lines within the attractor, cycling around the lobes, with apparently random behavior. This is a strange attractor having a non-integer dimension of about 2.06. Mark one point to represent today. Mark another point close to the first to represent another day of almost identical weather conditions. As the weather conditions progress from these two points, they may

(a) move together in one wing,

(b) move together in the other wing, or

(c) move apart, one into each wing.

By examining the attractor closely, it is possible to see that there are some conditions in which all the nearby points will follow similar paths and other areas where they will diverge quickly. Imagining one wing to represent one weather pattern and the other to represent a very different pattern, this can lead to a reasonable degree of certainty in the future for some points and almost total unpredictability for others.

The climate can be imagined as the attractor – only certain states are acceptable. Ice in the tropics, sweltering sands in Antarctica are not represented by points on the attractor. So these weather pattern will not occur, except if there is some catastrophic event. For environmentalists, the sixty-four thousand dollar question is: Is the attractor slowly being changed? If it is, what is the long term effect on our weather, lifestyles and environment? The long-term implications of this have only recently begun to be understood with specialists in the earth's history of the past two million years (the Quaternary period) having records of the planet's history.

The three differential equations for the Lorenz attractor are given below:-

$dx/dt = ay - ax$

$dy/dt = rx - y - xz$

$dz/dt = xy + bz$

x = a value proportional to the speed of motion of the fluid due to convection in the system being modelled.

y = a measure of the temperature difference between the warmer, rising fluid in the system and the cooler, falling fluid – that is, it is a measure of the horizontal temperature difference in the system.

z = a measure of the temperature difference as we move vertically through the system – that is, it is a measure of the vertical temperature difference in the system.

The constants are a, b and r.

a = a value proportional to the Prandtl number. It is based upon the physical parameters of the fluid involved such as its density and how efficiently it conducts heat.

b = a measure of the size of the area being represented by the equations.

r = the Rayleigh number for the system. It indicates the point at which convection starts for a particular system. Below this critical value we have steady convection and above it unstable convection starts up.

The set of three differential equations for Earth's magnetic field put forward by Chillingworth and Holmes was very similar to those for the Lorenz attractor. Without going into any length at all on the meaning or finer details, here are the equations:-

$dx/dt = ay - ax$

$dy/dt = zx - y$

$dz/dt = b - xy - zr.$

By plotting a zy plot of this strange attractor, it was not surprising that the Earth's magnetic field has an attractor very similar to the Lorenz attractor. This is the sort of behavior that the Earth's magnetic field exhibits over thousands of years, with the north and south magnetic poles flipping over every now and again, which is analogous to the trajectory flipping between the lobes of the attractor.

There are many other attractors with useful practical implications, such as Henon attractor (using iteration), Rossler equations, forced Duffing oscillator, Van der Pohl oscillator, strange attractors from real data (e.g. real pendulum, dripping water tap or faucet, stock market prices and other economic indicators, cardiac rhythms and arrhythmia), models for arms races and disease epidemics (SIR model) and so on. Being able to model disease epidemics by infections, such as tuberculosis (TB) and HIV (AIDS), are extremely important not just in developing third world countries but also in the modern societies of developed nations where immunocompromised AIDS patients can be infected with opportunistic infections such as Pneumocystis carinii pneumonia and virulent forms of multiple drug resistance TB.

Microclimate refers to small areas of an isolated nature which may exhibit specific climate conditions that will not be revealed by a nearby mesoclimate. Given vineyards on a particular hillside are examples. A microclimate also may exist in a particular field or pasture that may or may not enjoy the shelter of a windbreak or shading. Mesoclimate refers to a climate that is intermediate between a macroclimate and a microclimate. Mesoclimate may refer to a small valley, a forest clearing, a frost hollow, and open spaces in towns and villages. Macroclimate refers to the averaged climate of a large geographic area.

Climate change is part of the natural system and ecosystems have adjusted to natural changes in the past. However, the vast number of humans on our fragile planet has added a dimension never experienced before and this is producing change in the atmosphere, hydrosphere and oceans, which is irreversible in human time scales. The El Nino effect causing prolonged drought conditions in country areas of Australia, resulting in hardships to farmers, is a global phenomenon. It is thought to be mainly due to the greenhouse effect. But this is only part of the equation. Almost 71% of the earth's atmosphere is in contact with oceanic surfaces. The ocean-air interface, therefore, plays a dominant role in determining the water content and temperature of the lower levels of the atmosphere and perhaps of the total atmosphere.

The Southern Oscillation is defined as a massive seesawing of atmospheric pressure between the southeastern and the western tropical

Pacific. It involves a periodic weakening or disappearance of the trade winds, which triggers a complex chain of atmosphere-ocean interactions – here is a feedback between the temperature gradient of the equatorial Pacific, its eastern end normally being colder than its western end, and the east-to-west winds normally blowing along the equator. A warming in the east would weaken the gradient and thus the winds. This would lead to further warming, that is, positive feedback, a part of El Nino.

What is El Nino? For well over a century, Ecuadorian and Peruvian fishermen have referred to the annual appearance of warm water in the Pacific off their shores at Christmas time as El Nino (Spanish for the Christ child). It is a normal yearly event, but in some years is of much greater intensity, covers a larger area of water, and is prolonged. For instance, in the major 1982-83 El Nino event, the annual Peruvian anchovy fisheries catch dropped to less than one-half million ton (normally the catch will exceed 12 million tons). Also, unusually severe Pacific storms struck the California, Oregon, and Washington coasts; extreme droughts hits many parts of the world including the western Pacific and Mexico, and torrential rains and flooding drenched parts of South America and the southern United States.

The prolonged and intensified El Nino events is thought to come in seven-year cycles. Current consensus is that in recent times this is of a more frequent basis due to global warming. In 1983, John Geisler (University of Utah) and his colleagues used the high-speed CRAY-1 supercomputer to model El Nino (needing more than 100 hours of computer time), showing the complex and chaotic dynamics of El Nino, which is also a primary signature of weather prediction in general. Long-term weather prediction, on scales of one month to 2 or 3 years tends to be almost impossible, due to the inherent nature of chaotic dynamics "sensitivity to initial conditions".

The El Nino is, in simple terms, when the water surface temperatures off South America are higher than average and the water temperatures off the eastern Australian coast are lower. Because the hot air rises above South America in these conditions, the rains head away from Australia. But weather prediction is a complicated business, and rainfall is also influenced by many other factors, such as season changes, geography, rainforest cover - the microclimate effect, and so on.

In fact during each El Nino, some parts of the drought affected areas may even receive higher than normal rainfall. In other parts of the world, severe storms such as cyclone, hurricane (or known as tornado in the US) occur. One thing is certain, each individual El Nino can be of different intensity and duration, and affect different areas with resulting storms and flooding or drought - that is, causing weather extremes. In other words, it increase the climate variability, and this is of a regional, as well as, a global nature. Traditional agricultural farmers, sheep and cattle farmers must all learn to live with El Nino and better manage their lands. This entails the weatherman helping them to statistically predict when the next El Nino is forthcoming using the Southern Oscillation Index.

The El Nino is just beginning to be better understood by man over recent times, and therefore humans have a limited time to adjust to it. However, ecosystems have millions of years to do so, and Australian animals like the Red kangaroo has learn to adapt to the dry El Nino periods by "slowing down" their reproduction. The young are born in a very early stage; and in the pouch is where the young complete their development. The adult female has fewer young ones in her pouch, and the growth of tiny babies can be "delayed" until the wet condition arrives.

Climatologists base their long-term forecasts on satellite measurements of sea surface temperatures and a measurement of air pressure. The difference in atmospheric pressure between Darwin and Tahiti is known as the Southern Oscillation Index, being either positive or negative. A negative index is a strong indicator for El Nino; and this would also be associated with a higher temperature in the eastern Pacific. The index had records comparing air temperature in Darwin and Tahiti, representing the eastern and western ends of the temperature gradient of the equatorial Pacific respectively, going back 120 years. The variables vary in a reciprocal fashion in the two places, and has some predictive value for the El Nino. But despite that statistical record, long-term forecasters are hampered by insufficient records in tracking long-term trends. This can be explained by one of the hallmark scientific concept of Chaos, that is, "sensitive dependence on initial conditions" – system behavior that depends so sensitively on the system's precise initial conditions that it is, in effect, unpredictable and cannot be distinguished from a random process, eventhough it is deterministic in a mathematical sense. This is exemplified by the so-called famous "butterfly effect" – as someone with a bent for whimsy had put it, a butterfly flapping its wings in a South American jungle, it is said, can lead to a hurricane in China – popularized by people like James Gleick in his book "Chaos: Making a New Science", London: Penguin Books (1988). This book is regarded by some to be the "Holy Grail" book on the historical aspects of Chaos.

72

Ozone (O_3) has a concentration of about 10 ppm in the stratosphere. It acts as a protective shield selectively absorbing most of the harmful ultraviolet radiation from 2000 to 3000 Å from the sun. (The angstrom (Å) is equal to 10-10 m or 0.1 nm). Certain chlorofluoromethanes, such as CFC, are mainly used as refrigerants and propellants in aerosol spray cans. These compounds are highly volatile and essentially inert to chemical reaction. They slowly diffuse up to the stratosphere where they are decomposed by ultraviolet radiation between 1750 and 2200 Å to produce free chlorine atoms, which can then undergo a chain reaction with ozone, thus depleting the ozone layer.

The greenhouse effect and other phenomena like the progressive depletion of earth's ozone layer over the Antarctic, as shown by satellite studies, are areas needing the cooperation of world leaders, for example, in Earth Summit meetings, to reduce global climatic changes. Ironically, reducing greenhouse gases emission among the nations of the world to a lower target level proposed in the South Pacific Forum in September 1997, and in the Japanese city Kyoto in December 1997 would mean the loss of jobs to a country like Australia rich in natural fossil fuel coal resources, as the Australian Bureau of Agricultural and Resource Economics estimate that it could cost about 1.5% of Gross Domestic Product (GDP). (GDP is the total value of goods and services produced by an economy or group of economies, excluding income from foreign investment).

In actual fact, there are six greenhouse gases under debate in the international climate talks in Kyoto. Only three are included in the negotiations – carbon dioxide, methane and nitrous oxide. The effect of leaving three other "artificial trace gases" from the deal, such as the HFCs used in air-conditioners, is to make it more difficult for some countries, including the US and Australia, to achieve their targets. But Europe expects its own emissions of the other three artificial gases to rise strongly in coming years and so wants them out of the deal.

In September 1997, I was visiting the tiny oil-rich nation of Brunei in Southeast Asia. When the sun has appeared at all, it has been reduced to a blood-orange orb, barely visible through a dense pall of smoke. An environmental and ecological disaster is in the making. A number of deaths has been recorded in people suffering from respiratory ailments like asthma. Most people were wearing face masks to protect themselves from the dangerously polluted air (pollution index level way above 300 ppm) that they breathed, estimated to be equivalent to smoking forty cigarettes per day. The Orang-utan primate apes of Borneo and Sumatra were in grave danger of being extinct as they are very sensitive to smoke.

On September 26, 1997 an Indonesian Garuda Airbus A-300 crashes in the thick smog-hit island of Sumatra while approaching an airport to land, killing all 234 people on board. Although it was later revealed that simple communication misunderstanding between the air-traffic controller and the pilot regarding the aircraft turning left versus right is to blame, the thick smog may well have contributed to Garuda Airline's worst aeroplane crash disaster. Many flights in Malaysia and Indonesia have been suspended due to poor visibility.

Satellite pictures showed that smog caused by the out-of-control burning of perhaps over 200,000 hectares of scrubland, grassland and forest in the Indonesian archipelago, particularly in Sumatra and Kalimantan, is to blame. The smog haze is covering much of Southeast Asia including Indonesia, Malaysia, Brunei, Singapore, the southern Philippines and southern Thailand, resulting in people praying for the monsoon rain that is expected to arrive at this time of the year, to put out the fire; but due to the El Nino effect, this is not forth-coming for a few months. Logging and plantation companies are mostly to blame – once the primary rainforest timber trees are logged off, the remaining secondary forest growths, which is highly combustible are deliberately burned off to provide land for the lucratic palm tree and other types of commercial plantations. Small "slash-and-burn" farmers contributed only to a small proportion of the overall problem. Environmentalists are gasping at the political, legal and diplomatic failures which allowed it to happen. During this same period, the El Nino effect is also to blame for the drought conditions causing starvation and outbreaks of cholera and dysentery, especially in the highlands of Papua New Guinea.

In February 1999, my wife and I went back visiting parts of the eastern Malaysian states of Sarawak and Sabah during the festive Chinese New Year celebrations – "Gong Xi Fa Cai" (Happy & Prosperous New Year) – good food, family reunion, open house visits, the giving & receiving of ang pows (money in red packets), lion dances, pedestrian parades, and noisy ear-shattering fireworks galore! We toured the famous Niah Caves of Niah National Park (the Great Cave is one of the largest in the world) and the limestone cave chambers of Gunung Mulu National Park and finally climbed Mount Kinabalu (the highest mountain in Southeast Asia).

For your information (FYI), the Niah Caves of Sarawak are also well known as a source for that famous Chinese dish, bird's nest soup. Countless tiny swifts build their nests in crevices in the walls and ceilings of the caves, constructing them out of hardened sticky saliva. Collecting the nests is a dangerous occupation - making bird's nest soup a very expensive dish indeed. Mulu's Sarawak Chamber is the largest natural chamber in the world, and Deer Cave is the largest cave passage known to man. It has two huge entrances at either end of the mountain it penetrates. Perhaps the most popular attraction of this cavern, however, is the daily exodus of its colony of nearly a million bats. Every evening the bats stream from the cave to spend the night dining on Gunung's rich insect population. Fantastic visual images of these flying bats triggered nostalgic memories of having previously watched a massive flock of certain bird species flying together (or a massive shoal of certain fish species swimming together) in harmony and the entire flock of birds (or shoal of fishes) simultaneously changing direction in a smooth, graceful and coordinated manner (synchronization) without the birds (or fishes) colliding with one another. How is this "safety-in-numbers" complex emergent collective behavior or phenomenon possible? My quest to find out explanations to phenomena such as just mentioned has also provided me with aspirations to write this book.

Cybernetics is a term coined by American mathematician Norbert Wiener to refer to the general analysis of control systems and communication systems in living organisms and machines. In cybernetics, analogies are drawn between the functioning of the brain and nervous system and the computer and other electronic systems. The science overlaps the fields of neurophysiology, information theory, computing machinery, and automation. Principia Cybernetica Project (see "Welcome to the Principia Cybernetica Web" http://pespmc1.vub.ac.be/) is about philosophy. What is philosophy? Philosophy intends to answer the eternal questions: Who am I? Where do I come from? Where am I going to? What is knowledge? What is truth? What are good and evil? What is the meaning of life? Understandably, there is a huge literature on philosophy covered by this project, including direct and indirect answers to the question posted above on complex emergent collective behavior.

The complex emergent collective behavior of birds and fishes mentioned above is partially explained by the complex adaptive systems of birds or fishes exhibiting self-organizing behavior with positive/negative feedback and cooperation for the evolutionary purpose of improved survival of the species as a whole – the massive gathering of birds or fishes confer some protection through "safety in numbers" (which confuses potential predators nearby). The growth of complexity during evolution seems obvious to most observers. This issue can be objectively illustrated by analyzing the concept of complexity as a combination of variety and dependency. It is argued that variation and selection automatically produce differentiation (variety) and integration (dependency), for living as well as non-living systems. Structural complexification is produced by spatial differentiation and the selection of fit linkages between components. Functional complexification follows from the need to increase the variety of actions in order to cope with more diverse environmental perturbations, and the need to integrate actions into higher-order complexes in order to minimize the difficulty of decision-making. Both processes produce a hierarchy of nested supersystems or metasystems, and tend to be self-reinforcing. Though simplicity is a selective factor, it does not tend to arrest or reverse overall complexification. Increase in the absolute components of fitness, which is associated with complexification, defines a preferred direction for evolution, although the process remains wholly unpredictable.

Models have been proposed to explain how an evolutionary process interact with a decentralized, distributed system in order to produce globally coordinated behavior (the "evolving cellular automata" framework). Using a genetic algorithm (GA) to evolve cellular automata (CAs), it has been shown that the evolution of spontaneous synchronization, one type of emergent coordination, takes advantage of the underlying medium's potential to form embedded particles and their interactions. The particles, typically phase defects between synchronous regions, are designed by the evolutionary process to resolve frustrations in the global phase. Solutions (synchronization algorithm in terms of embedded particles and their interactions) to these models have implications both for understanding emergent collective behavior in natural systems and for the automatic programming of decentralized spatially extended multiprocessor systems. Therefore, the "evolving cellular automata" framework is an idealized means for studying how evolution (natural or computational) can create systems that perform emergent computation, in which the actions of simple components with local information and communication give rise to coordinated global information processing. Importantly, the results demonstrate, via a generally close quantitative agreement between the CAs and the embedded particle models (using certain computational mechanics framework in which a CA's information processing is described in terms of regular domains, embedded particles, and their interactions) that this new model class captures the significant functional features in the CAs' space-time behavior that underlie the CAs' computational capability and evolutionary fitness.

At the heart of most physiological adaptations is a system of regulation and feedback that helps maintain homeostasis. This "self-maintaining" aspect of physiology is considered to be a unifying principle, the same way that evolution serves as a unifying principle in biology. These two principles may each be treated as a separate question but they also meet (and are benefitted from being treated) on the common ground of evolution. Devising plausible biochemical models describing the evolution of, for instance, biochemical physiology is an active ongoing area of future research. It is bound to be a complicated affair involving providing answers to questions such as "How are the network of biochemical pathways in the cell coordinated as an integrated system?" and "How does this coordination come to exist?"

While climbing the gruesome 8.5 km trek to Mt Kinabalu's 4101-metre summit in Sabah, Malaysia; my wife and I would sometimes encounter the sturdy figures of Dusun and Kadazan men and women, working as porters carrying heavy loads of goods up and down the mountain. These native people belong to the two largest tribal groups in Sabah. They are extremely fit and well acclimatized to the effects of high altitude.

The Mt Kinabalu International Climbathon is dubbed "The World's toughest mountain race." At the Timpohon Gate (entry to the mountain trek), there is a signboard that refers to the annual winner results:

11th Mt Kinabalu International Climbathon (3 - 4 October 1998)

Women Category

1. Danny anak Kuilin Gongot, Malaysia, 02:03:47

Men Category

1. Ian Holmes, Great Britain, 02:42:07

Therefore, the 1998 winner (from amongst an assembled international field of elite competing athletes) for the women section was a local native girl. Wondering about the shorter time taken for the women? This is due to the fact that the men have to complete 21 km (Race route: Timpohon Gate - Summit - Administration Building), whereas the women's total distance for the race was only 15 km (Race route: Timpohon Gate - Laban Rata - Administration Building). After absorbing the breath-taking view at the top, we descended down the mountain to dip our tired aching bodies in the nearby soothing Poring Hot Spings, 43 km away.

Stock exchange, or stockmarket, is a market place in which members of the market buy and sell stocks and shares, and investments in companies or in governments. Stocks and shares fall into two basic categories: shares, or equities, which give the buyer part-ownership of the company in which he or she has invested; and stocks, or bonds, which lend money to a government or large company without giving a right of ownership. Both stocks and shares are called securities.

Some stock exchange terminology: A bear is one who has sold a security in the hope of buying it back at a lower price. A bear market is one in which bear investors would make profits – when prices on the market are falling. A bull is an investor who has bought a security in the hope of selling it at a higher price. Similarly, a bull market is one where prices are rising and in which bulls would prosper. A stag is an investor who applies for new shares when they are issued, hoping that when the shares are first traded on the open market, the price will rise. The stag would then sell the shares quickly to make a profit.

Compounding all the above problem is the recent sharp devaluation of many ASEAN (Association of South-East Asian Nations) countries' currencies, including the Malaysian ringgit, the Thai baht, the Indonesian rupiah, and the Filipino peso; needing the intervention of the International Monetary Fund and the World Bank with multi-billion dollar bail-out packages in Thailand, Indonesia, and the Philippines. This is proclaimed by Malaysian Prime Minister Dr Mahathir Mohamad to be due to the "immoral" currency trading by US financier and currency trader George Soros; and also due to a Jewish conspiracy to destabilize the Southeast Asian economy!

Of the three Asian countries needing these bail-out packages, Indonesia was the worst affected and slowest to recover. The long-standing corrupt practices of the former dethroned (in May 1998) Suharto government regime was to leave long lasting economic hardship for the Indonesian people. There was political instability with wide-spread riots, lootings and killings across the country. The minority ethnic Chinese population, holding the majority of the country's wealth, was especially targeted by the majority Malay group.

Although billionaire George Soros is blamed by Dr Mahathir for the currency crises, and associated with it, poor Asian foreign exchange and stockmarket performance; a deeper explanation can be provided by the relentless phenomenal unchecked annual economic growth rate of the Southeast Asian "tiger" economies, their main problem being the economies running at unsustainable debt levels. To stop the rot requires, for instance, many ASEAN nations to slash spending on major infrastructure projects, reduce current trade account deficit, lift depleted national reserves, bring inflation down, increase deregulation, improve tax collection, and other further stringencies in national budgets. Increasing exports and decreasing imports will reduce current trade deficit of a country to a lower percentage of its GDP.

Trading between countries are becoming part of an increasingly competitive and sophisticated global economy. By the end of October 1997, reverberation of the economic setback in Southeast Asia was felt in stockmarkets of countries from all continents. Major countries affected include the United States, Japan, Hong Kong, the United Kingdom, Germany, Singapore, Korea, Taiwan, and Australia. Northeast Asian country, Hong Kong, was the first country outside Southeast Asia to be hit hard - the Hang Seng index took a battering. Hong Kong, being recently handed over to China in 1997, may in turn affect the economy of the China, the most populous nation on earth with an incredible economic growth rate. Most economists felt that all these represent only sharp corrections, or knee-jerk reactions; and not a prolonged global economic slump.

The Dow Jones averages published include the most commonly quoted Dow Jones Industrial Average (DJIA). It also include one based on 20 transportation stocks, one based on 15 utility stocks, a composite based on all 65 stocks, and several bond averages. Depending on future economic trends, these averages are subjected to possible future alterations. They are among the most commonly used indicators of general trends in the prices of stocks and bonds in the US. Another example is the Nasdaq composite index, which is based heavily on technology stocks.

The much publicized DJIA index, based on the prices of 30 industrial stocks (stocks from the IBM computer company is also included), has its averages based on not just straight arithmetic means of the number (30) of listed stocks, but on averages where the divisor has been adjusted to compensate for stock splits, stock substitutions, and significant dividend changes. Despite this, analysts felt that the DJIA is not a good indicator of general market price trends as compared to, say, the Standard & Poor's 500 (S&P 500) Index, which is more broadly representative of the market stocks, and has more exposure to the important computer industry sector. Then you also have the S&P 100. Confused? If not, S&P is also the name of the US credit agency providing credit ratings (for example) for Asia's electric power utilities in the international arena.

On Tuesday 16 March 1999, the DJIA rose to over 10,000 points (a high of 10,001.78 points) for the first time in its 114-year history, 20 minutes into the trading session. This was due to a high demand for technology stock. There was a loud cheer as this psychological barrier was crossed. However, it lasted for only about 30 seconds on that day before closing at 9930.47. It was not to be until Monday 29 March 1999 before the Dow reach the all-time closing high of 10,006.78 points. Being just a mathematically derived index, there was no magic in reaching this number; or for that matter in reaching 20,000 or 100,000 in the distant future. (In a similar manner in Australia, the All Ordinaries index briefly crossed the "magic" 3,000 points barrier on 23 March 1999. The index surged to a record high of 3005.2 around midday before late selling saw it close at 2987.6 for the day. On 6 April 1999, it closes over the 3000 mark for the first time on 3032.9)

On the subject of economic collapse, let us now look at the Wall Street collapse of 1987. In October of that year, Wall Street crashed - the Dow Jones index fell 508 points in a single trading session, wiping billions of dollars off the stockmarket. It was called "Black Monday". Mathematicians investigating this new domain of Chaos theory applied their iterative processes to the economic data. They used methods such as the fake observable technique, and found that economic data had a limited degree of predictability. This is despite the fact that stockmarket is deterministic – there are rules which govern the behavior of stock prices.

If a group of people all have the same information and act in the same way, problems can arise. With modern computers, the players in the stockmarket had the same information. Many decided at the same moment that they wanted to buy the same stock – it was going to go way up, and make them all very wealthy. Except the people who owned that stock had also access to the same information and did not want to sell. The stockmarket became unstable.

In the stockmarket crash, worldwide stock prices plummeted. There is a computer loop arrangement called portfolio insurance. Once share prices in their share portfolios slip below a certain level, the computer automatically sells the shares. As share prices tumbled through these thresholds, the selling started to feed on itself and everybody wants to sell creating panic.

The computers around the world are linked and so a minor bit of bad news can become magnified very quickly. It seems that on one day all the investors went the same way, randomness was lost and the market crashed – "a sensitivity to initial conditions". Programmed computer trading emerged in the 1980s. It was designed to protect the large institutional investors from being caught out by any sharp corrections. The rippling market crashes around the globe were the first manifestation of a significant movement of hot money, high-velocity money that can be shifted from one market to another, always seeking a higher return.

In America today, investors have trillions of dollars invested in the stockmarket. A repeat of the 1987 crash may happen again if another war, such as the Gulf war were to occur, resulting in destabilizing oil prices; and also if inflation with rapid interest rise were to occur, resulting in more money being moved from shares into bonds. To safeguard against this, the New York Stock Exchange has introduced "cooling off" periods ("circuit-breaker" rule); for instance, if the Dow Jones were to fall more than 350 points in a day, trading is suspended for 1/2 hour; and if 550 points fall, 60 minutes trading halt. Some analysts believe that the circuit-breaker has ruled out the chance of a one-day fall as big as the 1987 crash. Future "crash" would take the form of a series of falls over days, weeks or even months. Other Dow Jones safety valves includes a trading system expanded to be able to handle five times daily trading volume.

Now, the Dow Jones Industrial Average fell an unprecedented 554.26 points to 7161.15, slicing $US600 billion off the value of stocks in its heaviest trading day on Monday 27 October 1997. There was panic with many investors including major institutions withdrawing from many US mutual funds, which control shares worth more than $US4000 billion, or almost half the $US9100 billion capitalization of the New York exchange – this Wall Street's sell-off was triggered by Monday's 5.8% plunge in Hong Kong's Hang Seng index, which set off a chain reaction in Asian and European markets, culminating in New York. This exceeded the 508-point drop just over a decade ago in the Black Monday crash of October 1987 – *deja vu*! However, this was far smaller in percentage terms, with Dow dropping only 7.2% compared with 22.6% in 1987, a fair way to go before a major crash.

As the New York stock exchange is the "epicenter" of world stockmarket trend, the eventual quick rallies from Wall Street has prevented a global economy melt-down, and led to most stockmarkets recovering. Towards the end of November 1997, financial crisis in Korea has necessitated financial bail-out of many billions of dollars by the International Monetary Fund. Giant financial institutions and companies in Japan have faced either severe financial crisis or suffered bankrupcies. However, Korea and Japan, whose problems are home-grown, were primed for trouble anyway.

If you are confused about the Asian economic crises, here is a simplier version of the story in a nutshell. It started in Thailand with an overvalued currency pegged to a rising US dollar, a property bubble and a forced devaluation. The ASEAN currencies mentioned above have to go the same way because they trade in similar products as Thailand. It "spread" to

86

northeast Asia, specifically South Korea and Japan, because the markets, which had wildly boomed Asia, were adding an anti-Asia sentiment to their buying and selling decisions.

The whole crisis moved through three distinct phases. First it was a currency crisis, which led to devaluations that will actually make the countries more competitive. Then it became a stockmarket crises. With stocks and currencies so devalued, the assets that back loans have massively declined in value, so now it is a banking crisis. Banking crisis is a much nastier beast than a currency or stockmarket crises, but we have experience on how to deal with it in the 1980s. To restore a banking system that has been devastated, it is necessary to let the bad banks go bust (while protecting their depositors), to increase transparency greatly, to increase capital reserve requirements, to heavily tighten lending criteria and to allow foreign participation in the domestic banking sector, thus importing best international practice into the banks. From the Latin American banking crisis in the early 1990s, we know that rapid economic growth can be reattained within two years.

The Southeast Asian economies have enormous residual strengths that the Latin Americans did not have – namely, high domestic savings rates, sound budgetary positions, lower current account deficits and comparative advantage, and solid track records in exporting products with high international demand. With any luck at all, the Southeast Asian tigers should be (and have been) roaring again within a few years. Even though South Korea, the 11th largest economy in the world at that time, has even greater residual strengths than the Southeast Asians, poor political decisions and upcoming presidential election has unfortunately temporarily delayed the country grasping the depth of action needed to solve its version of the banking crisis.

Japan is an entirely different kettle of fish altogether. Its assets, international and domestic are vast. It is the world's second largest economy at that time and commands 30% of global savings. In some ways the collapse of banks and finance companies towards the end of 1997 is encouraging. Japan has been dithering over its bad loans for six years prior to this, waiting for growth to come back and solve the problem painlessly. Long term macroeconomic infrastructure changes are required. No doubt, the economic recovery in Japan will be painful with many people losing their jobs.

For those of you interested in (some of) the how and why aspects of economic collapses, there are two interesting articles by the same authors Tad Hogg, Bernardo A. Huberman and Michael Youssefmir from the Dynamics of Computation Group, Xerox Palo Alto Research Center, Palo Alto, CA 94304:

(1) "The Instability of Markets" - available in postcript (125K) at [ftp://parcftp.xerox.com/pub/dynamics/markets.ps].
Abstract:
"Recent developments in the global liberalization of equity and currency markets, coupled to advances in trading technologies, are making markets increasingly interdependent. This increased fluidity raises questions about the stability of the international financial system. In this paper, we show that as couplings between stable markets grow, the likelihood of instabilities is increased, leading to a loss of general equilibrium as the system becomes

increasingly large and diverse."

(2) "Bubbles and Market Crashes" - available in postscript (587K) at [ftp://parcftp.xerox.com/pub/dynamics/bubbles.ps].

Abstract:

"We present a dynamical theory of asset price bubbles that exhibits the appearance of bubbles and their subsequent crashes. We show that when speculative trends dominate over fundamental beliefs, bubbles form, leading to the growth of asset prices away from their fundamental value. This growth makes the system increasingly susceptible to any exogenous shock, thus eventually precipitating a crash. We also present computer experiments which in their aggregate behavior confirm the predictions of the theory."

Two Americans, Robert Merton, of Harvard University, and Myron Scholes, of Stanford University were awarded the $US1 million 1997 Nobel prize (for economics) for their complex Black-Scholes (BS) model of call option pricing. What they did back in 1973 was to put a price on risk, namely, how to price financial options, carried out with the late Fischer Black. It turned risk management from a guessing game into a science; and its subsequent evolutions led to explosive growth in stock options and other financial derivatives. It also enabled shareholders to calculate exactly how much chief executives were getting paid, as compensation packages include more and more share options. Another by-product was that it opened the era of the Wall Street rocket scientist, a strategist schooled in physics or mathematics who makes money crunching numbers rather than playing hunches. In other words, by breaking down assets into constituent parts, it is possible to get rid of precisely those risks you do not want to keep and take on precisely those that you do. Emboldened by the risk management ideas contained within the BS equation, financial institutions use derivatives to offer cheaper and more efficient products such as mortgages and savings accounts. In addition, the BS equation instructs traders as to how they should juggle the proportions of options, cash and assets in their portfolios, so as to "hedge" their risks, immunize themselves against movements in the markets, and earn profits from the charges they put on trades. This sort of arbitrage is precisely what Merton and Scholes are doing through Long-Term Capital Management, a hedge fund with impressive results of high returns, with low volatility – every investor's dream!

Derivatives play an important role in protecting investors against the uncertainty or volatility of the markets. Although derivatives often come dressed up in fancy names, such as swaptions and quantos, they really boiled down to two basic sort of financial instrument: forward contracts (or futures) and options. Any asset, from a government bond to a bar of gold, is essentially a mixture of forward contracts and options. A forward contract commits the user to buying or selling an asset – say a treasury bill, or dollars – at a specific price on a specific date in the future. It is rather easy to price. The main difficulty is working out the cost of carrying the asset until it changes hand.

An option gives the buyer the right, but not the obligation, to sell or buy a particular asset at a particular price on or before a specified date. Pricing one is a trickier affair, as it involves putting a number on the probability that a buyer will exercise his option. This guesswork was taken out by the BS equation, published in May 1973. It has a "closed-form solution", meaning that sellers of options could bung in a number of variables and the formula would churn out a price. The big advantage of the formula is that it does not require option sellers to take a view on which way the price of the underlying asset will move; but this is not entirely foolproof, as some of the variables, such as the risk-free interest rate and the violatility of the underlying asset, may change over time. For the record, the following is a more technical run down on the BS model of call option pricing.

In this famous article (Black, F. and Scholes, M. (1973). The Pricing of Options and Corporate Liabilities. Journal of Political Economy, May-June 1973, pp. 673 - 654), Black and Scholes presented a model that determines the price of a call is a function of five variables: the current price of the underlying share, the exercise price of the call, the call's term to expiry, the volatility of the share price (as measured by the variance of the distribution of returns on the share), and the risk-free interest rate. First, they noted that the effect of an option – the movement of goods and funds that it engenders – can be replicated, or mimicked, by buying and selling various amounts of those goods and funds each day. That is, an option can be "built" out of other more familiar products, the values of which are closer to hand. Second, they assumed that arbitrageurs in the market are extraordinarily effective. This supposition of "perfect" arbitrage implies that the value of an option has to match that of its replicating portfolio. For arbitrageurs would quickly exploit any imbalance, buying and selling for profit, and would cause the forces of supply and demand to bring the prices back into balance.

The Black-Scholes-Merton model, sometimes just called the Black-Scholes model, is a mathematical model of financial derivative markets from which the Black-Scholes formula can be derived. This formula estimates the prices of call and put options. Originally, it priced European options and was the first widely adopted mathematical formula for pricing options. Some credit this model for the significant increase in options trading, and name it a significant influence in modern financial pricing. Prior to the invention of this formula and model, options traders didn't all use a consistent mathematical way to value options, and empirical analysis has shown that price estimates produced by this formula are close to observed prices.

In their initial formulation of the model, Fischer Black and Myron Scholes (the economists who originally formulated the model) came up with a partial differential equation known as the Black-Scholes equation, and later Robert Merton published a mathematical understanding of their model, using stochastic calculus that helped to formulate what became known as the Black-Scholes-Merton formula. Both Myron Scholes and Robert Merton split the 1997 Nobel Prize in Economists, listing Fischer Black as a contributor, though he was ineligible for the prize as he had passed away before it was awarded.

Roughly, their model determines the price of an option by calculating the return an investor gets less the amount that investor has to pay, using log-normal distribution probabilities to account for volatility in the underlying asset. The log-normal distribution of returns used in the model is based on theories of Brownian motion, with asset prices exhibiting similar behavior to the organic movement in Brownian motion.

The formula helped to legitimize options trading, making it seem less like gambling and more like science. Today, the Black-Scholes-Merton formula is widely used, though in individually modified ways, by traders and investors, as it is the fundamental strategy of hedging to best control, or "eliminate", risks associated with volatility in the assets that underlie the option.

Reiterating, the Black-Scholes-Merton formula is an estimate of the prices of European call and put options, with the core difference between American and European options being that European options can only be exercised on their one exercise date versus American call options that can be exercised any time up to that expiration date. It is also used only to determine prices of non-dividend paying assets.

Let us suppose that the Black-Scholes theory acts like a set of Newton's laws for the derivatives markets. In 1997, two Russian physicists, Kirill Ilinski and Gleb Kalinin from the University of St Petersburg, applied quantum theory (specifically through ideas from the simplest gauge theory, quantum electrodynamics (QED)) to global finance and came up with the more powerful version of BS theory. Impressively, they derived the entire BS theory from the heart of their gauge theory. Just as quantum theory "corrects" Newton's laws of classical physics, so Ilinski and Kalinin's calculations correct the BS theory. All of these is well explained in a feature article by Nicholas Dunbar entitled "Market forces" (New Scientist, vol 158 issue 2128, 04/04/1998, page 42 - 45). The following information are acknowledged to be either modified from, or directly copied from, this article by Dunbar.

Gauge theory is a set of mathematical tricks that lets physicists describe fundamental particles and the fields through which they interact. Modern gauge theory has its roots in the 1920s, when physicists noticed that the quantum description of a particle – its wave function – has a strange mathematical redundancy. It always contains an unimportant number called a "phase".

Fields come into the picture because quantum particles have wave-like properties, and tend to smear themselves out in space. So a particle doesn't have just one phase, but many – one for every point in space. Changing all these phases in exactly the same way, known as making a gauge transformation, has no effect on the basic equations of quantum theory. This seems like a trivial matter until you realize that physicists believe that such "gauge invariance" is a fundamental property of the laws of nature.

But if different points in space receive different changes, then the equations do change. Roughly speaking, gauge theory says that the fundamental fields of the Universe exist in order to maintain the status quo of gauge invariance, and to prevent changes in these unimportant phases from having any effect. If the fields also change, and in just the right way, they can compensate for any phase changes that are different.

The simplest gauge theory, QED, deals with the way electrons and positrons interact via the electromagnetic field, and it is one of the most accurate theories ever devised. In QED, the close connection between the fields and the phases has some peculiar consequences. Like most things quantum, the phases of particles fluctuate randomly. To stop these quantum fluctuations causing brief violations of gauge invariance, the field has to fluctuate in a corresponding way. It does this by stirring up "virtual" photons which transmit forces between charges, causing them to move and restore the balance. On large scales, the virtual photons become unobservable, and classical electromagnetism emerges.

Ilinski and Kalinin has translated this arcane conceptual framework into the financial world. Instead of electric charge, think about cash sitting in trading accounts ("electric cash"). If the account is in surplus, it's a positive charge, while if it's in debit, it's a negative charge. Cash can be invested in US dollars, yen, Euro dollars, whatever, because at any given time, the exchange rates are fixed, so the choice is unimportant.

This, Ilinski and Kalinin claim, is the financial equivalent of gauge invariance, with the choice of currency corresponding to a choice in phase for an electron's wave function. If every currency suddenly doubled in value, it shouldn't affect the value of our account because the exchange rates stay the same. However, if just one rate (say Euro dollar to yen) suddenly changes randomly, it creates an opportunity for what the financial world knows as "arbitrage". In other words, someone, somewhere sees the imbalance and can profit by making a clever exchange of currencies.

The imbalance stirs up an "arbitrage field" that acts to restore the balance, just like the electromagnetic field in QED. This field serves as a force between trading accounts, causing cash to flow between them as traders eliminate the arbitrage opportunity by taking advantage of it. The following three paragraphs illustrate further what we meant by arbitrage.

The exchange rates between US dollars, Euro dollars and yen fluctuate constantly, buffeted by supply and demand and other economic factors. You might think the US dollar – Euro dollar rate will go up tomorrow, and exchange your US dollars for Euro dollars now in anticipation. But you have no way of knowing this – all you can do is bet on the market, and your hunch might be wrong.

If, however, you have the capacity to trade US dollars, Euro dollars and yen all at the same time, there is a way of making a risk-free profit. It is called arbitrage. Suppose US\$1 = Euro\$2, US\$1 = 200 yen, and Euro\$1 = 100 yen, then by converting a US dollar into Euro dollars, then into yen, then back to US dollars, you wouldn't make any profit. But if the Euro dollar to yen rate suddenly changed so that Euro\$1 was the equivalent of 125 yen, then the same round trip transaction would yield a profit of 50 yen, or 25p. If this could be done instantaneously, it would amount to risk-free arbitrage.

In the language of Ilinski and Kalinin, the blip in rates would create an arbitrage field which would create trading that would restore the balance between currencies, by changing the US dollar – Euro dollar and Euro dollar – yen rates, for example.

Ilinski and Kalinin offer a more powerful version of the traditional BS theory of derivatives because they took into account the effects of arbitrage fluctuations. This is important because the BS assumption of "perfect" arbitrage isn't really true. First, arbitrageurs are fallible. And second, when an arbitrage opportunity occurs in the market, it persists for a short time before being washed out. Just as in QED electrons interact with vacuum

96

fluctuations of electromagnetic field, which causes them to skitter to and fro, so cash flows through markets in response to momentary imbalances that produce arbitrage opportunities. This is an effect BS model doesn't include.

Ilinski and Kalinin also discussed the "rationality" of traders – this is a number analogous to the charge of the electron – which measures how quickly and strongly traders respond to exploit arbitrage opportunities. Volatility is a measure of how much market prices fluctuate. It plays a role similar to Planck's constant, h, in QED. Because h is exceedingly small, quantum fluctuations normally only reveal themselves in the atomic world. If it were much larger, then our world would be a much stranger place. Similarly, as the volatility in a market becomes larger, the quantum-like fluctuations in it grow as well. On the other hand, and in line with QED, the traditional BS theory of derivatives emerges as a "classical limit" when volatility is small. Could traders make more money than before using their new model? Probably not, as efficient arbitrage should quickly render useless any model that gives one market player an advantage over the others.

As each new generation of derivatives hits the market, competition between banks forces traders to charge less for their services and drives the creation of new, more sophisticated products. This puts derivative traders in a tough spot. Unlike their clients, they try to avoid betting directly on the markets. Nor do they want to become victims of arbitrage by underpricing their derivatives. Therefore, the corrections offered by their new model (by estimating the effects of imperfect hedging) would probably provide some practical tools for traders working in real markets.

Emanuel Derman, head of quantitative strategies from Goldman Sachs, acknowledges a profound difference between the worlds of physics and finance. "In physics, you're playing against God," he says, "who doesn't change his mind very often. But in finance, you're playing against God's creatures, whose feelings are ephemeral, at best unstable, and the news on which they are based keeps streaming in." Unlike the natural world with its fixed and immutable laws, the financial universe seems to evolve mischievously to confound new ideas that help us understand it. In my opinion, the Grand Unified Theory of Finance is unlikely to exist; but if there ever is one, it may remain someone's closely guarded secret.

(5) Doomsday volcano. Our planet Earth is constructed with a crust, a mantle and an iron core. The motion of liquid metal in the Earth's (molten) core is what creates the magnetic field forming a 'shield' around our planet. This shield serves to deflect most of the solar wind whose charged particles would otherwise strip away the ozone layer that protects Earth from harmful ultraviolet radiation. Interestingly enough, Mars Pathfinger's radio signals indicate that Mars is constructed like the Earth's, except that scientists are not sure whether the iron core is solid or molten. Could global geological changes deep in the interior of our planet cause widespread major volcanic eruptions, especially near major cities causing heavy casualties? If so, the ashes from major volcanic eruptions may block sun ray from reaching the surface of Earth thus causing catastrophic global climatic changes.

7 Premature Babies

 My youngest daughter Jelena was born 13 weeks (3¼ months)

premature in City X, Australia on May 14, 2012 thus becoming a 27 weeks gestational age premature (Premi) baby classified as a 'very low birth weight infant' (VLBWI). My wife Jocelyn and I were grateful to all volunteers at Ronald McDonald House in City X providing invaluable help to families with their hospitalized babies and children, and to neonatal intensive care unit (NICU) health care staffs in Hospital A.

Discussions on premature babies will concentrate mainly on the medical subject of Human Physiology and Human Pathophysiology based on various scientific materials which incorporate advancement in human comprehension and knowledge of science. As a health professional, I often muse what life would be like being an engineer or scientist instead of a doctor. I would dream of retiring one day in my older years with the goal to be sitting in front of a computer in my spare time to browse through interesting websites, or watch relevant videos shared on YouTube, on scientific subjects of my choice. As the Internet is such a powerful modern communication tool to disseminate information, I would undoubtedly implore, with a burning passion, relevant people and organizations from all over the world to maintain and update the contents of useful websites for the educational benefit of present and future generations.

Jelena was an extreme premature baby, [to-be-exact] 27+2 weeks (viz. 27

weeks + 2 days) gestation 'Premi' baby, with birth weight of 1010 grams (2.2 pounds) – very low birth weight infants (VLBWI) – she was tiny enough to easily fit onto the palms of an adult hand. Her expected time of birth calculated using 40 weeks (9 months) term gestation was to be on Friday August 10, 2012 – this important date is utilized to calculate her 'corrected gestational age' and 'corrected age'. These two terms are important to use for Premi babies until the chronological age of 2 to 2½ years when they developmentally catch up with their term baby counterparts. For instance, Jelena will only transition from being a neonate to an infant at (by definition) 4 weeks (1 month) old corrected age [biological age] on September 7, 2012. This equates to the chronological age of 17 weeks (4 ¼ months).

Miraculously, Jelena survive without any consequential mental and physical impairment during subsequent follow-up baby checks, thanks largely to the dedicated professional medical team and allied health care providers involved in looking after her. At the time of her birth, Jelena already had four older siblings (three brothers and one sister) born in Australia: Jonah, Joelle, Jethro, and Jonty aged 12, 8, 7 and 4 years old.

My wife and I are devote Christians – the deep roots for this lie in my grandmother [in the remote past before she passed away], having been missionary in the Fujian province in China (with a history of early and prominent Wesleyan Methodist Christianity influence from the Americans). The large Foochow dialect speaking group from this province are especially renowned for being ambitious with seeking better life in many parts of developed countries like Australia, Canada, and USA. Being Foochow people, my grandparents originally migrated out of the poverty-stricken

100

Fujian province to seek better fortunes initially in Malaysia during the 1930's. They endured much hardship and gave birth to 9 children (which included my father) in Malaysia before the whole clan eventually migrating again to Australia during the 1980's, thus always seeking a better life. My 3 older sister & 2 younger brother siblings and I also joined this immigration – we studied hard in Australian universities to become successful engineer, dentist, optometrist & doctors by profession, and raise our own family. Therefore, with this religious background, it is not unexpected when I would often reflect on Jelena premature life beginning: It is truly an amazing grace that so many things that could have gone wrong for her did not do so because of God's divine interventions. Jelena could potentially be inflicted with some of the nasty conditions associated with prematurity affecting brain, eyes, lungs, and gastrointestinal systems; viz. intraventricular hemorrhage, retinopathy of prematurity, bronchopulmonary dysplasia, and necrotizing enterocolitis.

My wife and I sent all our children for education in Christian-based schools. We encourage our kids to grow up as God-fearing gracious people with good education and jobs, having happily married family life, spread the word of God at home and overseas through opportunistic organizations such as OMF International (formerly the China Inland Mission and Overseas Missionary Fellowship founded by James Hudson Taylor in 1865), offer generous gifts to charities & regular church tithes, and be child sponsors for the poor in other countries.

In late evening of a beautiful autumn day on Sunday May 13, 2012 in a regional Town Y in State Z, Australia is the location where my family was residing previously. My wife Jocelyn was 27+1 weeks pregnant with Jelena on that day. I was at work approximately 600 km away. Just before midnight, Jocelyn rang me on my mobile phone telling me that she was having heavy vagina bleeding. Jocelyn was terribly upset that she may have already lost Jelena by miscarriage. I rang the town ambulance [so that they can rush her to the local Hospital B] and the local Church's Pastor Richard and his lovely wife Kirsten [for their invaluable help to accompany/support and help look after the other kids at home]. Then I started to pray earnestly to uphold Jocelyn onto God's hand.

The early pregnancy stage ultrasounds, such as one performed on Friday February 24, 2012 for amniocentesis at 20 weeks gestation, had shown low lying placenta near the internal cervical os of the uterus. It was felt that there will be a good chance of placenta moving away from os as the pregnancy progress and not turns out to be placenta previa [which is implantation of placenta in the lower uterine segment over or near the os in late pregnancy stage – the placenta is thus positioned in front of the fetal presenting parts with likely occurrence of painless antepartum bleeding associated with high risk maternal/fetal morbidity and mortality]. The Hospital B's obstetrician rapidly assessed Jocelyn clinically and performed an obstetric ultrasound. Massive maternal bleeding from placenta previa was diagnosed and timely life-saving blood transfusion was commenced. Intramuscular steroid injections were given to Jocelyn in order to help mature Jelena's two underdeveloped tiny lungs. Knowing that Jelena will be born an extreme Premi baby from the urgently required emergency life-saving caesarean section, a professional informed decision was made for my pregnant wife [with Jelena transferred in-utero] to be expediently escorted by a medical retrieval team via plane to Hospital A whereby there is a Level III Neonatal

102

ICU (NICU) equipped to look after Jelena when she was born. The operation was performed under general anesthesia at Hospital A resulting in Jelena birth at 6:20am on Monday May 14, 2012 – what a dramatic and eventful day!

With an initial APGAR score 2 (at 1 minute) and 6 (at 5 minutes), Jelena was intubated and ventilated for 2 days with one dose of surfactant given to improve her lung compliance. She stayed for almost 7 weeks total in NICU needing ongoing caffeine treatment (until 34 weeks corrected gestational age), 35 days nasal CPAP therapy, and supplemental oxygen for total 45 days from birth for Respiratory distress syndrome and apnoea of prematurity. Other treatments included initial total parental nutrition (TPN) supplementation via peripherally inserted central catheter (PICC) line. We enrolled Jelena in the Premi-Remi Study for this PICC line insertion procedure in her left ankle saphenous vein on May 21, 2012 when she was 7 days old (corrected gestational age 28+2 weeks). This was intended to be a randomized double-blind controlled clinical trial to determine whether Remifentanil is effective for treating procedural pain in neonates. Remifentanil is an esterase metabolized [thus, not dependent on the immature liver enzymes for metabolism] opioid analgesic with rapid onset and an effective biological half-life of 3 to 10 minutes. Therefore, its theoretical advantage is that it may provide the superior analgesia of an opioid without causing prolonged respiratory depression.

Jelena packaged NICU care also encompass 24-hour breast and special formula milk feeding through oral- and naso-gastric tubes, Indomethacin drug to successfully close her patent ductus arteriosus, two separate blood transfusions, and brief phototherapy for jaundice. On Saturday June 30, 2012 (weight 1992 grams), she was transferred by NETS (Newborn & pediatric Emergency Transport Service) from Hospital A to the special care nursery at Hospital B. She stayed there for a further 4 weeks until Friday July 27, 2012 (weight 2585 grams) with I jokingly quoting "Jelena got discharged home just in time to watch the opening ceremony of London XXX 30th Olympia". Being a keen supporter of the sporting underdog, I was thrilled for Andy Murray, the long awaited British male tennis player to 'break the drought' by winning his maiden grand slam title at the 2012 US Tennis Open (in addition to winning the coveted Olympic gold for men single tennis at the Olympic Games prior to this). By that time, Jelena was completely weaned off nasal prong oxygen and can suck well enough to avoid all supplemental tube feeding. On Friday July 13, 2012, two weeks before discharge, Jelena weigh 2345 grams (with a feeding tube up her nose). When I gratefully updated the Premi-Remi Study researcher nurse at Hospital A on Friday October 26, 2012, Jelena was 24 weeks (6 months) chronological age but only 11 weeks (2½ months) corrected age, and was fantastically healthy weighing in at just over 5000 grams (5 kg).

The Ronald McDonald House (RMH), located in the compound of Hospital A, with its indoor games room and playground was absolutely magnificent in providing great fun-time unit accommodations for entire families with their sick kids in hospital. My family was provided with one of those units to stay while Jelena was dependent on critical care for precious physical / mental growth and nourishment. It was a bitter sweet time for me. The Director of Anesthetic Department at work would generously grant me prolonged periods of extra leave from work to stay at RMH to help out my wife. I would worry about all the possible complications that Jelena may develop arising from being such a Premi baby. However, these were also special times with abundant lasting nostalgic memories for my family as we were able to visit Jelena at NICU many times daily to provide her with extra 'kangaroo care' (in addition to that obtained from Jocelyn & my other children), and to walk my older three children Jonah, Joelle and Jethro to attend the hospital school. Kangaroo care for Jelena as a Premi baby is invaluable not just for physiological & psychological warmth and bonding, but also for holistic brain development. The daily to-and-fro medical helicopter flights at the hospital helipad were especially keenly watched by my youngest son Jonty. The few weekend trips for my family to visit City X and nearby area, and the surprise visits from Pastor Richard, his wife Kirsten, church members, & friends were all extremely welcome and pleasant events.

On top of everything that had happened during those 7 weeks of high turmoil, here is a final mis-adventurous note: My accident-prone eldest daughter Joelle broke her right elbow again on Wednesday June 27, 2012 while playing on the playground slide in RMH. This meant that she had to play the 'sick-role child' until her elbow was fixed at Hospital A – this being the second operation on the same elbow after having previously breaking it (and needing operation) 3 years prior to this episode.

In summary, the above materials could nicely constitute an insightful albeit sketchy portrait of my entire family's life stories at that point in time. These stories would roughly reflect a segment of my personal life story and my autobiography.

8 Modern Physiology

This chapter concentrates on "Advanced Human Physiology" which includes areas of discussion on Human Physiology (and Pathophysiology) not usually included in mainstream "traditional" basic Human Physiology from many current physiology textbooks and articles. Human Physiology is a vast subject and whether we are talking about basic or advanced Human Physiology in this book, we can only barely scratch the surface of both these areas. Even so, we will be able to illustrate many fundamental principles and provide you with a flavor of "Modern Physiology".

In this modern age of automatic devices, engineering systems such as automatic pilots, chemical-process control and speed governors are widely used. Economic theory prescribes ways of responding to inflation, unemployment and other crucial indicators by adjusting taxes and interest rates. All these (as well as the human body alluded to earlier) involves many feedback control systems. A mathematical model appropriate to many of the feedback control systems is a linear differential equation with constant coefficients connecting an input and an output. The Laplace transform provides a simple language for studying these equations and especially for highlighting important physical concepts, such as stability. The closely related Fourier transform brings out the frequency aspects of the process. In essence it allows us to predict behavior under general conditions from knowledge of how the system responds to a sinusoidal input of arbitrary frequency. This important feedback mechanism occurs in both biological living systems and non-biological non-living systems, and the scale of the systems varying from the very smallest to the very largest.

Of course we expect there will be "surprising" parallelism between biological negative feedback, general psychological negative feedback, and negative feedback in your love life. Homeostasis: a term used in systems thinking to describe the action of negative feedback processes in maintaining the system at a constant dynamic equilibrium state.

In the stability analysis of a homeostatic system, while it is generally true that by adding enough negative feedback one can stabilize any given system, and that by adding enough positive feedback one can destabilize any given system, the behavior between extremes is more complicated. It could make sense biologically to have an inducible expression if it was costly to produce, or if it interfered not only with the production of but also with physiologically desirable reactions. Feedback mechanisms can be combined. We have considered so far autonomous differential systems, without delays. Delays are however known to be involved in biological systems, because for example mRNA synthesis and transport (in eukaryotic cells) are certainly not instantaneous. Delays can be a major source of instability in negative feedback circuits (the intuitive situation being that of a correction to an offending variable which is applied too late, and has increased too much by the time it becomes effective, thus causing a swing in the opposite direction of the offending variable), and have been shown to effectively cause oscillations in biological systems. Intuitively, the corrections to the variations of a variable come "too late", and give rise to an ever-expanding series of "over-corrections", a phenomenon commonly known as hunting. Of course, real-world systems oscillations do no keep expanding, because these systems are not linear and there is a saturation in synthesis rates (such non-linearities are actually necessary for the occurrence of asymptotically-stable limit cycles). Such long negative-feedback circuits seem to be the basis for cicardian clocks and mitotic oscillations, and it has been shown that they could generate oscillations in MAP-kinase cascades; a model system for a biological clock, which has been implemented in the prokaroyte E. coli, oscillates only when the weight of the negative feedback circuit is strong enough

Stability analysis

These examples show that, while it is generally true that by adding enough negative feedback one can stabilize any given system (see below), and that by adding enough positive feedback one can destabilize any given system, the behavior between extremes is more complicated (and it should be because the stability of a variable can depend on a coefficient very far away in the interaction graph). This fact becomes even more obvious when one calculates the Routh-Hurwitz equations for a given system. These equations provide necessary and sufficient inequalities on the coefficients of the Jacobian matrix of a system for this system to be stable. These inequalities, which with current mathematical knowledge can only marginally be generically simplified, are quite complex even if the underlying system is highly structured in a simple fashion.

Relationships between Jacobian matrix structures and biological or chemical system behaviors have already been investigated. The problem of the stability of qualitative matrices (matrices grouped according to the signs of their coefficients) has also received extensive attention in the past, before

108

computers made their way into everyday scientific life. If a system comprises a positive autocatalysis loop, a positive circuit of length 2, or any circuit, positive or negative, of length 3 or more, then the system can be made unstable by altering weights of interactions while strictly conserving their signs. To make system stability a property robust against changes in values of its interaction weights with an interesting goal from evolutionary standpoint because it frees the system from certain particular kinetic values, it seems necessary to limit to a strict minimum the number of negative feedback circuits of length greater than 2. Positive feedback, if present, can always destabilize the system if its amplitude is sufficiently augmented; however, the existence of positive feedback is a necessary condition for certain useful behaviors of regulation systems, and thus often cannot be avoided. Furthermore, it is remarkable that if one modifies a stable system by introducing positive autocatalysis on variables previously not autocatalyzed, the resulting system can always be brought back to stability by altering interaction weights without changing their sign.

From an evolutionary standpoint, it is less costly to provide each system variable with local, negative feedback, rather than to ensure long feedback circuits are maintained in such a way as to stabilize variables with no local feedback. Local, negative feedback could be provided by specific & non-specific proteolysis [which would not only serve purposes of automatically removing partially degraded proteins or transducing certain signals].

To conclude this discussion of the respective relationships of positive and negative feedback to stability, we point out that autocatalysis has a special role in that it must be negative on average for the system to be stable (the trace of a matrix is the sum of the real parts of its eigenvalues), and that the theorem of the "dominant diagonal" (a well-known theorem from linear algebra has the intuitive interpretation that if for each variable the autocatalysis loop has a negative (respectively positive) weight sufficiently higher in absolute value than the sum of other interactions affecting the variable, then the system is stable (respectively unstable).

In cellular physiology, to avoid delays between triggering of a signal and mRNA transcription, and then mRNA translation, are very noticeable in cells, especially eukaryotic cells; it could therefore be that positive feedback has a role in stabilizing systems which involve long-range negative feedback loops, that length being bounded from below by structural, biological constraints. Systems in which a variable is under positive auto-catalysis, and under control of a long negative feedback circuit, could also be relevant to the modelling of the immunological system.

Negative feedback mechanism

Negative feedback mechanism is a universal ubiquitous mechanism occurring essentially in biological processes in all living organisms and non-biological processes in all non-living systems. Variability – in Health vs Disease X – is change in set point and variable which can be self-limiting (non-repetitive) or chronic (repetitive).

Analysis of Health & Disease X effects using Parameters from Same (Intra-organ) Organ A. Negative feedback mechanism is not involved.

Analysis of Health & Disease X effects using Parameters from two or

110

more Different (Inter-organ) Organ A, B, C, etc. with "one lagging behind the other" and/or "out of phase with each other", with same vs different polarity e.g. Primary vs Secondary hyperthyroidism or hypothyroidism. Negative feedback mechanism is involved.

.

.Doctors and intensivists working in Intensive Care Unit (ICU) have traditionally conducted many invasive procedures for close monitoring of the patient's clinical parameters to enable many drastic therapeutic interventions to be carried out to optimize patient survival outcome. I have personally noted that in dealing with extremely sick ICU patients that even in extreme dire-strait pathophysiological circumstances, the sick human body is still instinctively geared and adapted to survive for as long as possible through evolutionary defence mechanisms. Therefore, in the modern ICU setting, one of our goals is that we must be careful in so carrying out our therapeutic interventions (thus creating an "artificial" pathophysiological homeostasis which is likely to be suboptimal) that we do not tamper too much or must be selectively wise in tinkering with this optimal "natural" pathophysiological homeostasis created by the body own survival mechanisms (unless it is directed at correcting the underlying cause or problem). In the trauma life support situation, acronyms such as AMPLE history Allergy Medication Past (medical and surgical) history Last eat/drink, Event/Environment; ABCDE; Primary survey ABCDE Airway management and cervical spine control, Breathing (ventilation) Circulation and bleeding Disability Expose and protect from the environment; In a hypovolemic shock situation, the body vascular system can be artificially divided into three physiological, not anatomic, parts. These are Part I – the part that serves the heart, brain, and lungs which receives the highest priority, Part II the part that serves the abdominal contents and retroperitoneum which receives the next highest priority and Part III – the part that serves the extremities which receives the lowest priority. When the circulation of oxygenated blood is reduced for any reason, the system of priorities among these three physiological components of the vascular system takes effect. Blood is shunted from the lowest priority area to those areas that are more sensitive to the loss of oxygenated blood and are essential to maintaining life.

Liver (hepatic) failure can be acute or chronic. The nomenclature for the subgroups of acute liver failure (ALF) remains non-standardized but usually consist of fulminant hepatic failure (FHF) and late-onset hepatic failure (LOHF). The term FHF should be reserved for patients who develop hepatic encephalopathy within 8 weeks of first symptoms, whereas LOHF is used when encephalopathy appears between 8 and 26 weeks. Patients with LOHF have a lower incidence of cerebral edema, develop renal failure more frequently and have a worst prognosis without liver transplantation. The hepatic encephalopathy of ALF is classified into four grades of increasing severity ranging from Grade I (general apathy), Grade II (lethargy, drowsiness, variable orientation and asterixis), Grade III (stupor with hyperreflexia and extensor plantar reflexes) to Grade IV (coma with cerebral edema being present in over 80% of cases). The clinical course of patients developing severe ALF can be seen to rapidly progress through the four grades of encephalopathy with repeated clinical examination.

Hepatic encephalopathy in patients with chronic liver disease is termed portal-systemic encephalopathy and it is rarely associated with cerebral edema. Mild to moderate hypoxemia caused by of 20 - 30% intrapulmonary

112

shunting (the hepatopulmonary syndrome) occurs in some patients. Pulmonary hypertension is a less common but well-recognized accompaniment of chronic liver disease. The Child-Pugh's classification below is the most important prognostic factor for early variceal upper GI bleeding and survival. Patients in Child Class A have a mortality with each bleed of less than 10%, in Class C of at least 70%, and in Class B of intermediate mortality.

Indications

Evaluating prognosis in Cirrhosis

Criteria

Total Serum Bilirubin

Bilirubin <2 mg/dl (or <35 umol/l): 1 point

Bilirubin 2 to 3 mg/dl (or 35 to 50 umol/l): 2 points

Bilirubin >3 mg/dl (or >50umol/l): 3 points

Serum Albumin

Albumin >35 g/l: 1 point

Albumin 28 to 3.5 g/l: 2 point

Albumin <28 g/l: 3 point

INR

INR <1.70 (or PT<4 sec): 1 point

INR 1.71 to 2.20 (or PT 4 to 6 sec): 2 point

INR >2.20 (or PT>6 sec): 3 point

Ascites

No ascites: 1 point

Mild ascites controlled medically: 2 point

Severe ascites poorly controlled: 3 point

Encephalopathy

No Encephalopathy: 1 point

Grade I-II Encephalopathy controlled medically: 2 point

Grade III-IV Encephalopathy poorly controlled: 3 point

Interpretation

Child Class A: 5 to 6 points

Life expectancy: 15 to 20 years

Abdominal surgery peri-operative mortality: 10%

Child Class B: 7 to 9 points

Indicated for liver transplantation evaluation

Abdominal surgery peri-operative mortality: 30%

Child Class C: 10 to 15 points

Life expectancy: 1 to 3 years

Abdominal surgery peri-operative mortality: 82%

Let us look at a case scenario of a FHF. Patient X is a 53 year old woman who spent 10 days in Intensive Care Unit Y. She is comatose on ventilator and inotropic support with acute irreversible end-stage terminal (alcohol induced) liver failure and severe hepatic encephalopathy, hepatorenal and hepatopulmonary syndrome, sepsis from E. coli spontaneous bacterial peritonitis, persistent high fever above 39degree Celsius, and progressive cerebral edema as the terminal event. Increasing intracranial pressure (ICP) in cerebral edema has threefold phenomenon, referred to as Cushing's triad of (1) rising blood pressure, (2) change in respiratory pattern, and (3) decrese in pulse rate. Three different levels or stages have been identified with increasing intracranial pressure and increasing brainstem involvement.

Level One ICP Decorticate posturing (flexion of upper extremities and extension of torso and legs), Pupils med-size and reactive, Cheyne-Stokes breathing Level Two ICP Decerebrate posturing (extension of upper extremities, extension of torso and legs) Pupils mid-sized and fixed, central neurogenic hyperventilation; Level Three ICP Flaccid, does not react to pain, Pupils dilated and fixed, Ataxic medullary) breathing or Apnea.

CONCEPTS: Most, if not all, clinical parameters would show Chaos and fractals, variability in health and disease. Let us look at just two of them – variability of breathing and the sodium electrolytes osmostat level setting accelerated or ever faster cyclical Cheyne Stoking breathing (respiratory center) and cyclical electrolyte variability seemingly tending to occur, interestingly enough, "one lagging behind the other" and/or "out of phase with each other"! These concepts have different connotations to one other. In particular, the first "one lagging behind the other" concept is used to refer to Process A occurring before Process B but Process A terminating before Process B. This is exemplified by the clinical caveat that the appearance of radiological signs in a developing pneumonia often lag behind that of corresponding clinical signs but the disappearance of radiological signs in a resolving pneumonia often lag behind that of corresponding clinical signs. The second "out of phase with each other" concept is used to refer to the idea that rise and fall of Process A and Process B are at opposing polarity to each other. Then, we also have the "mixed" state where both the lagging and out of phase concept occur simultaneously. Finally, processes involving all these concepts occurring in the healthy and diseased state are also highly likely to be qualitatively different from each other.

Variability in both the healthy physiological and the diseased pathophysiological case is due to negative feedback mechanism, and can easily be understood with what I call the "Mini-tug-of-war" concept below:

Any dynamic parameters (e.g. breathing rate, blood sodium concentration) can only increase or decrease above the inherent or designated normal "set-point" value(s) for that parameter. One (or more) factors will increase the parameter under scrutiny, which I shall call the "up factors" and, similarly, one (or more) factors will decrease the same parameter, which I shall call the "down factors". With perpetual opposing motions from negative feedback mechanism, the overall combined effect (in terms of randomness [or "chaotiness"], intensity and duration) from the up factor(s) will roughly be equal to, and countered by, that from the down factor(s) resulting in the healthy physiological variability – this is analyzed on the shortest time scale interval possible. [On a larger time scale interval (which is a manifestation of fractional geometry), physiologists have long known about the healthy variation in parameters e.g. diurnal (circadian) rhythm, cortisol, biphasic temperature with a drop variation in a ovulating woman's monthly menstrual cycle. Needless to say, in the diseased state, this variation will also be affected, for instance, in Cushing's disease (hypercortisolism), this diurnal variation is frequently found to be opposite in nature – e.g. the reversed diurnal variation.]

However, in the diseased state, this same overall combined effect from the now pathological up factor(s) {or down factor(s)} will be relatively more powerful than that from the non-pathological down factor(s) {or up factor(s)} resulting in a disturbed less healthy variability. This less healthy

116

variability with a now higher (or lower) set point value(s) will also have a lesser amplitude, frequency and chaoticness to it, thus looking "less variable" due to the dominant effect from the pathological up factor(s) {or down factor(s)}.

For a given total body water, intravascular volume is determined by [Na+] (controlling distribution between ICF and ECF) and plasma oncotic pressure (controlling distribution between ISF & the intravascular compartments).

Conclusion: Under normal circumstances, the mechanisms controlling total body water by the thirst-ADH mechanism maintain a constant ECF [Na+]. This controls the distribution of the TBW between the ECF & the ICF. The net result is to maintain a constant cell volume. This is advantageous for maintaining optimum function of most cells but is particularly important for the brain. (It is interesting to note that the osmoreceptors themselves are brain cells.) The kidney is the major regulatory organ in the system.

The system that maintains the blood volume is a negative feedback homeostatic mechanism which consists of two components: an intrinsic system & an extrinsic system. The intrinsic system is the slow but very powerful system that is very effective at maintaining a constant intravascular fluid volume.

Sudden loss of blood can seriously impair tissue oxygen delivery and the intrinsic system is not able to deal with this situation. The extrinsic system consists of all the neurohumoral factors that are together extremely important in the rapid response to acute hypovolemia. This system includes the sympathetic response & several hormones (angiotensin II, aldosterone & ADH). Together these cause vasoconstriction, venoconstriction & renal retention of Na+ & water. This is an appropriate short term response.

However, over longer periods, the intrinsic system is dominant in maintenance of a constant intravascular volume. The system responds to an increase in blood volume by causing a pressure diuresis and a pressure natriuresis. Blood volume returns to normal.

The mechanism involves control of the amount of water and sodium in the circulation. ECF volume must necessarily be controlled if intravascular volume is controlled. Setting a certain blood volume automatically sets the volume of the ISF at a particular level. The actual distribution of fluid between these two compartments is determined by the balance of oncotic & hydrostatic pressures as defined in Starling's hypothesis.

It follows that this system is effectively determining the volume of the ECF. As sodium is essentially restricted to the ECF, then the sodium content of the body must also be determined by this system. Alternatively, as the excretion of Na+ by the kidneys is controlled by this system (eg the intrinsic and the neurohumoral factors determine the GFR & aldosterone levels), then this system may be viewed as regulating total body sodium. (In contrast to the thirst-ADH mechanism which controls total body water).

Summary: The osmoreceptor-thirst-ADH mechanism regulates total body water and maintains a constant ECF tonicity (i.e. constant [Na+]). The purpose of this is to maintain a constant ICF volume (i.e. a constant cell volume). This control is rapid and precise.

The blood volume control system maintains a constant blood volume and consequently a constant ECF volume. This control has a slow but powerful component and a more rapid neurohumoral component.

Managing Hyponatraemia (low sodium):

Step 1: Exclude 'factitious' hyponatraemia

This is usually easily excluded on history. It is suspected if other measured electrolytes seem diluted also. A repeat blood collection and electrolyte profile should be obtained to check the clinical suspicion. No treatment is required.

Step 2: Consider isotonic hyponatraemia

Isotonic hyponatraemia suggests pseudohyponatraemia due to hyperlipidaemia or hyperproteinaemia. If the lipid level is high enough to cause depression of the measured sodium level, then the serum will be obviously lipaemic and this is easily assessed by inspection. The [Na+] in the plasma water is normal. No treatment is required.

Step 3: Consider hypertonic hyponatraemia

This can occur with high levels of poorly permeable osmotically active substances present in the ECF. Such substances include glucose, mannitol, sorbitol, ethanol, methanol, sodium diatrizoate and ether. The diagnosis is usually obvious with hyperglycemia or mannitol infusion. Occasionally assessment of the osmolar gap may be useful.

Transfer of water from the intracellular to the extracellular fluid in response to the extracellular hypertonicity causes a form of dilutional hyponatremia. The plasma [Na+] decreases by roughly one millimole per liter for every 4 millimoles per liter rise in blood glucose above normal; this allows calculation of a 'corrected' [Na+] but this is not particularly useful clinically.

Cellular dehydration occurs in all hypertonic states. This may be clinically desired (e.g. mannitol 20% infusion in head injury patients) or pathological (e.g. diabetic ketoacidosis, hyperosmolar non-ketotic coma).

Step 4: Assessment of hypotonic hyponatraemia

This group accounts for the majority of clinical cases of hyponatraemia. The previous three steps exclude 'factitious' hyponatraemia, pseudohyponatraemia and hypertonic hyponatraemia. This is usually not difficult on clinical assessment & inspection of results.

Hypotonic hyponatraemia may be divided into 3 groups based on a clinical assessment of the ECF volume of the patient and these groups represent changes in total body sodium content. These groups are:

Hypotonic hyponatraemia with hypovolaemia
Hypotonic hyponatraemia with normovolaemia
Hypotonic hyponatraemia with increased ECF volume

Causes of hypernatremia can be distinguished. These are (i) Pure water

120

loss, (ii) Hypotonic fluid loss, and (iii) Salt gain.

The fluid loss can be considered as consisting of two components – an isotonic fraction and a pure water fraction. This is useful because the loss of pure water is loss of fluid from all compartments in proportion to the volume; however, the isotonic loss is loss from the ECF only. The ECF is decreased more than if all the loss had been of pure water. The rise in serum sodium is not as marked as with pure water loss but hypertension and shock are more likely to develop. Cerebral edema could occur during treatment phase. Pure water loss is associated with less decrease in blood volume and less hypotension than with loss of the same volume of a hypotonic fluid.

Pure water loss must be extreme before shock occurs. The reasons for this are:

- The fluid loss is distributed throughout all the compartments in proportion to their volume.
- Plasma is only 7.5% of total body water.
- Altered Starling forces (especially increased oncotic pressure & decreased hydrostatic pressures) tend to maintain fluid intravascularly.
- Excessive salt gain causes an increase in ECF volume at the expense of ICF. This may be marked enough to cause acute pulmonary edema.

9 Body Fluids and Electrolytes

Traditional medical teaching recommend a doctor seeing a patient to follow clinical steps in chronological order from: History (Hx) -> Examination (Exam) -> Investigation (Ix) -> Diagnosis (Dx) -> Treatment (Rx). In practice, this chain of event is generally followed with proviso that the "->" sign be replaced with the "<->" sign which is taken to meant that both forward and backward directions between two steps can (and usually do) occur. Certain step(s) may have to be repeated more than once for instance, test results are routinely used to assist with the diagnosis of a patient, assess the effect of Rx or the progress of a disease. Sometimes, a step(s) may be bypassed or be followed out of sequence for instance, Ix may not be justified in clear-cut cases, or Ix may necessarily and prudently be delayed following provisional Dx and Rx in a critically ill patient.

Let us analyze the theories and concepts behind the intimately related acid base system and the body fluid & electrolytes system, which is commonly integrated into the Body fluid, electrolytes and acid base system. These areas can be very confusing and to master them takes time for veterinary students and human medical students, and even for the cleverest doctors. Understanding the data in an arterial blood gas (ABG) panel requires an appreciation for not only acids and bases, but also ventilation, gas exchange, dynamics of electrolyte and water movement, plasma composition, respiratory control, and renal mechanisms of hydrogen ion, electrolyte, and water excretion. We must also develop an understanding of a host of other organ, metabolic, and structural dysfunctions that can potentially contribute acid or base loads to the extracellular fluid. A complete treatment of the topics in these areas is a massive undertaking and, therefore, I will only outline what is essential for the purpose of the book. We may eventually or one day apply them to create the grand finale of this subject – a computer program using "basic" (pseudolanguage) programming language on "Management flow chart for practical approach to Body fluid, electrolytes and acid base system" which we trust will be useful to help clinicians and, hopefully, lay people to better comprehend this subject.

Theories on Acid base system

Bare protons or hydrogen ions (H+) do not exist in solution. Protons are associated and react with surrounding water molecules to form H3O+ (the hydronium ion) but this is most accurately represented by {H:(H2O)n}+. However, we will continue to use H+ simply out of convenience. It is usual in discussions of acid-base balance to assume the activity coefficient of solutes is equal to one and use (the approximate) concentration instead of the more precise "effective concentration" called activities. This is obviously not correct but the errors introduced are usually small and not clinically relevant. The acid-base regulation in the body is extremely important and the hydrogen ion concentration within the body is one of the most tightly regulated of all physiological variables. The body always tries to maintain this concentration in the range of 36 to 44 nanomol/l. This concentration, usually described in terms of pH (where, loosely, $[H+] = 10^{-pH}$ & $pH = \log (1/[H+])$ = -log [H+]; strictly, hydrogen ion activity instead of hydrogen ion concentration should be used instead), correspond to the pH in the range of

7.36 to 7.44. The limits of survival in humans cover a ten-fold range of H+ from 160 to 16nmol/l, which corresponds to pH of 6.8 to 7.8. The pH is a contrived symbol, which represents a double non-linear transformation of [H+]. The pH is typically 7.4 in plasma ([H+] about 40nmol/l) but different lower intracellular values of pH. In theory, values of pH could range from negative infinity to positive infinity but the practical limits in aqueous solutions are from -1.2 to +15 reflecting [H+] varying from 15 to 10^{-15} moles/l. Concentrated hydrochloric acid used by chemists has a pH of -1.1. By convention, pH is just a number and does not have any units. However, there is a loose use of the term 'pH units' as a tool to assist explanation of certain concepts. In chemistry, treatment for the exact thermodynamic basis of pH measurement and temperature is quite a complex area. For example, it includes a term for the 'ground state potential' that must be arbitrarily defined at every temperature which means that the absolute value of measured potential at any particular temperature cannot be precisely determined and thus that pH values obtained at different temperatures, strictly speaking, cannot be compared. However, this is not an issue for the clinician to worry about.

Why is maintaining a normal pH so important? Mainly because of the following: -

(A) Even though hydrogen ions concentration is extremely low, an alteration in this concentration has major effects on the relative concentrations of every conjugate acid and base of all the weak electrolytes.

(B) The function of proteins (large molecule effect) is dependent on the net protein charge because it determines the 3-D shape of the molecule and its binding characteristics. The activity and function of all intracellular and extracellular proteins (including enzymes) is optimized because their charge (which is dependent on pH) is kept constant.

(C) The Davis hypothesis (1958) – intracellular trapping of metabolite intermediates (small molecule effect) is maximized at an intracellular pH of neutrality (pN). Neutrality is defined, for aqueous systems, as the state when $[H+] = [OH-]$. By the Law of Mass Action applied to the dissociation of water, then $pN = 0.5 \times pKw'$ (where pKw' is the ion product for water). All the known low molecular weight and water soluble biosynthetic intermediates possess groups that are essentially completely ionized at neutral pH. These groups are phosphate, ammonium and carboxylic acid groups. This hypothesis implies that the advantage to the cell was the efficient trapping of these ionized compounds within the cell and its organelles.

(D) The "imidazole alpha-stat hypothesis" (1972) – proposed by Reeves and Rahn is an extension of the conclusions from the Davis hypothesis by considering the dissociation constants (pK) for these metabolic intermediates. Maximal ionization with consequent intracellular trapping of metabolic intermediates occur at pN. In the body, there is strong evidence that ideal intracellular pH change with temperature such that this pH remains at or close to pN. This is achieved by appropriate temperature induced changes in the pK of the imidazole group of histidine. To achieve this requires that the degree of dissociation (known as alpha) of imidazole remain constant despite changes in temperature – it is this idea or theory about the constancy of the imidazole alpha that is referred to as the imidazole alpha-stat hypothesis. Alpha is 0.55 in the intracellular compartment and it remains constant despite changes in temperature i.e. pK is changing with change in temperature. Measurements have confirmed that the mean intracellular pH of man is 6.8 at 37°C, which is indeed the pN at that temperature. Other

125

measurements of intracellular pH have also shown that it change with temperature in order to remain equal to pN at each temperature – how is this achieved? It has been calculated that for the body to maintain this temperature-pH relationship requires certain things. There must be a buffer system with a pK which is approximately one-half that of water (because a buffer is most effective close to its pK) and which changes its pK so that it maintains this relationship as temperature changes. The buffer must be present in sufficient concentration and have certain chemical properties (eg. delta H° = 7 kcals/mol). This buffer system turns out to be from protein buffering, largely due to the imidazole group of histidine (the only protein-dissociable group that has the correct pK and whose pK changes with temperature in the appropriate way), and it is aided by phosphate and bicarbonate buffering. For this system to work optimally via maintaining the imidazole alpha constant also requires a constant CO_2 content at different body temperatures. This means that respiratory ventilation must be regulated to maintain the imidazole alpha in the blood. This regulation to maintain imidazole alpha constant in blood will then also result in imidazole alpha being maintained in other compartments such as intracellular fluid as well. It is speculated that the respiratory control that adjusts ventilation probably involves proteins whose activity is altered in an appropriate direction by an alpha-stat mechanism. Adjustment of ECF pCO_2 is necessary as this maintains a constant relative alkalinity of the ECF relative to the ICF so there is constancy of the gradient for H+ across the cell membrane. In reality this does not mean that ventilation has to increase markedly with decrease in temperature because the reduced metabolic rate in hypothermia state will automatically result in decreased CO_2 production. As we shall see later on in the Treatment (Rx) segment, the alpha-stat hypothesis approach which is not

126

to temperature correct blood gas results but use values reported (or only ask for values to be reported) as measured in the machine at 37°C and interprets this values against the reference range relevant to 37°C, has implications for clinical practice e.g. management of hypothermia during cardiopulmonary bypass.

(E) The "pH-stat hypothesis" – generally less well accepted alternative theory to the alpha-stat hypothesis. The pH-stat hypothesis approach argues that the pH should be kept constant despite changes in temperature. This approach is used by anyone who temperature corrects blood gas results to the patient's temperature but then interprets this values against the reference range relevant to 37°C. No reference range is available for temperatures other than 37°C. From the intellectual attraction of theoretical arguments, the best approach is probably to never temperature correct blood gas results and use the values as measured in the machine at 37°C and interprets these values against the reference range relevant to 37°C – this is the widely accepted alpha-stat approach. Major differences in outcome between groups of patients managed by the pH-stat or the alpha-stat technique have not been clear. Cells are capable of functioning despite the presence of minor perturbations. Clinical studies have concentrated on which is best for the heart (myocardial outcome) and which is best for the brain (neurological outcome). The lower pH in the pH-stat technique is maintained by having a higher pCO_2 level. This increases cerebral blood flow.

Hydrogen ion balance consist of hydrogen ion turnover (Internal Balance) and net H+ production/excretion (External Balance). The net production of hydrogen ions requires excretion from the body to maintain a stable body pH and the turnover of hydrogen ions refers to H+ being produced and consumed in chemical reactions without any net production. The turnover of hydrogen ions in the body is huge (1.5 moles/day from lactic acid, 80 moles/day from adenine dinucleotide, 120 moles/day from ATP, and at least 360 moles/day involved in mitochondrial membrane H+ movements) and very much larger than net acid production (12 moles of CO_2 and 0.1 moles of fixed acids). However, for acid-base equilibrium, the net acid production by the body must be excreted. This external acid-base balance also includes any acids or bases ingested or infused into the body. Acid base balance means that the net production of acid is excreted from the body each day.

The pH, bicarbonate, and the dissolved (dis) & combined (com) gas content of blood (in mL/dL of blood containing 15g of Hemoglobin) are shown below:

	Arterial Blood	Venous Blood
pH	7.40	7.36
[HCO3-] mmol/l	24.1	26.1
pO2 mm Hg	95	40
pCO2 mm Hg	40	46
sO2 %	97	75
O2 (dis, com)	0.29, 19.5	0.12, 15.1
CO2 (dis, com)	2.62, 46.4	2.98, 49.7
N2 (dis, com)	0.98, 0	0.98, 0

The pCO_2 is 7 - 8 mmHg higher and the pH is 0.03 - 0.04 unit lower in venous than arterial plasma because venous blood contains the CO_2 being carried from the tissues to the lungs. Therefore the calculated venous

128

[HCO3-] is about 2 mmol/l higher than that from arterial blood. We note that the hemoglobin of blood at the ends of the pulmonary capillaries is about 97.5% saturated with O_2 (pO_2 = 97mm Hg). But because of a slight admixture with venous blood that bypasses the pulmonary capillaries (physiological shunt), the hemoglobin in systemic arterial blood is only 97% saturated (total O_2 of about 19.8 mL/dL: 0.29mL in solution and 19.5mL bound to hemoglobin).

The Bicarbonate Buffer System

The major buffer system in the ECF is the bicarbonate buffer system. This is responsible for about 80% of extracellular buffering. It is the most important ECF buffer for metabolic acids but it cannot buffer respiratory acid-base disorders.

The components are easily measured and are related to each other by the Henderson-Hasselbalch equation.

Henderson-Hasselbalch Equation

pH = pK'a + log10 ([HCO3] / 0.03 x pCO_2)

The pK'a value is dependent on the temperature, [H+] and the ionic concentration of the solution. It has a value of 6.099 at a temperature of 37C and a plasma pH of 7.4. At a temperature of 30C and pH of 7.0, it has a value of 6.148. For practical purposes, a value of 6.1 is generally assumed and corrections for temperature, pH of plasma and ionic strength are not used except in precise experimental work.

The pK'a is derived from the Ka value of the following reaction:

CO_2 + H2O <=> H2CO3 <= H+ + HCO3- (where CO_2 refers to

129

dissolved CO_2). Note that the first reaction is slow, and exceedingly small amounts of H2CO3- are formed unless the enzyme carbonic anhydrase is present. This enzyme is especially abundant in the walls of the lung alveoli, where CO_2 is released; and it is also present in the epithelial cells of the renal tubules, where CO_2 reacts with H2O to form H2CO3. The second component of the system is predominantly shifted to the left where H2CO3 ionizes weakly to form small amounts of H+ and HCO3-. In the ECF, the HCO3- occurs predominantly as sodium bicarbonate which ionizes almost completely to form Na+ and HCO3- as follows: NaHCO3 --> Na+ + HCO3-.

The concentration of carbonic acid is very low compared to the other components so the above equation is usually simplified to:

CO_2 + H2O <=> H+ + HCO3-

By the Law of Mass Action:

Ka = [H+] . [HCO3-] / [CO_2] . [H20]

The concentration of H2O is so large (55.5M) compared to the other components, the small loss of water due to this reaction changes its concentration by only an extremely small amount. This means that [H2O] is effectively constant. This allows further simplification as the two constants (Ka and [H2O]) can be combined into a new constant K'a.

K'a = Ka x [H2O] = [H+] . [HCO3-] / [CO_2]

Substituting:

K'a = 800 nmol/l (value for plasma at 37C) and [CO_2] = 0.03 x pCO2 (by

130

Henry's Law) [where 0.03 is the solubility coefficient] into the equation yields the Henderson Equation:

$$[H+] = (800 \times 0.03) \times pCO_2 / [HCO3-] = 24 \times pCO_2 / [HCO3-] \text{ nmol/l}$$

Taking the logs (to base 10) of both sides yields the Henderson-Hasselbalch equation:

$$pH = \log10(800) - \log (0.03\ pCO_2 / [HCO3-])$$

$$pH = 6.1 + \log ([HCO3] / 0.03\ pCO_2)$$

On chemical grounds, a substance with a pKa of 6.1 should not be a good buffer at a pH of 7.4 if it were a simple buffer. The system is more complex as it is 'open at both ends' (meaning both [HCO3] and pCO_2 can be adjusted) and this greatly increases the buffering effectiveness of this system. The excretion of CO_2 via the lungs is particularly important because of the rapidity of the response. The adjustment of pCO_2 by change in alveolar ventilation has been referred to as physiological buffering.

Key Fact: The bicarbonate buffer system is an effective buffer system despite having a low pKa because the body also controls pCO_2

The human body response to an acid-base perturbation (or frequently referred to as the way the body 'defends' itself against acid-base disturbances) has three components. This response, utilizing negative feedback mechanisms, can be considered by looking at how they affect the ([HCO3]/pCO_2) ratio in the Henderson-Hasselbalch equation.

First defence – The Immediate Response: Buffering

The Major Body Buffer Systems

Site	Buffer System	Comment
ISF	Bicarbonate	For metabolic acids
	Phosphate	Not important - concentration too low
	Protein	Not important - concentration too low
Blood	Bicarbonate	Important for metabolic acids
	Hemoglobin	Important for carbon dioxide
	Plasma protein	Minor buffer
	Phosphate	Concentration too low
ICF	Proteins	Important buffer
	Phosphates	Important buffer
Urine	Phosphate	Responsible for most of 'Titratable Acidity'
	Ammonia	Important - formation of $NH4+$
Bone	Ca carbonate	In prolonged metabolic acidosis

The Bicarbonate Buffer System: Processes involved in Buffering

ECF

43% (by bicarbonate & protein buffers)

ICF

57% (by protein, phosphate and bicarbonate buffers) due to entry of $H+$ via:

Na+-H+ exchange 36%

K+-H+ exchange 15%

Other 6%

Second defence – The Rapid Response: Respiratory Response via alteration in ventilation (arterial pCO_2)

Third defence – The Slow Response: Renal Response via alteration in bicarbonate (HCO3-)

Buffering is a rapid physicochemical phenomenon. The body has a large buffer capacity. The buffering of fixed acids by bicarbonate changes the [HCO3] numerator in the ratio (in the Henderson-Hasselbalch equation).

The Respiratory Response: Alteration in Ventilation

Adjustment of the denominator pCO2 (in the Henderson-Hasselbalch equation) by alterations in ventilation is relatively rapid. An increased CO2 excretion due to hyperventilation will result in one of three acid-base outcomes:

- correction of a respiratory acidosis

- production of a respiratory alkalosis

- compensation for a metabolic acidosis.

Which of these three circumstances is present cannot be deduced merely from the observation of the presence of hyperventilation in a patient.

This respiratory response is particularly useful physiologically because of its effect on intracellular pH as well as extracellular pH. Carbon dioxide crosses cell membranes easily so changes in pCO_2 affect intracellular pH rapidly & in a predictable direction.

The Renal Response: Alteration in Bicarbonate Excretion

This much slower process involves adjustment of bicarbonate excretion by the kidney. This system is responsible for the excretion of the fixed acids and for compensatory changes in plasma [HCO3] in the presence of respiratory acid-base disorders.

Consider the management of a patient who is cooled during open heart surgery: Alphastat Management during Induced Hypothermia (an Example)

A patient cooled to 20degree C for cardiac surgery. Assume an arterial sample was drawn and analyzed at 20degree C and showed pH 7.65 and pCO_2 18 mmHg. Now if this same sample was analyzed at 37degree C then at that temperature, the values would be pH 7.4 and pCO_2 40 mmHg. So which value do you want reported to you:

The values for 37degree C can be interpreted against known reference values for 37degree C and they would be considered to be normal. This is the alphastat approach equivalent to assessing results against appropriate reference range for 20degree C but without having to know what it is.

The values for 20degree C could also be interpreted against the reference values for 37degree C. [Actually the blood gas machine would measure at

134

37degree C then apply the correction formulae and report what the values are at 20degree C]. This is the pH-stat approach (i.e. the pH must be kept at 7.4 at every temperature). So it would be decided that the patient had a significant respiratory alkalosis and measures would be taken to correct this.

Clearly the two approaches can result in quite different therapies being applied.

Summary of important aspects

The acid-base behavior of blood has been fairly well characterized. All descriptions of this behavior (the "traditional approach" to acid-base) have however been very approximate, until the advent of modern computers, which allow a more accurate assessment of the complex acid-base characteristics of human blood. The first person to spend time with a computer and describe such behavior (termed the "physicochemical approach" to acid-base) was Peter Stewart, in 1981. The Stewart approach to the acid-base balance in the blood is complex and numerically intractable without the use of a computer.

Pre-Copernicus Kepler's Laws for planetary motion may be derived from Newton's Law of Gravitation. Calculations of Earth and Lunar orbits for space travel by rockets to reach the moon required only classical Newtonian physics (Newton's Law of Gravitation and Newton's Three Laws of Motion) to accomplish this mission. If humanity aspires to go deep into or beyond our solar system, the limitations of Newtonian physics is quickly illuminated – Special and General theories of Relativity by Einstein is required. Some would even argue that due to limitations of the early interpretations of special and general relativity and the conflicts with quantum mechanics, an even more accurate or advanced theory viz. Quantum relativistic (Quantum gravity) theory, incorporating quantum theory to the two relativity theories, is required. As discussed in other parts of this book, in the limit from the (high) relativistic speed/mass to the (low) classical speed/mass scenario, all three theories are thought to be correct and consistent with each other in the following manners:

Quantum relativistic theory -> General theory of relativity -> Newton's Law of Gravitation -> Kepler's Laws for planetary motion

Pre-Copernicus -> Copernicus-> Special theory of relativity -> Newton's Three Laws of Motion

To use another astrophysics analogy, the precession (not completing a closed ellipse) of Mercury could not be explained by classical physics, but Einstein's General Theory of Relativity gave a precise explanation for it. This can inspire us to provide the rationale behind saline induced hyperchloremic acidosis and furthermore, to provide a platform for the use of a more physiological crystalloid during emergency resuscitation.

Stewart's methods will have to be tested in the clinical scenario over the next few decades, even though it is more mathematically sound than prior methods. In the same way, even Einstein's theory of relativity had to be tested by confirming that light bent during the solar eclipse.

"Post-Copernicus era": The two approaches are very similar in the way

136

that acid-base disorders are classified and measured. The major difference is in the explanation and interpretation of acid-base disorders and control mechanisms.

In a similar manner, two slightly different analogies for acid base system can be advocated for the self-consistency and "in the limit" The Van Slyke equation may hold in normals or near-normals or mildly ill patients but we should use it with caution in the critically ill and provide a slightly better model of how acid-base works than does the conventional approach. I believe that Stewart provides a refinement of the conventional approach. Under many, perhaps most circumstances, the 'old-fashioned' approach works fine, but we should be aware of the exceptions (gross volume dilution with fluids which have a low SID; hypoalbuminemia in association with metabolic acidosis) and invoke the physicochemical approach in these circumstances. The traditional approach to understanding acid-base is based on use of Henderson-Hasselbalch equation. This approach is only a partial solution to the problems of acid-base, and therefore breaks down in certain circumstances.

The Stewart's methods have previously been recognized that there was a close connection between acid-base balance and electrolyte balance. Where Stewart is revolutionary is that he mathematically proves that strong electrolytes act as the control (independent variable) in determining the pH. Hitherto, a mere correlation (not causation) between electrolytes and acid-base balance had been recognized.

Use both Stewart and conventional approaches with caution! In addition,

137

each clinician who makes therapeutic decisions should appreciate the limitations of the model they are using. He or she should also relate the model to the limitations in laboratory estimation of the numbers that go into the model. For example, in the hospital where I was work, the standard deviation of the estimates of serum sodium concentration is 3 mmol/l. I don't believe I can trust a serum sodium of "170 mmol/l" as I have seen a repeat estimate on the same specimen come out as "177 mmol/l"! We have also known since 1977 that small variations in sampling technique may have profound effects on arterial blood gas analysis – e.g. substantial reductions in PCO_2 related to heparinization of arterial blood gas samples. Be careful when you plug the numbers you obtain into any model, and then make dramatic alterations in clinical management based on small numbers, especially where there may be multiple sources of error!

This new approach also helps us explain how our therapeutic interventions work. Both the traditional and physicochemical approaches are influenced by the following.

Temperature STEWART: Temperature influences the constants in most of the equations that govern the complex equilibria in blood. Any calculation of a dependent variable will be influenced by a temperature that differs from 37degrees Celsius. In order to calculate the resultant effect on dependent variables, we need to know all the constants for that temperature.

Whole blood – STEWART: If we look at whole blood - a mix of red cells and plasma – then things become more complex. If we take a sample of blood with everything at equilibrium then we can look at the red cell and plasma as two separate compartments, which interact. In each compartment,

Stewart's principles must apply – if we know the independent variables, we can calculate the dependent ones!

It is thus possible to work out what is happening in the red cell compartment, just as we did for the plasma compartment. Things are more tricky in the red cell compartment, as there is an enormous amount of weak acid (in the form of hemoglobin), and the behavior of this weak acid is, to put it mildly, very complex. I'm not aware of anyone who has sat down and accurately applied Stewart's approach to the red cell.

How do the two compartments interact? As Stewart pointed out, there are only two ways that the compartments can interact:

1. By diffusion of CO_2 across the red cell membrane (this is rapid, so the PCO_2 in both compartments rapidly equilibrates);

2. By movement of strong ions (notably the chloride ion).

Let us now compare the two systems below. The traditional approach to acid-base concentrates on the Henderson-Hasselbalch (H-H) equation. This is simply a modification of one of the six equations we use in describing the relationship between the dependent variables we wish to calculate, and the independent variables that govern them. The H-H equation will of course always hold, but it cannot be used to explain the behavior of dependent variables, which will be influenced by all of the independent variables in the acid-base system. There appear to be at least two "subdivisions" of the traditional approach – those who concentrate on plasma bicarbonate concentration as a measure of metabolic acid-base disturbance, and those who look at base excess as a measure of this disturbance. This is starkly illustrated by the lack of consensus in the Acid-base terminology document published in the Lancet way back in 1965.

STEWART APPROACH: In order to determine the pH of human plasma at 37 degrees Celsius we need to have in our possession:

1.   The partial pressure of carbon dioxide, PCO_2.

2.   The concentration of weak acid in the plasma. Weak acids in plasma that interact with hydrogen ions – the main significant acid that does so under normal circumstances is the protein albumin, with a small contribution from phosphate.

3.   The strong ion difference (SID) of that plasma; simply the difference in activities of the various completely dissociated positive and negative ions in blood. The main contributors are sodium, potassium and chloride. (Note that lactate is also a strong ion).

4.   A system of equations that describes the relationship between these three independent variables.

5.   A means of solving these equations – need a computer and an appropriate program.

We say that pH is a dependent variable and there are many others, for example concentration of bicarbonate (often abbreviated to [HCO3-]), concentration of the carbonate ion, hydroxyl ion concentration, and so on. The pH depends on the first three items listed above, and we talk about each

140

of these three items as independent variables. We will usually have the first three in our sweaty little hands (at least, for most ICU patients); Stewart has given us the last two.

If required, modern blood gas machines will report the pH value for actual patient temperature but this 'corrected value' is calculated mathematically from the pH measured at 37°C in the machine. The change in pH with temperature is almost linear and 'anaerobic cooling' of a blood sample (i.e. cooling in a closed system) causes the pH to rise. The Rosenthal correction factor is recommended for clinical use.

Rosenthal Correction Factor

Change in pH/ change in T = 0.015 pH units/°C

Example

If the measured pH is 7.360 at a blood gas electrode temperature of 37°C, then the pH at a patient temperature of 34°C is [7.360 + (37-34)(0.015)] = 7.405.

10 CROP Effects

Many seemingly different or opposing dynamical processes (the temporalness or "time-like") and geometrical shapes (the spatialness or "space-like"), be it from actual systems occurring in nature or from theoretical mathematical models, bear some format or pattern of "resemblance" within itself (or to one another) by being "repetitive" or "opportunistic". [In the context here, the word "systems" is taken to imply systems and/or subsystems; and they may also consist of a mixture of processes and shapes (the spatio-temporalness or "spacetime-like").] CROP effect is a term coined to depict this "entanglement" property of repetition or opportunism. CROP is pseudonym for Complex Repetitive Opportunistic Phenomenon. Both the repetitive and opportunistic phenomena are complex or complicated (some may, however, choose to describe it as subtle or vague). The opportunistic phenomena is symbolized by oCROP and the repetitive phenomenon can be either of a finite (symbolized by fCROP effect) or infinite (symbolized by iCROP effect) nature. Examples of various terms or closely-related terms or combinations of terms there-off that are (or maybe) used in the scientific literature to describe this effect of entanglement include overlapping, blurring of the boundary or distinction, ubiquitous, generalization, analogy, similarities, subset of, repetitive, cyclical, recurring, parallel, opposite, converse, reciprocal, paradoxical, flip-flop effect, corollary, interesting, surprising, contrary to popular belief, remarkable, beautiful, close or deep or hidden or rich or subtle connection/relationship, intricate network/web of connections, intricate figure of extraordinary beauty with

142

infinite repetitive elements [referring to fractals], common fundamental principles, layered network of interconnections, converges, diverges, in the limit as variable x approaches zero/infinity, and so forth.

As I have just alluded to, the CROP effect is nothing more than a resemblance among the same (intra-CROP effect) or different (inter-CROP effect) processes, shapes, models or systems. Although it is an important concept, albeit of limited value (in concretely explaining many "things" in the proper scientific sense), it is not a scientific theory per se. Nonetheless, this loose resemblance via repetition and opportunism could be viewed either objectively, semi-objectively or subjectively to be of varying depths when used in different situations, viz. intuitively viewed as having varying degree of the presence (or absence) of the repetitive and/or opportunistic components; and this "varying depths" obviously depend to a large extent on the quantitative, semi-quantitative or qualitative nature of the "thing" being analyzed. In other words, the CROP effect, be it of a superficial or deep nature, may be seen as a crude yardstick measure of "Complexity" of a particular "thing" being analyzed; and as such, it belongs more to a Qualitative complexity measure. We see various examples of the CROP effect being mentioned in other parts of this book.

Invented number systems and base notation

Theoretically, bases for number systems can consist of all positive integral numbers greater than one (ie, 2, 3, 4, 5, 6,.....). For instance, systems are called binary (or dyadic), ternary, quaternary, quinary, senary, septenary, octenary (or octal), nonary, denary (or decimal), undenary, duodecimal, hexadecimal, vigesimal, and sexagesimal, corresponding to base values of 2,

143

3, 4, 5, 6, 7, 8, 9, 10, 11, 12, 16, 20, and 60, respectively. The importance of the binary system to information theory and computer technology derives mainly from its convenience in representing systems that are of two-state or bistable nature such as "on – off", "open – closed", "go – no go". The use of base-2 numbers has also greatly simplified the advances in digital computer capabilities. No smaller number system can be used to represent information, since there must be at least two symbols to distinguish meaning. Using only two symbols (1 and 0) eliminates the need to recognize the 10 symbols (0, 1, 2,....., 9) needed in base 10. However, the binary system's use of only two distinct symbols is partly offset by the requirement of many more digits to represent a number. Hence, apart from the bit (contraction of binary digit) commonly used in computing, the octal (a byte = 8 bits in computer jargon) and hexadecimal modes also find frequent application in electronic computing devices. As a historical example, the Babylonians developed (2000 - 3000 BC) a positional system with base 60 – a sexagesimal system – requiring 60 symbols (0, 1, 2,....., 59). The reasons for the choice of 60 are obscure, but one good mathematical reason might have been the existence of so many divisors (2, 3, 4, and 5, and some multiples) of the base, which have greatly facilitated the operation of division.

Some immediate properties of the base notation system are: -
(1) a rational number (expressible as a ratio of integers) remains a rational number when expressed in any integral base.
(2) any rational number can be represented by a terminating or repeating basimal (decimal in base 10). But how about if the same rational number is represented by a base other than 10?
The answer is as follows: Regular numbers are those whose prime factors

144

divide the base; the reciprocals of such numbers thus have only a finite number of places (by contrast, the reciprocals of non-regular numbers produce an infinitely repeating numeral). In base 10, for example, only numbers with factors of 2 and 5 (e.g. 8 or 50) are regular, and the reciprocals ($1/8 = 0.125$, $1/50 = 0.02$) have finite expression [an example of the fCrop effect]; but the reciprocals of other numbers (such as 3, 7 and 54) repeat infinitely ($1/3 = 0.\underline{3}$ and $1/54 = 0.0\underline{185}$, respectively, where the underline indicates the digits that continually repeat) [an example of the iCROP effect]. In base 60, only numbers with factors of 2, 3, and 5 are regular; for example, 6 and 54 are regular, so that their reciprocals ($1/6 = 10/60 = 0.\ 10$ and $1/54 = 1/60 + 6/60^2 + 40/60^3 = 0.\ 1\ 6\ 40$) are finite. Note that the reciprocal of 54 is represented by an infinitely repeating numeral in base 10 yet in base 60, the same number, 54, is represented by a finitely repeating numeral [an example of the oCROP effect]. In other words, we can choose to represent the reciprocal of a given number by either a finite or infinite numeral via selecting the appropriate number base systems.

Here, we take the opportunity to also explain how base 60 is able to facilitate the operation of division: The entries in the multiplication table for (example) $0.\ 1\ 6\ 40$ are thus simultaneously multiples of its reciprocal $1/54$. To divide a number by any regular number, then, one can consult the table of multiples for its reciprocal e.g. 3 divided by 54 is equal to 3 multiplied by $1/54$. Confused about how the fractions in base 60 are written? That is because we are not familiar with using the base 60 system. Not to worry – for our purpose, all we want to do is to decide whether there is any "overlap" between systems of different bases. We won't delve for great details into the logic behind the following, but looking at the three sets of even, odd and prime numbers, the three points below can be easily derived: -

(1) For two even-number bases, if the smaller base has all its prime factors that are part of the (larger group of) prime factors for the larger base, then all regular numbers contained in the smaller base are also part of the (larger group of) regular numbers for the larger base, an example is the base 10 and 60 as illustrated above. If there is only a partial overlap (some common prime factors) between the two, then there are common regular numbers between the two (plus the regular numbers that are unique to each base).

(2) For two odd-number bases, in addition to the two scenario for even-number bases, there is a third scenario with no common prime factor (and thus also no common regular number) between the two. The reason is simple: unlike the "universal" number, 2, where all even numbers are divisible by it with no remainders, there is no such number existing for all odd numbers. We will ignore the (irrelevant) case of the "universal" number, 1, where all numbers are divisible by it – the division resulting in answers being simply the original numbers being divided by 1.

(3) For an odd-number and an even-number base, there can either be some common prime factors, or no common prime factor between the two. This translates to having common regular numbers for the first case. In the second case with no common prime factors, there is no common regular number between them. This result is indifferent to whether the odd-number or even-number base is bigger.

It should be noted that there are many useful methods of obtaining or classifying numbers, each accompanied by separate nomenclature and theory, such as those displayed under the following number systems terminology: even numbers, odd numbers, counting (or integral) numbers, prime numbers, natural numbers, integers (whole numbers), surds, algebraic numbers, transcendental numbers, etc. For examples:

even numbers: 2, 4, 6, 8, 10,.....

odd numbers: 1, 3, 5, 7, 9,.....

prime numbers: 2, 3, 5, 7, 11,.....

integral numbers: 1, 2, 3, 4, 5,.....

natural numbers: 0, 1, 2, 3, 4,.....

integers: .....-2, -1, 0, 1, 2,.....

The four arithmetic operations are addition, subtraction, multiplication and division. The fundamental laws of these operations are: -

The commutative law of addition: $a+b = b+a$

The associative law of addition: $a+(b+c) = (a+b)+c$

The commutative law of multiplication: $ab = ba$

The associative law of multiplication: $a(bc) = (ab)c$

The distributive law: $(a+b)c = ac+bc$

For the purpose of this book, the invented number systems can be conveniently grouped into the following elementary or rudimentary kinds of numbers:

(1) rational and irrational numbers, together making up the real number system.

(2) pure imaginary numbers (a pure imaginary number is a number whose square is negative).

(3) complex numbers (a complex number is the sum of a real number and a pure imaginary number) and hypercomplex numbers (e.g. quaternions – they can be further subdivided into the real and complex types). This third and

final group can also be lumped together under what I have loosely termed the "unpure" imaginary numbers.

It is clear that any extension of the concept of a complex number beyond the system of complex numbers itself is possible only at the cost of some of the usual properties of numbers. Historically, the first such number system was the quaternions, discovered by Sir William Rowan Hamilton in 1843. The ordinary rules of algebra hold for quaternions except that multiplication is noncommutative. In 1870, Benjamin Peirce invented a whole new branch of mathematics – the study of linear associative algebras, consisting of a host of new algebras in which associative or commutative properties are abandoned. Hundreds of algebras were subsequently developed, including the algebras of Cartan, Dedekind, Dickson, Frobenius, Kronecker, Peirce, Scheffers, Sylvester and Weierstrass that are part of the (larger group of) prime factors for the larger base, then all regular number, 1, where all numbers are divisible by it – the division resulting in answers being simply the original number.

In the Big Bang theory, scientists have advocated three possible scenario for the universe: steady-state, open or closed. Both the steady state and open cases assume that the universe is expanding forever. In the open case all the cosmic material comes into being more or less in one go at the outset. As the universe expands and the galaxies fly apart, so the average density of matter declines; and, furthermore, this expansion may occur at either a constant or accelerated rate (due to the postulated dark energy derived from the cosmological constant (vacuum energy) – if it is positive or negative, it can act to speed up or slow down the expansion of the universe).

148

By contrast, in the steady-state case, the average density remains constant as matter is continually created, forming into new galaxies that occupy the widening spaces between the old ones. On a large scale, the universe stays much the same from epoch to epoch, like an ever-replenished reservoir dam. Finally, in the closed case, the universe will expand (due to the repelling force from the heat energy generated from the initial big bang "explosion") for a very long time until this ever decreasing inflationary tendency is overcome by the collapsing tendency (due to the attraction force from the total gravitational pull generated from all matter in the universe) – this would require a certain critical amount of average matter density to be present in our universe – hence, the frentic search by astronomers for possible missing or invisible matter, such as the dark matters, etc. After the expansion phase, the universe will start to contract back to its "Big Crunch" after an equally very long time.

Some interesting philosophical issues immediately arise from all these: Because a steady-state universe has infinite age, life might be imagined to have existed forever. Eternal existence of life is not possible in the open universe case, as energy, required to sustain life, will very slowly run out due to the ever decreasing gravitational force resulting in all stars eventually burning out (even massive black holes will slowly "evaporate" away to nothing through quantum mechanical processes). This corresponds to a state of dark, vast and cold empty space with lifeless celestial objects, with an ever-decreasing order (which equates with an ever-increasing entropy) in the universe – a final state of thermodynamic equilibrium will never (somewhat paradoxically) be reached. Physically and psychologically, the point on whether the arrow of time reverses and runs backward in the contraction phase of the closed universe can be rigorously argued in a scientific manner to be not possible. However, if the expanding and contracting phases of the closed universe is thought to repeat forever, with resulting cycles of big bangs and big crunches, then life in the universe might be imagined to have began and ended repeatedly. This is somewhat analogous to the Hindu belief of reincarnation, although the Hindu religious doctrine of reincarnation of life occurs much faster, with the soul entering into another (human or animal) body straight after death.

Scientific evidences point to the earliest signs of life on earth being present 3.45 billion years ago, 300 million years after the Earth's crust cooled sufficiently to support liquid water. The hominid (human) evolution occurred around 3 to 5 million years ago. Modern humans or homo sapiens are, intelligent-wise, the most advanced of all living creatures on Earth. As a source of conflict and debate, people with Christianity and some other religions believe that the Earth and universe is created by God, or a similar omnipotent being. Furthermore, if you look up the Bible and estimate the time elapsed [through the use of "time-line" for biblical characters in the book of Genesis from the Old Testament] since Adam (and the Earth, universe and all other living plants, land, sea and flying creatures) was created by God, you will arrive at a figure of the order of six thousand years only for the current age of our universe. This "time-line" is constructed by chronologically adding and plotting the life-spans of the biblical characters against time; and using the historical assumption that Abraham came to Canaan around 1900 BC and that Jesus was actually born around 6 years before the commonly accepted birth date of 1 AD. According to the Bible, Adam and Eve (created by God from one of Adam's ribs while he was in a

150

deep sleep) were the ancestors of all mankind. As the first human on Earth, Adam lived for 930 years. In Genesis 5:27, it was mentioned that [the oldest human ever was] Methuselah lived to an extraordinary ripe old age of 969 years, whereas the longest possible life-span of modern human today is known to be only between 100 to 110 years at the most, with life-span between 110 to 120 extraordinarily rare. Interestingly enough, Genesis 6:3 stated: Then the Lord said, "I will not allow people to live forever; they are mortal. From now on they will live no longer than 120 years." All these events occurred in the "beginnings or pre-history" where people spoke only one language. Then people of the world built Tower of Babel and God decided to mix up the language of all people and scattered them all over Earth. This is the biblical explanation for the origin of all languages (and perhaps also all different racial groups) in this world. But then the Bible also mentioned about the 12 sons of Jacob, who become ancestors of the 12 tribes of Israel.

Is immortality possible? If the ageing process were to be eliminated, we may live on average about 600 years. Perhaps the only thing that would kill us would be accidents such as being struck by lightning, a meteor, or a shark attack. Life expectancy of up to 150 might now be a realistic goal to achieve given our current scientific research and knowledge.

Working with mice, Dr. Simon Melov at the Buck Institute for Age Research in San Francisco, has found the gene that affected the way in which the animals aged. After months of study, Melov isolated genes that produced chemicals (such as in the form of a compound known as a Synthetic Catalytic Scavenger (SCS) that slowed down the ageing process. He then decided to breed mice without one of these genes. The results were incredible – the geneless mice were dying quickly of heart disease, liver dysfunction, and a whole range of other pathological abnormalities. Injecting these geneless mice – who were dying quickly – with SCS, Melov discovered that the cells of their brains stopped dying. In Britain, at Manchester University, Dr Gordon Lithgow gave one group of Nematode worms (a multi-cellular organism) the SCS and left the other group untreated. With a lifespan of 20 days, the untreated worms were all dying, and the worms with SCS were still alive, some living for 40 days – it was a major scientific breakthrough. Reference: "Extension of Life-Span with Superoxide Dismutase/Catalase Mimetics" by Simon Melov, Joanne Ravenscroft, Sarwatt Malik, Matt S. Gill, David W. Walker, Peter E. Clayton, Douglas C. Wallace, Bernard Malfroy, Susan R. Doctrow, and Gordon J. Lithgow Science Vol 289 nos 5484 Sep 1 2000: 1567 - 1569.

The main culprit contributing to the ageing process is the oxygen that we breath in for survival. Oxygen is a double-edged sword – it is absolutely required for life, but it can also be very toxic through the free radicals in oxygen that cause aging. Sunlight is a major source of oxygen radical damage – the wrinkling of our skin is almost 95 percent the result of exposure to UV radiation. Cigarette smoke is another major source of oxygen radicals. Therefore, avoiding the sun, not smoking, eating foods that naturally contain anti-oxidants (fruits and vegetables), and finally, taking anti-ageing drugs (such as anti-oxidant drugs powerful enough to kill off all the free radicals) is likely to prolong life. Caloric reduction, eating fewer calories by eating less food, may prolong life since restricting the supply of calories limits the amount of oxygen that is consumed and hence the production of these damaging radicals.

There are many anti-ageing drugs currently on the market. Hormone replacement therapy (HRT) is now widely used to combat the symptoms of menopause. It also helps with osteoporosis and can protect against heart attacks. Scientists have discovered the adrenal hormone DHEA (dehydroepiandrosterone) reduces the deterioration of the immune system in rats. Injections of the human growth hormone HGH have reversed some of the ageing symptoms in elderly men. ATP (Adenosine Triphosphate) can help with mobility in older people. The anti-oxidant, Coenzyme Q10 (CoQ10), may play a role in fighting heart disease by improving the symptoms of heart-failure patients. Melatonin, the hormone secreted from the brain's pineal gland, is triggered by darkness and makes humans feel sleepy. The pineal gland calcifies with age and melatonin level may decrease. As melatonin is the body's way of rejuvenating itself, giving melatonin is a potential means of slowing down the ageing process. However, this issue is controversial as some studies have failed to detect a difference in the melatonin levels of elderly people compared to that with young people, and also, the potential carcinogenicity of melatonin in rats and mice has been shown to occur.

Telomeres are the strands of DNA that tie up chromosomes. In the 1980s, discovery was made that the telomeres in reproductive cells were much longer than those in skin, muscle and nerve cells (known as somatic cells). It was suggested that if the somatic cells could not repair their telomeres, they would shrink. Telomerase is an enzyme found in reproductive and cancer cells that repairs telomeres. Without telomerase, a cell will gradually age. Subsequently, researchers have proven that the

153

telomerase gene can be activated in somatic cells – this may decrease cellular aging and thus increase our lifespan.

It is thought that damage to DNA is the driving force behind the ageing process and a major factor in the development of cancers. A specialized protein called PARP-1 works by mending damaged strands of the genetic material DNA. People who age rapidly have a deficiency of PARP-1 in certain types of tissue. Animals which have a short lifespan, such as mice, have a less effective form of the protein. Raising levels of PARP-1 can help cells to carry out repairs more effectively. The role of the vitamin, niacin, also plays a role in DNA repair.

Stem cells are either "multipotent" adult stem cells (adult bone marrow blood cells that are at their very earliest stage of development) or "pluripotent" embryonic stem cells (inner layer cells of the blastocyst stage of the embryo), able to differentiate into any of the 220 cell types that make up a human body, from the kidney, heart and liver to the skin, neuronal and pancreatic. Stem cell research is, of course, controversial and has been conducted on both embryonic and adult stem cells. Harvested adult stem cells from the brains of adult mice can be shown to grow into nerve and muscle cells. The potential for cell rejuvenation is endless, for instance, using stem cells to provide cure for insulin-dependent diabetes, Parkinson's disease, Alzheimer dementia, Amyotrophic lateral sclerosis (Lou Gehrig's disease); repair severe eye damage, damaged cardiac muscle and arteries from a heart attack, brain damage from stroke, spinal cord damage; provide skin for grafting to full-thickness skin burns; and ultimately, perhaps, growing human tissues and organs for transplantation purposes.

Gene therapy has been advocated for correcting life-threatening genetic disorders. In 2000, a 27 year old American volunteer, Jesse Gelsinger, died from overwhelming infection from a transgenic adenovirus being used to deliver an essential liver enzyme – this therapy is also, sadly, only a temporary fix. Using retroviruses that integrate their DNA permanently into the human chromosome runs the risk of triggering cancer or knocking out vital genes. In addition, viral genomes are typically too cramped to accept anything but the smallest human genes – so big genes must be pruned which may compromise function. To circumvent all these problems, a functional mini-chromosome has been re-engineered from an aberrant chromosome 10 (dubbed Cousin 47) by Dr Andy Choo and colleagues from the Murdoch Children's Research Institute in Melbourne, Australia during 2001 over a 4 year period – please refer to relevant Proceedings of the National Academy of Science for more information. This promising Human Engineered Chromosome has three essential features of, firstly, a centromeric "knot" providing attachment point for elastic protein fibres that pull newly duplicated chromosomes apart and hauling them to opposite ends of dividing cells; secondly, 2 telomeres which prevent chromosome's end unravelling; and, thirdly, origin-of-replication sites where DNA duplicating enzymes can key into chromosome to copy it segment by segment. Unwanted genes between centromere and telomeres are excised and an all-purpose slot on this chromosome then allows accommodation of new therapeutic genes – this finalized chromosome may then be inserted into stem cells and be versatilely progenited into specialized cells of tissues and organs.

Sports and exercises such as swimming, walking and jogging helps maintain cardiovascular health and fitness and have been proven as a tried and true method of living longer. Foods (such as tea, red wine and chocolate) containing a group of compounds known as polyphenols can contribute to our overall anti-oxidant defence, thus slowing the decay our bodies.

The human sexual drive and function is important for the well-being of our physical and psychological health. Viagra (Sildenafil) is recommended for treatment of erectile dysfunction in adult males. It restores impaired erectile function by increasing blood flow to penis, resulting in natural response to sexual stimulation. Its exact mechanism of action is by being a selective inhibitor of cyclic guanosine monophosphate specific phosphodiesterase type 5, which is the predominant phosphodiesterase isoenzyme in human corpora cavernosa (part of the penis). It is no wonder that, due to human nature as we know it, Viagra is a 'best-seller' drug, as are (or will be) anti-ageing drugs that can prolong life.

Many living creatures, including humans, have maternal (and paternal) instincts – a strong desire to have and raise their own off springs – in so doing propagate their own genes onto the next generation. Infertility is a disease or condition of the reproductive system often diagnosed after a couple has had one year of unprotected, well-timed intercourse, or if the woman has been unable to carry a pregnancy that results in a live birth. Approximately 35% of infertility is due to a female factor and 35% of infertility is due to a male factor. In the balance of cases, infertility results from problems in both.

Assisted reproductive technologies (ARTs) is any procedure that

156

uses high technology to combine sperm and eggs. In man and/or woman, examples of medical terms associated with ART procedures and techniques include In vitro fertilization (IVF), Assisted hatching, Microsurgery, Micromanipulation, Microsurgical tubal repair, Intracytoplasmic Sperm Injection (ICSI), Gamete intrafallopian tube transfer (GIFT), Zygotic intrafallopian tube transfer (ZIFT – sometimes called Tubal embryo transfer or TET), Micro Epididymal Sperm Aspiration (MESA), Percutaneous Epididymal Sperm Aspiration (PESA), and Testicular Sperm Biopsy (TESE).

Some examples of fertility drugs include the following:-

Clomiphine Citrate (Clomid) which stimulate increased output of pituitary gonadotrophins,

Nafarelin acetate (Synarel nasal spray) which is a potent agonistic gonadotrophin releasing hormone (GnRH) analogue, and

Recombinant human follicle stimulating hormone (rhFSH) (Puregon) which contains follitropin beta, a hormone known as follicle stimulating hormone (FSH). FSH belongs to the group of gonadotrophins, which play an important role in human fertility and reproduction. FSH is needed in women for the growth and development of follicles in the ovaries.

There will always be proponents of infertility treatment, especially couples desperate for children after many years of trying. However, undergoing infertility treatment is not without its problems. There is the psychological anguish and high monetary expenses if there were to be failures and multiple attempts to conceive with multiple courses of expensive infertility treatment required. Then there is the morbidity or mortality associated for both mother and babies if multiple pregnancies were to result

157

from infertility treatment – for instance, extreme premature infants are at a high risk of brain damage and blindness. Other alternatives for infertile couples are either to adopt children or have babies through a surrogate mother (with its moral and legal dilemma).

Genes encode proteins that perform all of the fundamental activities within cells. Proteins are the molecular machines that carry out genetic instructions. Proteins transmit messages, repair damage, provide the building blocks for tissues and carry out reactions essential for life. Abnormalities in protein production or function have been connected to many diseases and health conditions. To understand how best to treat a particular disease, it is necessary to identify the proteins associated with that disease and to understand how they function.

Proteomics is the systematic analysis of all protein sequences and protein expression patterns in tissues, which involves the isolation, separation, identification and functional characterization of all of the proteins in an organism. Proteomics involves the comprehensive functional annotation and validation of expressed proteins in health and disease. Understanding the functional characteristics of proteins and their activity requires a determination of cellular localization and quantitation, tissue distribution, post-translational modification state, domain modules and their effect on protein interactions, protein complexes and pathway information, ligand binding sites and structural representation as well as direct links to drug design and discovery. The development of functional proteomic maps of cellular activity allow us to create a bridge from genomic information to the discovery of new medicines.

158

There are two main approaches to proteomics: expression proteomics and functional proteomics. Expression proteomics involves the study of proteins in comparative tissue samples. Functional proteomics is the study of how proteins interact with other cellular components in order to determine protein function.

Genome sequence information, derived from works through the Humane Genome Project (see The Human DNA story below), is becoming available in unprecedented amounts. The absence of a direct functional correlation between gene transcripts and their corresponding proteins, however, represents a significant roadblock for improving the efficiency of drug discovery. This overabundance of unannotated genomic information may, paradoxically, contribute to an overall reduction in the efficiency of the biopharmaceutical industry today. With the endeavor to discover drugs through this method, proteomics is set to be one of the biggest fields in drug development in 21st Century, spawning an industry worth billions of dollars.

Is life the result of creation or evolution or some other process? In the taxonomy of living things, all life is thought to evolve from the earliest living organisms, as depicted in the "evolutionary tree of life". Is the apes from Africa our modern evolutionary ancestors? Where did life began? Is it from Earth, and if so, which place (or places) on Earth? Is it the hot and hostile deep volcanic ocean vents as proposed by Paul Davies in his book The Fifth Miracle: The search for the origin of life (Penguin Press 1998), the gentle "warm little pond" as suggested by Darwin in his theory of evolution, or some other place on Earth? Or is it extraterrestrially from another planet like Mars, or from both Earth and Mars (with or without cross-colonization through the interchange of hardy primitive living microbes or superbugs proposed to be able to survive on meteorites derived from, and exchanged between, the two planets)? What is the meaning or definition of life? Unless of a direct scientific nature, any indirect (scant and often weak) scientific evidences obtained ("unEarthed" or "unMarsed") from Earth or Mars is likely to be controversial, and people – scientists and ordinary people alike – could debate these issues "all day till the cows come home". Current astronomical data collected support the fact that even if a tiny proportion of all the stars in our universe have some sort of planetary systems similar to our solar system, this will equate to an immense number of stars with planets in our universe – thus, arguably, there is a non-zero probability of intelligent life ("green man") &/or microbes ("green slime") existing outside Earth (if the right conditions for life to florish were to be present on that planet).

As a Christian, I have come across some interesting but mystical concepts during some of the church camps in my early life. They are derived from interpretations by theologians, based mainly on the final biblical chapter of Revelation in the New Testament. This is the so-called "God's outline of History", with three slightly different versions (which I won't go into details): Pre-Millennialist, A-Millennialist or Post-Millennialist. This "history", as understood by me, is as follows: Eternity past -> Israel Old Testament -> Crucifixion of Christ -> Resurrection of Christ -> New Testament -> 7 years of Tribulation [the period between Rapture (Christ Returns for His Church, then the Church in Heaven) and the Second Coming of Christ] -> 1000 years of Messiah's Kingdom -> Present World destroyed -> Last Judgement -> Eternity New Heavens and New Earth.

Genesis 1:1-2 said: "1 In the beginning, when God created the universe, the earth was formless and desolate. 2 The raging ocean that covered everything was engulfed in total darkness, and the power of God was moving over the water." Hypothetically, sitting on the fence, so to speak – one may propose that after these two verses, the Bible did not mention how long God waited before He endowed Earth with living things, etc. Perhaps our all-patient God created the universe some 15 billion years ago as per Big Bang theory, confidently knowing that Earth will subsequently form on its own accord (through our universe taking the "natural cosmological evolution" as proposed by scientists today). Darwin's theory of evolution then "kick-started" the most primitive living organisms some 3.45 billion years ago. After another long wait (but just a blink of the eye for God), at about 6000 years ago, starting from the first day, God decided to create "Day" and "Night"; then progressively, the more advanced living creatures, etc; and, finally, on the sixth day, God created Adam and, later on, Eve. The seventh day (Sabbath – which is Sunday for Christians) is a "rest day" for God when he completed his creation. Let this idea be dubbed Sitting-on-the-fence Theory!

There is a closely related theory called the Gap Theory, so named because of the so-called gap it leaves between Genesis 1:1 and 1:2. According to this theory, God actually created the earth many billions of years ago as stated by the evolutionists. This earth was the site of previous life forms, including the dinosaurs. Eventually, however, this creation was destroyed, and God remade the earth about 6000 years ago using the same raw materials.

However, there is a whole host of other controversial issues to be contemplated with, such as the hominid evolution – deemed to have occurred 3 to 5 million years ago. Another theory to explain how the earth could age only about 6000 years while the universe around us aged about 15 billion years uses the fact that time slows down in strong gravitational fields. This theory was put forward by D. Russell Humphreys in his book Starlight and Time (Humphreys, D. R. 1994. Starlight and Time. Master Books. Colorado Springs, Colorado). Humphreys rejected the notion that the earth is part of a boundless universe called a hypersphere and used the notion that the solar system is located near the center of the universe. This central location allows for a strong concentration of gravitational forces at the time of creation. In the so-called White Hole Cosmology, the universe is assumed to have finite boundaries with our earth situated near the center. Putting those boundary conditions into Einstein's equations of General Relativity, it is postulated that God created the entire universe in a very dense format - so dense that it formed a white hole. A white hole, with a shrinking event horizon and matter expanding out of it, is essentially black holes running in reverse. White holes eventually spew enough material out to lose their white hole status. As the universe began expanding (in a similar manner to the Big Bang theory), the outgoing material passed through the event horizon, which is a place where clocks would be momentarily stopped relative to clocks further out. At one critical moment of the expansion, the event horizon would reach the earth, and clocks there would also momentarily stop. In this way where clocks (and all physical processes) tick at different rates in different parts of the expanding universe, the entire universe that crossed the event horizon could have aged billions of years, while the age of the earth still retains its basic 6,000 year age. This theory is proposed to also explain

162

the red shifts of galaxies and the cosmic microwave background.

Of course, there will be many critics of this controversial theory. Without going into any great details, in arguments from "The Present Invalid Nature of Humphreys' White Hole Cosmology" by Robert A. Herrmann from the Mathematics Department, U.S. Naval Academy in Annapolis in June 1996, Humphreys' model may fail to achieve the goals claimed in, at the least, one instance. While investigating the cosmological constant "Lambda" in this model, a direct contradiction is obtained in the use of this constant by Humphreys.

All serious dating methods (radiometric age dating, dendrochronology, ice core analysis, varve deposition, and more) yield ages far older than Humphreys' methods. Thus D. Russell Humphreys allegedly broke all the rules of science. He uses flawed logic, overly simple models, and twisted data to sell his young Earth. On the other hand, in "A Review of Dr. Russ Humphreys A Young-Earth Relativistic Cosmology" by David J. Tyler in 1994 Copyright 1996 The Biblical Creation Society, Tyler wrote "Even if Humphreys is wrong in his biblical interpretation, he has contributed significantly to cosmological studies. We have known that presuppositions are important for the Big-Bang theory – but Humphreys has worked this through in some detail. Furthermore, he has demonstrated that with different presuppositions, different conclusions are possible. A door has opened – Christian students of cosmology will find this research a great stimulus to their own thinking."

Freedom to practice one's beliefs and religion is a fundamental human right. We must all have tolerance and humility in our hearts in order to live in harmony with one another. Being a Christian, and in spite of certain conflicts between Science and Christianity, I will always worship God and uphold the fundamental Christian teachings – the crust of which is epitomized by Bible verse John 3:16 "For God loved the world so much that he gave his only Son [Jesus Christ to die for us], so that everyone who believes in him may not die but have eternal life." The same should apply for another different religion that a person may hold dear to his or her heart.

The Human DNA story

Human DNA consists of approximately 30 billion bases. The traditional Watson-Crick double helix DNA model (by James Watson and Francis Crick in 1953) takes the form of a long twisted ladder. The two legs of the ladder swirl round one another, connected by rungs made of "base pairs", the carriers of genetic information. Each rung is made of either the molecule adenine (A) paired up with Thymine (T), or Guanine (G) paired up with Cytosine (C). The long twisting legs of the ladder are chemically built up of sugar and phosphate groups linked together into a strong "sugar phosphate backbone". Therefore, the primary structure of DNA is a helix. The DNA molecule also twist itself up into complex knots – its secondary and tertiary structure. Because each of the 30 billion bases can be of either A, T, G or C, the total number of possible combinations and permutations is astonishingly large – this translates to the ability to store a gigantic amount of information. This is the principle behind the proposal of advocating DNA material as a powerful organic data storage medium ("DNA chips") in future computers.

164

"Deconstructing DNA" is the title of an article by freelance correspondent, Ayala Ochert, in New Scientist Vol. 158 No. 2134 pp 32-35, 16 May 1998. This interesting scientific report is about an imaginative British artist, Mark Curtis, who has created a possible astonishing alternative to Watson-Crick model. Curtis found that the two and three-dimensional (2D and 3D) representations of Watson-Crick model in textbooks does not make sense. Using what he refers to as "geometric first principle", he associated a pentagon (5-sided figure) with each base pair. Because DNA has ten base pairs to each turn of helix, he then placed ten pentagons (each representing a base pair) around a decagon (10-sided figure). Geometrically, this is aesthetically pleasing as the decagon is the only regular shape that can fit round a decagon in this way without leaving any space.

In real life, the DNA is a structurally sound 3D double helix. Therefore, each pentagon is given some thickness and two (instead of just one) pentagons are now joined together to represent a base pair. In this way, ten paired-pentagons stacked up and arranged around a decagon gives a helix with the known dimensions of DNA. Also, Curtis has changed the Watson-Crick's base pairing, from its traditional way with edges that face outwards, to edges that face inwards. In other words, Curtis is constructing his model of double helix solely from the base pairs themselves – whereas the Watson-Crick structure gets its shape from the sugar phosphate backbone. However, X-ray crystallography results show that the bases pair up Watson-Crick style, not Curtis style. Does this mean the death knell for Curtis's creation? Perhaps not. His hypothetical structure may mirror something that could in principle exist in the real world.

Since the 1950s, DNA has been found to be a far more complex beast than scientists first imagined. Many unusual forms of DNA have been found, such as the circular DNA in bacteria, the left-handed zig-zag helix of Z-DNA, triple-helical DNA, parallel-stranded DNA and P-DNA. In in vitro experiments, it was discovered that when the twist on a DNA molecule exceeded its natural number of turns (natural twist) by 3%, the molecule will change suddenly in some regions – the base pairs became unpaired, sticking out to the sides instead of sitting inside the external scaffolding of the sugar-phosphate backbone. This changed DNA form is known as P-DNA, and its structure is thought to occur in DNA transcription, when the base pairs need to open up for easy reading. In viral DNA, the effect of a coat of proteins glued to the outside of the DNA strand serves to give the same effect as the mechanical twisting.

Finally, Curtis's DNA has another intriguing feature. Although its overall structure is independent of the sequence of bases, the details of the molecule's surface are sequence-specific. This fundamental sequence-specific property of DNA's surface is essential for proteins to be able to recognize sequences of bases and bind to them, switching genes on or off. Curtis's DNA structure seems to allow for this recognition process.

The Human Genome Project (HGP) is an international effort formally begun in October 1990 to discover all the 60,000 to 80,000 human genes and make them accessible for further biological study. (A gene is a segment of DNA that codes for a particular function.) Another project goal is to determine the complete sequence of the 3 billion DNA subunits (bases in the human genome). It uses many ideas originally developed in information theory and cryptography to sequence DNA; namely, algorithmic complexity developed through the work by Solomonoff, Kolmogorov, and Chaitin (the SKC theory). However, this truly monumental coding task relates mainly to the spatialness or "structural" aspect of the human genome. The temporalness (dynamic) aspect of the human genome consists of many characteristics, such as the actual processes of genes coding the protein synthesis, controlling the operation of other genes, etc. For instance, the process of protein synthesis consists of the unwinding of the DNA helix, then the copying of RNA from the DNA template, and finally the attachment of ribosomes to mRNA and its subsequent interaction with the chain to serve as a template for the manufacturing of proteins from amino acids. (The four types of base present in RNA are identical in nature to DNA bases, except that it substitutes Uracil for Thymine.)

As part of the HGP, parallel studies are being carried out on selected model organisms such as the bacterium E. coli to help develop the technology and interpret human gene function. The DOE Human Genome Program and the NIH National Human Genome Research Institute (NHGRI) together make up the U.S. Human Genome Project.

A codon is a triplet of bases coding a single amino acid. There are four bases at each of the three positions within a codon – so there are 64 (= 4^3) codons, and three of these are "punctuation" codons. The following logical "conclusions" may now be drawn: There are 61 codons possible (plus 3 "punctuation" codons), yet only 20 amino acids are produced. Thus given a gene, coding for a small protein comprising 100 amino acids, we could have (on average) 3 different codons for the same amino acid, and 5×10^{47}

different genetic arrangements coding for the same protein! Given that many proteins will have similar selective or functional effects (due to similar shape – their folding characteristics), we can see that genetic variation, in itself, does not guarantee any phenotypic variation on which selection may then act, there may be long periods of stasis when nothing happens, before a mutation occurs that actually causes an effect. This redundancy of the code allows some freedom for bases (especially the third position) to change (silent mutation) without altering the resulting amino acid sequence. It also allows the code for different genes to overlap – this is seen to be beneficial for organisms, such as viruses, with relatively little DNA. In most cases, the third base in a codon is irrelevant; with the resulting amino acid determined by the first two. This led to the speculation that the present code may well be derived from an earlier form in which codons consisted of pairs of bases.

Answers to some burning questions such as: "Why is there the enormous amount of junk DNA (as only about 20% of DNA strand actually contains genetic codes)?" requires the combined spatio-temporalness aspect of human genome for clues. (In fact, a "thorough" study of any part of the human genome system must involve, one way or another, both the spatialness and temporalness aspects.) Perhaps the junk DNA is needed to enable the reading of gene sequences. Perhaps it provides space for writing new genes. Perhaps it serves both (and maybe even other) purposes.

Extinctions

Extinction has played an important role in the development of life on the Earth. Of all the species that have lived here since life first appeared, only somewhere between one-in-a-thousand and one-in-a-hundred is still living today. The vast majority became extinct typically within one to ten million years of their first appearance. It is thought that the distribution of the sizes of extinction events approximately follows a power-law form, taken by some to be indicative of criticality in the processes giving rise to extinction. It has been suggested that the power spectrum of extinction intensity during one of the prehistoric mass extinction has a $1/f$ form. The distribution of the lifetimes of genera may also be a power law.

There are evolutionary and environmental processes which drive extinction. The effects of large-scale coevolution can be modelled on ecosystems. Organisms coevolve because of interactions between them – predation, competition, parasitism and other effects can force one species to evolve as the result of the evolution of another. It has been argued that co-evolution can give rise to a "domino'" effect, in which the initial mutation of one species causes a wave, or avalanche of evolution to travel across the ecosystem. It is possible that such collective evolution effects, because of their size, could be the dominant mode in the evolution of ecosystems. Large coevolutionary avalanches may be rare, but some theories suggest that they are still frequent enough that any particular species is more likely to be touched by one such than to undergo a spontaneous mutation of its own.

The possible existence of coevolutionary avalanches raises intriguing questions about the way in which ecosystems organize themselves and about the closely related issue of extinction rates. Chief amongst these questions is that of self-organized criticality in evolution – although controversial, theories based on this concept may help explain the nature of fossil record.

169

A contrasting view to the models based on self-organized criticality above is that extinction is caused by external perturbations on the ecosystem. The dynamics of this new model, in sharp contrast to the previous models, evolve solely in response to differing environmental stresses on species. Interestingly, this new model also shares with the other models the feature that it gives reasonable qualitative agreement with the power-law distribution of the data.

Therefore, it is not necessary to invoke critical processes or other internal dynamical effects to account for the observed fossil extinction record. Ongoing research is focusing on a search for characteristics in the fossil data that would constitute distinctive signatures of either internal co-evolutionary interactions or external environmental stresses. One possible signature for external environmental stress has been suggested to be the so-called "aftershock" extinctions.

Our sun is a middle-aged star of 4.6 billion years old. In about 5 billion years into the future, the center of the sun will shrink (which will increase the rate at which hydrogen changes into helium and thus giving off greater amount of energy) and become hotter with outer regions of the sun expanding about 48 to 64 billion km. The sun is thus a red giant, causing the earth's temperature to become too high for life to exist. Being a cataclysmic event, all life on earth (including humans) will become extinct unless our civilization are scientifically advanced enough to allow us to escape and settle on another suitable planet in our universe. After the sun has used up its thermonuclear energy as a red giant, the sun will shrink to about the size of the earth, becoming a white dwarf. While doing so, it will throw off huge amount of gases in a violent eruptions called a nova explosion. Then after billions of years as a white dwarf, the sun will have used up all its heat, thus becoming a black dwarf.

Supernova is an exploding star in its final death throe that can become billions of times as bright as the sun before gradually fading from view (from several weeks or months to a period of years). It can leave behind different type of objects such as a small dense neutron star, a rapidly rotating & highly magnetized neutron star called a pulsar, an invisible & immensely powerful gravitational object called a black hole, or no remaining object of any kind.

Several great (prehistoric) mass extinctions occurred during prehistoric times. The more publicized mass extinction occurred around 65 million years ago (postulated to be due to earth colliding with a large asteroid), when the dinosaurs and many species of marine life died out. The largest mass extinction was about 240 million years ago, at the end of the Permian Period, when possibly as much as 96% of all species disappeared. Two other mass extinctions took place during the Mesozoic Era – one around 205 million years ago, the other around 138 million years ago. It has been thought that the earth exposure to large amounts of lethal cosmic rays (originating from a supernova) may have contributed to mass extinctions if the position of earth along one the spiral arms of our rotating Milky Way galaxy was in close enough proximity to a supernova explosion (which coincide with the timing of the mass extinction).

In our current modern industrialized times, we are damaging our planet and experiencing another great wave of many species extinctions occurring over an unprecedented and relatively short period of time of over fifty to a hundred years caused by just one species – namely us, the humans. This will result in a decrease in the biodiversity of our planet and a impoverished world for our future children, and is occurring by 5 ways – overharvesting, introducing alien species, destroying places where species live, creating small islands of habitat, and polluting the atmosphere. However, through collective effort and cooperation between countries, we humans, as the most intelligent species of all, also have the capacity to slow or reverse this trend of destruction.

There may also be an association between some species and other life-form extinctions and Ice Ages. About 13,000 years ago, not long after the first humans arrived in the New World (North America), all but a few of the remarkable creatures – collectively known as the Pleistocene megafauna (such as the wooly Mammoth, mastodons, Saber-toothed cat, Prehistoric horses, giant sloths and glyptodonts) – had vanished. Three competing hypotheses – climate change, overhunting and hyperdisease – acting in combination in a nonlinear chaotic dynamical interactions may be theorized to explain this event – this may subsequently be "put to the test" in a computer stimulation. As the Ice Age wound down, the forests and savannas in which the megafauna foraged would have dwindled in size and become less diverse. The Stone Age hunters armed with primitive weapons may have killed slightly more animals than were born over about a thousand years in a gradual attrition. Some disease-causing microbe – one to which New World organisms lacked resistance – tagged along with the Stone Age hunters running amuck may have killed off many of these creatures.

Ice Ages are intervals of time when large areas of the surface of the globe are covered with ice sheets (large continental glaciers). The term is used to describe time intervals on two very different scales. It describes long, generally cool intervals of Earth history (tens to hundreds of millions of years) during which glaciers waxed and waned. The term also describes shorter time periods (tens of thousands of years) during which glaciers were near their maximum extent (also known as "glaciations").

Many glacial advances and retreats have occurred during the last billion years of Earth history. These glaciations, not randomly distributed in time, are concentrated into four time intervals – during the late Proterozoic (between about 800 and 600 million years ago), during the Pennsylvanian and Permian (between about 350 and 250 million years ago), during the late Neogene to Quaternary (the last 4 million years) and during parts of the Ordovician and Silurian (between about 460 and 430 million years ago). Once ice sheets start to grow, they probably contribute to their own further development. This positive feedback occurs because ice sheets reflect more sunlight back into space than does ground not covered by ice. The reflected sunlight would otherwise warm the Earth's surface.

Glaciations are affected by a number of factors interacting to produce conditions favoring the formation of ice such as:
(1) Plate tectonics (changing of continental positions and uplifting of continental blocks). The presence of large land masses at high latitude appears to be a prerequisite for the development of extensive ice sheets.
(2) Reduction of CO_2 in the atmosphere. A general reduction in amount of

CO_2, an important greenhouse gas, in the atmosphere may contribute to the development of ice ages. Decreases in the amount of CO_2 in the atmosphere may lead to global cooling. Many processes (including many complex interactions among organisms, ocean currents, erosion, and volcanism) can cause a long-term decrease in the amount of CO_2 in the atmosphere.

(3) Changes in the Earth's orbit (Milankovitch Theory, first popularized in about 1920). The Serbian astrophysicist Milutin Milankovitch is best commemorated for developing one of the most significant theories relating Earth motions and long-term climate change. The Earth's orbit varies through time. Important parameters that vary include the eccentricity of the orbit around the sun, the tilt of the Earth's axis, and the precession of the equinoxes (the direction the north pole points). Variation in these three factors changes the amount and distribution of incoming solar radiation, thus affecting glaciations.

The Earth's orbit around the sun is not a circle, but rather it is an ellipse. The shape of the elliptical orbit, which is measured by its eccentricity, varies from between one and five percent through time in a periodic manner (periodicity is approximately 100,000 years). The eccentricity affects the difference in the amounts of radiation the Earth's surface receives at aphelion (the point on its orbit when the Earth is farthest from the sun) and at perihelion (the point on its orbit when the Earth is closest to the sun). The effect of the radiation variation is to change the seasonal contrast in the northern and southern hemispheres. When the orbit is highly elliptical, one hemisphere will have hot summers and cold winters; the other hemisphere will have warm summers and cool winters. When the orbit is nearly circular, both hemispheres will have similar seasonal contrasts in temperature.

Although the amount of change in radiation is very small (less than 0.2%), it is apparently extremely important in the expansion and melting of ice sheets.

The Earth's axis is tilted with respect to its orbit around the sun. Today the tilt is approximately 23.5 degrees. The tilt varies from between 22.1 and 24.5 degrees in a periodic manner (averages about 40,000 years). Because this tilt changes, the seasons as we know them can become exaggerated. More tilt means more severe seasons – warmer summers and colder winters; less tilt means less severe seasons – cooler summers and milder winters.

Twice a year, the equinoxes, the sun is positioned directly over the equator. Currently the equinoxes occur on approximately March 21 and September 21. However, because the Earth's axis of rotation "wobbles" (like a spinning top), the timing of the equinoxes changes. The change in the timing of the equinoxes is known as precession. Although the timing of the equinoxes is not in itself important in determining climate, the timing of the Earth's aphelion and perihelion also changes. Like the timing of the equinoxes, the timing of the aphelion and perihelion is also affected by the wobble of the axis of rotation (axial precession). This increase the seasonal contrast in one hemisphere and decrease the seasonal contrast in the other hemisphere. The aphelion and perihelion change position on the orbit through a cycle of 360 degrees. The cycle has two periods of approximately 19,000 and 23,000 years. Together these combine to produce a generalized periodicity of about 22,000 years. The effects of precession on the amount of solar radiation reaching the Earth are closely linked to the effects of tilt. Variation in these two factors cause radiation changes of up to 15% at high

175

latitude, thus greatly influences the growth and melting of ice sheets.

The periods of these orbital motions above are together known as Milankovitch cycles. Using these three orbital variations, Milankovitch was able to formulate a comprehensive mathematical model that calculated latitudinal differences in insolation and the corresponding surface temperature for 600,000 years prior to the year 1800. He attempted to correlate these changes with the growth and retreat of the Ice Ages. A study published in the journal Science in 1976 examined deep-sea sediment cores and discovered that Milankovitch's theory did in fact correspond to periods of climate change. Reference: J.D Hays, John Imbrie, and N.J. Shackleton, "Variations in the Earth's Orbit: Pacemaker of the Ice Ages," Science, 194, no. 4270 (1976), 1121 - 1132.

The intuitive notion of complexity is well expressed by the common definition: "a complex system is an arrangement of parts, so intricate as to be hard to understand or deal with." For our purpose, by complex system, it is meant a system comprised of a (usually large) number of (usually strongly) interacting entities, processes, or agents, the understanding of which requires the development, or the use of, new scientific tools, nonlinear models, out-of equilibrium descriptions and computer simulations. Understanding the behavior of the whole from the behavior of its constituent units is a recurring theme in modern science, and is one of the central topic on the study of complex systems. The term "complexity" is harder, but not impossible, to define narrowly in science. This book has attempted to do so under the innovative Spherical Model of Numbers and the Spherical Model of Science.

176

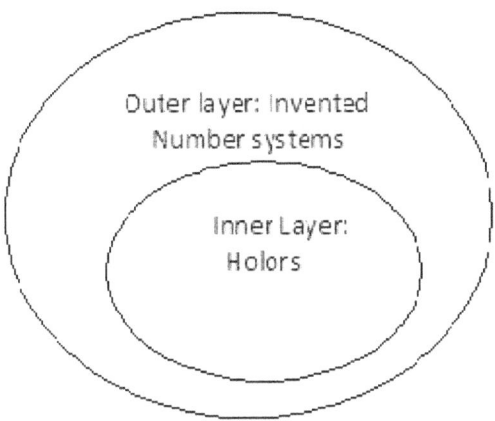

Spherical Model of Numbers: Holors form its inner layer "foundation". Invented Number systems & their Base (radix) notation form its outer layer. Note: Holors are the universal "Alphabet of Science". Holor algebra & Holor calculus are the universal "Language of Science". Using analogy, the various Invented Number systems represent arbitrarily selected different human languages used to enforce "Alphabet of Science" & "Language of Science".

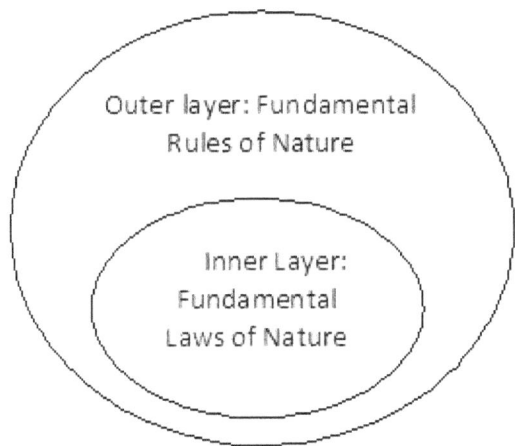

Spherical Model of Science: Fundamental Laws of Nature form its inner layer "foundation" (mainly applicable to Nonliving Things) and Fundamental Rules of Nature form its outer layer (mainly applicable to Living Things) with both layers always obeying all "principles" as derived from inner and outer layers of Spherical Model of Numbers. Note: The somewhat metaphoric use of "laws"

177

for Nonliving Thing and "rules" for Living Things in this context can be intuitively differentiated. To avoid subsequent confusion about any observed omission or discrepancy, I have to strongly stress here that with respect to predominantly the inner layer of Spherical Model of Science, Science of Complexity can be applied to three very different mechanical settings; namely, Quantum mechanical, Relativistic mechanical and Relativistic quantum mechanical setting. Perhaps a bit misleading, we could also include a fourth one, Classical (commonly known as Newtonian) mechanical setting.

Relativistic mechanics is an extension of Newtonian mechanics conforming to the principles of Special Relativity – in particular, in the low-velocity approximation, the ratio v^2/c^2 (where v is the velocity of the particle, c is the velocity of light) can be neglected in Special Relativity calculations, which then correctly reduce to the calculations using Newton's laws of motion. Also, in Einstein's field theory of gravitation (General Relativity), the relationship between mass-energy and space-time curvature is a relativistic generalization of Newtonian law of gravitation, and in the limit of weak gravitational field strength, Newton's law of gravity is an accurate theory for gravity. Roughly speaking, Classical mechanical setting is a crude approximation of its Relativistic mechanical counterpart. The use of Classical mechanics is more than sufficient for most of our everyday life purposes – even many calculations required, for instance, for our solar planetary system and space travel can still be safely and accurately done under the framework of Classical mechanics.

The statics and dynamics aspects of structures and processes in all four settings can be either of fractal or nonfractal geometries and chaotic or nonchaotic dynamics respectively. The resulting applications may, respectively, be thought of as the Classical complexity, Quantum complexity, Relativistic complexity and Relativistic quantum complexity. For all of these complexities, they can each be looked at in the broader sense, under Qualitative and Quantitative Complexities. Furthermore, depending on whether discrete or continuous type of mathematics is used to analyze these four domains, each can be further subdivided into either discrete or continuous subdomain using this criteria. The terms "discrete" and "continuous", with their respective analogy of "digital" and "analog" in computer science, have subtle but important differences in meaning when applied to different situations in nature. Not only that, but in relation to this matter, there is a deep hidden connection between the (classical and quantum) computing world and the natural world, as you shall see when you read through the rest of this book. But that is another story by itself, and I will have to leave it as that here.

In fact, more than all known structures and processes in the above mentioned four mechanical settings; everything from the most complicated thing to the simpliest (or most trivial) thing is amendable to both Qualitative and Quantitative Complexities analysis. The "everything" and "thing" refer to every conceivable abstract mathematical equation, formula, rule, law, theory, etc.; every conceivable mathematical model of nature; every conceivable living or nonliving, natural or man-made (artificial) and organic or inorganic system, structure, organization, process (whether biological, geological, chemical, physical, or nuclear, etc.), culture, ecology, etc. in the universe; and so forth. A system (whether closed, open or isolated; homogenous or heterogenous; and at, close to, or far from equilibrium), be it of a physical, chemical, nuclear, biological, social, or institutional nature, etc., can be a collection of organic or inorganic matter, or even institutional entities, etc. Therefore, when we refer to "everything" in the current context, we literally meant everything in the whole wide universe, from the most massive scale to the tiniest scale, from the slowest known processes to the fastest known processes, from the most complex system to the simpliest system, from the highest extreme energies to the lowest (zero or negative) energies, and finally, from the past, present or predicted future. The most distant past is, of course, the creation of universe about 15 billion years ago, as explained by Big Bang theory (or some other plausible scientific theories). The highest extreme energies is thought to occur in the earliest tiny fraction of a second in Big Bang and in celestial black holes, where gravitational forces in both cases are infinite, and mathematical singularities exist, at their centers.

When each of the "everything" is explicitly expressed in its appropriate scientific format, the two complexity measures can then be applied to it. Implicit in the previous statement are (some of) the controversial issues of: -

(1) Whether every single "thing" can be expressed in a satisfactory or accurate scientific manner. For instance, is a scientific model or theory a good enough approximation of a real world situation? (It may well be required to be an exact model or theory of the real world situation, as exemplified by the "criteria" for the Theory of Everything.) Are we able to quantify certain hard-to-quantifiable concepts (e.g. psychological concepts such as the human/animal consciousness level, intelligence level, complex behaviors, etc), as an obvious prerequisite for a "thing" to be expressible in its appropriate scientific format? (It may well be that we can only quantify those concepts in a semi-quantitative manner - this may perhaps be due to our intrinsic lack of knowledge or ignorance on the situation, or even perhaps be due to those concepts being intrinsically (and correctly) qualitative or semi-

quantitative in the first place.) I strongly suspect that my proposed Spherical Model of Science, itself a Semi-quantitative Complexity on theories, can only be intrinsically and correctly semi-quantitative in nature. However, proving its correctness or that it is the best model, if at all possible, is a different kettle of fish altogether.

(2) Whether the right design and/or decision on the types of Qualitative and Quantitative Complexity measures are being employed. Certain parts of this book are devoted to the expansion of this idea, and have exploited the subtle and useful relationships between these two measures.

Let us highlight what we meant by the previous two statements through the simple example of the arithmetic operation of addition: "1+1+1 = 3". The Qualitative Complexity consist of the arithmetic operation "+", the mathematical symbol "=" and single-digit integers; whereas the Quantitative Complexity consist of the numbers 1, 1, 1 and their resultant sum of 3. Let us now easily illustrate a possible conflict of definition arising from all this. A Semi-quantitative Complexity may be defined to consist of two "+" signs, one "=" sign and four single-digit integers (or more accurately, three "1s" and one "3"). But "1+1+1 = 3" is also equivalent to "1+2 = 3". Therefore, we may perhaps be justified into saying that the (simplified) Semi-quantitative Complexity is now one "+" sign, one "=" sign and three single-digit integers (or more specifically, one "1", one "2" and one "3"). Also, the numerical parts of the corresponding Quantitative Complexity is now altered to 1, 2 and 3.

How about using the binary number system instead of the decimal number system? Now, "1+1+1 = 3" and "1+2 = 3" in base 10 is equivalent to "1+1+1 = 11" and "1+10 = 11" in base 2. As you can see, there are now single- and double-digit integers (in base 2) to contend with. Does it matter whether two odd-number base systems are used instead of two even-number base systems, or one odd and one even? Well, the answer to that is: "Yes, but only if we are dealing with rational fractions." This is explained through the concepts of regular and nonregular numbers. If you need to refresh your memory on these concepts, let me remind you that you can refer to them in relevant section on "Invented number systems, base notation and holors". From our vantage point of view here, the important message is that, under certain circumstances, a rational fraction (reciprocal of a number) can have only a finite number of places (regular number) when expressed in one base, and yet the same rational fraction can have, simultaneously, an infinitely repeating numeral (nonregular number) when expressed in another base.

On the Spherical Model of Numbers, we could use slightly different Semi-quantitative Complexity consisting of a measure based on the number of ways in which all positive integer numbers can be added up in all possible combinations and permutations of the permitted number bases for those numbers. One could also look at those same numbers in terms of the number of digits, the "+" signs, etc., as alluded to above. The other important and fundamental issue dealt with in Spherical Model of Science is the (proposed) concepts involved in semi-quantifying complexities involved in the unification of theories (meta-theory, meta-meta-theory, and so forth), and also in theories by themselves.

The aspects behind the Quantitative, but not necessarily "purely" Quantitative Complexities (in other words, it could also be Semi-quantitative

182

Complexities closer in relationship to the purely Quantitative rather than the purely Qualitative Complexity), have previously been dealt with by other renowned scientists in the past. In particular, this involved the concepts behind complexity measures such as the length of the shortest computer program needed to print out a particular message string (variously known as the algorithmic complexity or algorithmic information content or algorithmic randomness of the string); and the shortest possible solution time for a computer to solve a certain kind of problem (the computational complexity of the problem). A significant "short-coming" of these two methods is that, individually, they only deal with their respective spatial and temporal aspect, thus neglecting the combined spatio-temporal aspect. However, in no way does this detract the practical value of those two methods as, for one thing, the "purer" quantitative (or semi-quantitative) nature and the definitions of the two methods dictates that either the time or length variable be looked at separately, but not both simultaneously.

A note of considerable import is that, a satisfactory complexity measure with the important spatio-temporal aspect taken into account has to be a Semi-quantitative Complexity, whereas those measures with either the spatial or temporal aspect taken into account are Quantitative (but can also be Semi-quantitative) Complexities. Of course, Qualitative Complexity can be looked at involving all three aspects viz. spatio-temporal aspect, spatial aspect and temporal aspect. A convenient 3-point summary of applicable complexity measures (derived using the final few concepts of preceding paragraph) is prepared below for easier visualization:-

#1 Spatio-temporal aspect -> COMPULSORY Semi-quantitative Complexity measure application

183

#2 Both Spatial and Temporal aspects -> EITHER Quantitative Complexity measure OR Semi-quantitative Complexity measure application
#3 Spatio-temporal, spatial and temporal aspects -> OPTIONAL Qualitative Complexity measure application

My gut feeling is that, speculatively, there is a spectrum of complexity measures, ranging from the (one) purely Qualitative Complexity (akin to "Descriptive Complexity") measure to the (one) purely Quantitative Complexity (akin to "Computational Complexity") measure with the (more than one) complexity measures falling in between these two belonging to the category of Semi-quantitative Complexity, for each given situation. All these would apply to most, if not all, given situations in nature. For a given situation, whether we can theoretically derive this comprehensive range of complexity measures, each useful in its own right, is another matter. Here, "useful in its own right" may mean that the definition of explicit (Semi-quantitative) complexity indicators (such as static, dynamic, potential, effective and crude complexities, power spectrum, degree of mixing, entropy, thermodynamic functions, automaton representation of a language, etc.) will allow the establishment of general criteria for the identification of analogies among seemingly unrelated fields and for the inference of effective mathematical models – this can perhaps be thought of as the complementarity of various complexity indicators in specification of a complex system. Indeed, this has the connotation that Semi-quantitative Complexity may compile the list for the most "useful" group of complexity measures; and intuitively, perhaps Quantitative Complexity is next in line, followed by Qualitative Complexity (that is, if we can quantify "usefulness" in the first instance if you get my pun!) There is a further gradient (gradation) of usefulness in Semi-quantitative Complexity. A schematic diagram on usefulness property of various Complexity measures is depicted below:-

Usefulness level   *Legends:-*

```
 ^              $         Qual. Com. measure (one) = @
 |         $    $         Semi-quant. Com. measures (many) = $
 |      $       $         Quant. Com. measure (one) = #
 |   $          $
 |   @          #
 ------------------------------>  Complexity measures
```

There are many faces that complexity can manifest itself through patterns of different origin and conditions under which various forms of complexity can arise and evolve. Therefore, we need to introduce a sequence of increasingly sharp mathematical methods for the classification of complexity. Currently, we are at a stage where a systematic and critical ordering of traditional and novel complexity measures, relating them to well-established physical theories (such as statistical mechanics and ergodic theory), and to mathematical models (such as measure preserving transformations and discrete automata), are possible.

One of the biggest concern is that the mathematics employed in many situations, with respect to the Science of Complexity, is of the discrete type; whereas many, if not all, of the processes and structures in nature are continuous. This is not a destructive criticism on discrete mathematics. On the contrary, apart from this theoretical limitation, this branch of mathematics has contributed wonders to our understanding on these scientific fields and their related areas. As the dynamics of a system evolves continuously with time, so too must the dynamical aspect of the system's complexity evolves in a continuous manner with time. To use the (often approximate form of) continuous mathematics, the object (usually a pattern generated by some unknown rule) is investigated in the infinite-time or in the infinite-resolution limit, or in both, as appropriate. The current difficulty of describing or reproducing its scaling properties in the correct and complete format of continuous mathematics is due to our lack of insight into the mathematical tools and structures required for the satisfactory scientific modelling of many given real-life situations in nature.

Complexity is an extremely rich area of research that draws together ideas, methods and phenomena from almost all traditional fields of knowledge. What does this area cover? The simple answer is, "Basically anything complex and having parts that interact with each other." This is also roughly equivalent to saying, "Basically all complex systems." There are already a significant number of papers, journals and books out there, written either in paper or electronic format, by many knowledgeable and renowned scientists on this subject and its related topics. These scientists often hailed from many seemingly unrelated scientific disciplines. Researchers from many areas are involved in this science; including from physics, evolutionary biology, neuroscience, economics, medicine, philosophy, sociology, computing, engineering, game theory, chemistry, mathematics, cognitive science, social sciences, artificial intelligence, information science, and many others. This list is not exhaustive. It is a true interdisciplinary endeavor. Many more papers, journals and books dealing with complexity from these areas, from an empirical or theoretical viewpoint, or both, will no doubt be published or compiled by this new breed of researcher on a regular basis in the near future. One obvious important component of any published material should be their cross-disciplinary approach or perspective so as to promote the cross-fertilization of ideas among all the scientific disciplines having to deal with their own complex systems.

Online materials published in the electronic format can be in the Hypertext Markup Language (HTML), Portable Document Format (PDF), TeX, HyperTeX, LaTeX, and PostScriptTM formats; and in Archive or Downloadable form. Big sized files are usually compressed (zipped) with a software program such as Winzip, for quicker transfer across the Internet. Once downloaded to your personal computer, those files can subsequently be uncompressed (unzipped) before they can be viewed. All of these could be considered as "Complexity on the Internet".

A final word on complexity: Until recently, science has made progress by breaking large systems down into smaller and simpler parts, studying and explaining how these parts operate and then putting them back together again. Now, there is a new science – complexity – that asks how the world is put together. For instance, knowing how a single neuron works does not explain how the brain works because of interactions of millions of individual neurons. We stand on the threshold of this new science that promises to have monumental and wide-ranging implications for our understanding of nature.

The Science of Complexity, or simply complexity, is the major new theory that unifies all sciences. Put in another way – it is a developing area of science with an all-encompassing theoretical and practical applications to all fields of science. Significantly, this will also, incidently and inevitably, put all fields of science in perspective with one other – this is done by being able to tease out the underlying connecting theories (based on the Science of Complexity), which is common among all fields of science. Naturally, being medically oriented, I have a somewhat greater predilection for the field of medicine. Under the wide umbrella of the Science of Complexity is its hallmark theory, the Complexity Theory. In the current context here, this Complexity Theory is in no way identical to Qualitative, Semi-quantitative and Quantitative Complexities.

A useful "classification" for the Science of Complexity is to think of it as consisting of the Complexity Theory and the Qualitative, Semi-quantitative and Quantitative Complexities; with complexity indicators, such as entropy and power spectrum, belonging to the Semi-quantitative Complexity; and concepts from (for example) chaos and fractals belonging to, or are incorporated in, Complexity Theory. A simplified but highly educational definition for Complexity Theory is that it is the study of complex system, and complex system is [as defined previously above "by complex system, it is meant a system comprised of a....."]. Traditionally, and also in the view used by this book, Complexity Theory is typified by four of its major distinguishing features: self-organization, emergent phenomena, nonlinearity and concepts from chaos-fractal.

To expand Complexity Theory a bit further: Complexity Theory consists of, roughly, the study of complex adaptive and nonadaptive systems, concepts from the (metaphoric) "edge of chaos", information gathering and utilizing system (IGUS), and more..... Complexity Theory is situated on the outer layer of the Spherical Model of Science; as opposed to Qualitative, Semi-quantitative and Quantitative Complexities, where they are applicable to both layers of this model (and also to "everything in the whole wide universe"). Symbolically, complex adaptive system is best characterized by its dynamic fitness landscape.

In contrast, complex nonadaptive system is best characterized by its static fitness landscape. The fitness landscapes and factors influencing them can either be endogenous (internal) or exogenous (external) to the underlying systems. The way the dynamic fitness landscape evolves or changes has to be of a continuous nature, and these changes are pseudo-random (only looks like, but is not, totally random, since, on closer analysis, they have underlying order or pattern in them) due to the effect of the underlying deterministic chaos. Many concepts from chaos and fractals are applicable to both types of complex systems. The idea of the fitness landscape being used in this manner is a new concept.

The very complex is one of the equally challenging, important and great frontiers of science. Is the Science of Complexity resulting in a true scientific revolution? I may be biased, but I truly believe so; and, perhaps, we should aptly name this the Complexity Revolution. We are on the doorstep of designing many new tools, concepts and fresh explanations for many age-old puzzles. This revolution will alter the way that scientists approach their subject matter. Mankind has always realized that the physical universe is complex – splendidly, exuberantly, richly complex – but only recently are we beginning to comprehend just how subtle and extensive that complexity may be. In many instances, the physical universe – be it of a cultural nature, an organic or inorganic nature, a living or nonliving nature, and so on – is exquisitely complex beyond our wildest imagination.

When I first realized the significance of using **fundamental ideas such as Spherical Model of Numbers and Spherical Model of Science as the most appropriate "representation" for the simplest model synthesis of complexity**, I shouted, "Eureka!", as (part of my proposals) all other "higher" forms of complexities may now be derived from, or based on, this basic model. (By the way, "Eureka" is Greek for "I have it!", and in this context, it signifies a cry of triumph at a discovery.) Three extremely important terms (which do not exist in English dictionary) coined by Yours truly & connected to Science of Complexity, are spatialness, temporalness & spatio-temporalness. They all belong to Complexity Theory.

Finally, the general theoretical framework and principles behind the Science of Complexity have been presented in this book. This core knowledge would act as the foundation and materials for future research and our quest for better understanding in this exciting area of science, as we journey through the remaining of an exciting new millennium, the 21st Century. Albert Einstein is arguably the greatest scientist and genius of all time. I am fond of Einstein's idea that science is a quest for the secrets of the Old One – his metaphor for the creator of the universe.

"There are inherent Black and White Laws with complete accuracy applicable to Nonliving Things but only Black and White–like Laws with incomplete accuracy applicable to Living Things." By Dr. Bernhard Helpful (June 1, 2019).

Above is my quote using Black and White (B&W) concept when applied to "Elementary" Nonliving Things and "Emergent" Living Things. Mathematical-based proofs for Nonliving Things with simple or complex Elementary problems must be B&W correct (with absolute 100% certainty). Probability-based proofs for Living Things with simple or complex Emergent problems can only be B&W–like correct (with arbitrarily chosen level of statistical significance of less than and thus never equal to 100% certainty).

My two main research papers in 2019 on Riemann hypothesis, Polignac's and Twin prime conjectures were submitted to a mathematical journal. These can be accessed as "original version" viXra papers listed as (Helpful, B., Solving Incompletely Predictable problem Riemann hypothesis with Dirichlet Sigma-Power Law, April 2019) and (Helpful B., Solving Incompletely Predictable problems Polignac's and Twin Prime conjectures using Information-Complexity conservation, April 26, 2019) [reproduced respectively in Appendix 1 and 2].

Errata Notice: There are mathematical errors present in the first and second paper as relevant equations which were [incorrectly] treated as 'equations' instead of being [correctly] treated as 'inequations'. These papers are corrected, refined and amalgamated together as (Helpful B., Solving Incompletely Predictable problem Riemann hypothesis with Dirichlet Sigma-Power Law http://vixra.org/pdf/1903.0483v6.pdf, April 12, 2019) with all equations, inequations and mathematical arguments peer reviewed to be correct and complete.

We compare for similarities and contrast for differences between Riemann hypothesis (RH) and Polignac's and Twin prime conjectures (P&TPC) below.

Firstly, RH and P&TPC are *sine qua nons* classified as Incompletely Predictable problems with their rigorous proofs only possible when acknowledged and treated as such.

Secondly, to solve Completely Predictable problems require the simplicity of deriving their rigorous proofs based on simple properties whereas to solve Incompletely Predictable problems require the complexity of deriving their rigorous proofs based on complex properties ("meta-properties"). For Incompletely Predictable problems, the [few] complex properties are derived from the [many] underlying simple properties. As opposed to Completely Predictable problems such as dealing with even number gaps (= 2) or odd number gaps (= 2) endowed with simple properties which are easy to solve, dealing with RH and P&TPC which are Incompletely Predictable problems endowed with complex properties are mind-boggling hard to solve. RH and P&TPC are respectively connected with Incompletely Predictable entities nontrivial zeros (directly) derived from Riemann zeta function and prime numbers (directly) derived from Sieve of Eratosthenes. The act of explaining closely related Incompletely Predictable entities of the two types of Gram points which are (directly) derived [dependently] from Riemann zeta function should logically be classified as Incompletely Predictable problems. The Incompletely Predictable entities of composite numbers which are (indirectly) derived [dependently] from Sieve of Eratosthenes should logically be incorporated together with prime numbers to help solve P&TPC.

Thirdly, the two sets of prime and composite numbers exist at 'Numerical relationship interface' with (solitary) "outlier" even prime number '2'; and the three sets of nontrivial zeros (or Gram[x=0,y=0] points), Gram[y=0] points (or 'traditional' Gram points) and Gram[x=0] points exist at 'Axes intercept relationship interface' with (solitary) "outlier" negative Gram[y=0] point.

Fourthly, deep seated connections exist between Riemann zeta function, $\zeta(s)$, and complete Set all prime numbers 2, 3, 5, 7, 11, 13,... [but not complete Subsets of prime numbers with each uniquely derived from prime gaps 1, 2, 4, 6, 8, 10,...]. The equivalent Euler product formula from Equation 1 in (Helpful J. , April 12, 2019) with product over all prime numbers [instead of summation over natural numbers] can also be used to represent $\zeta(s)$. Thus Set all prime numbers is intrinsically "inscribed" in $\zeta(s)$. Prime number theorem, fully delineated by prime counting function [denoted by $\pi(x)$], describes asymptotic distribution of all prime numbers among positive integers by formalizing intuitive idea that prime numbers become less common as they become larger through precisely quantifying rate at which this occurs using probability. Note: we must instead use Dirichlet eta function, $\eta(s)$, [which is the *proxy* function for $\zeta(s)$] to solve RH. Solving RH is instrumental in proving efficacy of techniques that estimate $\pi(x)$ efficiently thus confirming "best possible" bound for error ("smallest possible" error) of this theorem.

Fifthly, solving RH involves rigorously proving the complete Set nontrivial zeros [of known infinite magnitude] to be located on so-called critical line whereas solving P&TPC involves rigorously proving the existence of complete Set all prime numbers [of known infinite magnitude] to be constituted by Subsets of prime numbers [each proposed to be of infinite magnitude] uniquely derived from all even number prime gaps [proposed to be of infinite magnitude]. RH deals with 'Set' whereas P&TPC deals with 'Subsets'. Note that Polignac's conjecture concerns Subsets of prime numbers derived from all even number prime gaps 2, 4, 6, 8, 10,… but Twin prime conjecture concerns Subset of prime numbers derived only from even number prime gap 2. Thus the later conjecture is intrinsically just part of the former conjecture. From Set theory, the Sets and Subsets of prime numbers will always comply with 'well-ordering Principle' (which states that every non-empty set of positive integers contains a least element) and 'pigeonhole principle' (which states that if n items are put into m containers with n>m, then at least one container must contain more than one item).

We now provide a Hierarchical Classification for Elementary-Emergent Fundamental Laws (EEFL). Implied by the definition for 'Fundamental Laws', then EEFL must by default be perfectly applicable to Terrestrial human beings on planet Earth (endowed with advanced civilization) and also Extraterrestrial alien beings on some hypothetical remote planet (endowed with super-advanced civilization). Thus one could also appropriately coin our Fundamental Laws as the Extraterrestrial-Terrestrial EEFL.

In order of increasing complexity, we have the following Laws:

194

Law I	Simple Elementary Fundamental Law for "simple" Nonliving Things with simple properties
Law II	Complex Elementary Fundamental Law for "complex" Nonliving Things with complex properties
Law III	Simple Emergent Fundamental Law for "simple" Living Things with simple properties
Law IV	Complex Emergent Fundamental Law for "complex" Living Things with complex properties

Solving Completely Predictable and Inompletely Predictable problems:

Solving Completely Predictable problems in both Simple 'Nonliving' Elementary and 'Living' Emergent cases	*Many* Simple properties → [Simple Elementary and Emergent Solutions]
Solving Incompletely Predictable problems in both Complex 'Nonliving' Elementary and 'Living' Emergent cases	*Many* Simple properties → *Few* Complex properties → [Complex Elementary and Emergent Solutions]

The three types of entities:

Type I Entities	Completely Unpredictable entities
Type II Entities	Completely Predictable entities
Type III Entities	Incompletely Predictable entities

Type I Entities occur purely as totally random physical processes in nature e.g. radioactive decay is a stochastic (random) process occurring at level of single atoms. According to Quantum theory, it is impossible to predict when a particular atom will decay regardless of how long the atom has existed. For a collection of atoms, expected decay rate is characterized in terms of their measured decay constants or half-lives.

The location-based definitions for Type II Entities and Type III Entities are:

Completely Predictable (Type II) Entities	Locationally defined as entities whose position is **independently** determined by **simple** calculations using simple equation or algorithm **without** needing to know related positions of all preceding entities in neighborhood.
Incompletely Predictable (Type III) Entities	Locationally defined as entities whose position is **dependently** determined by **complex** calculations using complex equation or algorithm **with** needing to know related positions of all preceding entities in neighborhood.

Postulated association between Entities and Laws:

Law I is obeyed by Type II Entities	e.g. (simple Nonliving Thing) even and odd numbers with even number and odd number gaps.
Law II is obeyed by Type III Entities	e.g. (complex Nonliving Thing) nontrivial zeros related to RH, and prime numbers related to P&TPC.
Law III is obeyed by Type II + Type III Entities	e.g. (simple Living Thing) human heart as an organ manifesting hemodynamic and electrical properties.
Law IV is obeyed by Type I + Type II + Type III Entities	e.g. (complex Living Thing) human brain which is often dubbed "the most complex structure in the universe" manifesting a whole range of neuro-psychological and neuro-psychiatric properties.

Human heart can simplistically be thought of having a "plumbing system" consisting of heart muscle pump, coronary arteries and cardiac valves; and an "electrical system" consisting of specialized heart muscle cells giving rise to pacemakers and electrical conduction pathways & networks.

Human brain is the most complex organ in human body. It produces our every thought, action, memory, feeling and experience of the world. It consists of jelly-like mass of tissue weighing around 1.4 kilograms, and contains a staggering one hundred billion nerve cells (neurons).

The complexity of the connectivity between these cells is mind-blowing with each neuron making contact with thousands or even tens of thousands of others, via tiny structures called synapses. Our brains form about a million new connections per second. Our conscious mind commands and our subconscious mind obeys. Thus, our subconscious mind is an unquestioning servant that works day and night to make our behavior fits a pattern consistent with our emotionalized thoughts, hopes, and desires. The pattern and strength of the connections is constantly changing and no two brains are alike. It is in these changing connections that memories are stored, subconscious mind operate, habits learned and personalities shaped by reinforcing certain patterns of brain activity, and losing others.

Structurally, the human brain contains "grey matter" and "white matter". The grey matter is the cell bodies of the neurons, while the white matter is the branching network of thread-like tendrils called dendrites and axons that spread out from the cell bodies to connect to other neurons. However, the human brain also has another even more numerous type of cell called glial cells. These outnumber neurons about ten times over. Once thought to be support cells, they are now known to amplify neural signals and to be as important as neurons in mental calculations.

In summary, the human brain manifest Natural Intelligence, consciousness, self-awareness, memory; mental illness such as anxiety, depression, schizophrenia; "dark triad" of personality consisting of three negative traits [viz. the tendency to manipulate others (Machiavellianism), seek admiration and special treatment (narcissism), and to be callous and insensitive (psychopathy)]; and "light triad" of personality consisting of three positive traits [viz. the opposite of Machiavellianism (Kantianism), valuing dignity and worth of each individual person (humanism), and believing that people are fundamentally good (Faith in humanity)].

Artificial Intelligence (AI) in Nonliving Things can be regarded as human endeavor to simulate Natural Intelligence in Living Things using powerful computers such as super-computers or quantum computers. DNA is a double helix, while RNA is a single helix. Both have sets of nucleotides that contain genetic information. DNA is a molecule that contains instructions for Living Things to be born, mature, reproduce, and died.

One would commonly concur that there are 'Simple' Living Things such as bacteria without brain and 'Complex' Living Things such as intelligent human with highly developed brain. The dividing line between Living Things and Nonliving Things is that the former is "powered" by DNA with an important implication that Natural Intelligence, consciousness and self-awareness can only be "powered" by DNA [which are organic]. Then by reasonable assumption, properties such as consciousness and self-awareness can never be present in AI created using computers [which are inorganic].

Creationism versus Evolution debate for **Nonliving Things** (obeying Law I and Law II) giving rise to **Living Things** (obeying Law III and Law IV) is compared and contrasted below:

Process of Creationism	Process of Evolution
Associated with major religions e.g. Islam and Christianity. From the Bible, Adam and Eve was estimated to be created just over 6,000 years ago by world's leading young-earth creationist organizations.	Atheists usually believe the Big Bang (when our Universe was created) occur about 13.8 billion years ago. The first true man appeared 13.7998 billion years after the beginning (or about 200,000 years ago).

People from most western countries generally embrace Christianity which is symbolized by the Bible. This is reportedly the biggest bestseller of all time with historically the Bible Creation starting at Before Christ (B.C.) 2000 whereby the earliest Scriptures are handed down from generation to generation orally. Circa B.C. 2000 - 1500 is when the book of Job, perhaps the oldest book of the Bible, was written. The New King James Version (NKJV) of the Bible, as an example, was published in the modern era in [A.D.] 1982. The term anno Domini (A.D.) is Medieval Latin meaning "in the year of the Lord" used to label or number years in the Julian and Gregorian calendars.

To avoid conflicts, humanity must respect the freedom to practice [or not practice] all different religions. How can we reconcile the huge time discrepancy noted above between the process of creationism and evolution? A controversial thought is perhaps the process of "simplified" evolution with natural selection (survival of the fittest) and adaptation as plausible mechanisms occurs in both Simple and Complex Living Things on short, medium and long term scale in past, present and future. Complex Living Things with brains can only arise through creationism. In particular, the complex neuronal brain tissue can only be "created" by God and cannot "evolve" from simple living tissue over the four eons of geologic time scale.

The human genome is the complete set of nucleic acid sequences for humans, encoded as DNA within the 23 chromosome pairs in cell nuclei and in a small DNA molecule found within individual mitochondrion. Mitochondrion is an organelle found in large numbers in most human and non-human living cells in which the biochemical processes of respiration and energy production occur. It has a double membrane, the inner part being folded inwards to form layers (cristae). Population geneticists believe ancestral human population lived somewhere in Africa and started to split up some time after 144,000 years ago [give or take 10,000 years] – the inferred time at which both the mitochondrial and Y chromosome trees make their first branches.

Mitochondria [plural noun] can thus be used to study the detailed Human Family Tree whereby they live inside human cells but outside the nucleus, thus escaping the shuffling of genes that occurs between generations and are passed unchanged from mother to children. In other words, the tiny rings of genetic material in mitochondria are bequeathed only by the egg cell and thus through the maternal line. In principle, all people should have the same string of DNA letters in their mitochondria. In practice, mitochondrial DNA has steadily accumulated changes over the centuries because of copying errors and radiation damage. This has resulted in different set of lineages for mitochondria being descended in different racial groups from particular regions and continents. Example, Europeans belong to a different set of lineages designated H through K and T through X. The split between the two main branches in the European tree suggests that modern humans reached Europe 39,000 to 51,000 years ago.

The following are subjective comments. For Nonliving Things, we would intuitively associate performing calculation $2 + 3 = 3 + 2 = 5$ as a simple case of elementary Completely Predictable problem; and solving RH and P&TPC as complex cases of elementary Incompletely Predictable problems. For Living Things, we intuitively associate analyzing human heart as a simple case of emergent Completely Predictable problem; and analyzing human brain as a complex case of emergent Incompletely Predictable problem. Because mathematical language for describing complex Incompletely Predictable problems in Nonliving Things (such as weather forecasting) and Living Things (such as determining neurophysiology of human memory) are convoluted, we can only ever obtain approximate models of these problems.

Riemann hypothesis was proposed by famous German mathematician

Bernhard Riemann (17 September 17, 1826 – July 20, 1866) in 1859. Twin prime and Polignac's conjectures were proposed by French mathematician Alphonse de Polignac (1826 – 1863) in, respectively, 1846 and 1849. In this chapter, I offer my personal opinion with some bold statements on why intractable open [Incompletely Predictable] problems in Number theory of Riemann hypothesis (RH), Polignac's and Twin prime conjectures (P&TPC) have previously not been solved for over 150 years. Chapter 11 above outlining complicated similarities and differences between these open problems will already provide good insight why delay in solving them occur.

Perhaps most mathematicians have a weird sense of humor. My task to explain this delay is simplified using quirky terms 'Beautiful Mathematics' (BM) and 'Sexy Mathematics' (SM) to provide succinct mental pictures thus promoting optimal understanding by the general public.

Beautiful Mathematics (BM)	Mental picture for Completely Predictable problems which are **easy** to solve requiring BM which involves analyzing their (intrinsic) **simple** properties.
Sexy Mathematics (SM)	Mental picture for Incompletely Predictable problems which are **difficult** to solve requiring SM which involves analyzing their (intrinsic) **complex** properties

With the [few] complex properties derived from the underlying [many] simple properties in Incompletely Predictable problems, the caveat is that only through correctly analyzing these complex properties (or "meta-properties") will we ever obtain their complete solutions. Underlying simple properties in Incompletely Predictable problems can be falsely perceived to be complex properties. Complex properties in Incompletely Predictable problems can be hidden away in some subtle manner. Thus actual complex properties are notoriously difficult to correctly decipher with mathematicians frequently **barking up the wrong tree** in Incompletely Predictable problems. Another mistake is mathematicians utilizing "manifestations" of relevant complex equations or complex algorithms from current Incompletely Predictable problems to compare with near-identical "manifestations" derived from some other *seemingly-related* Incompletely Predictable problems that were successfully solved in the past; and subsequently [incorrectly] claiming successful proof – this is dubbed by me as **'pseudo-proof'**.

In other words for Incompletely Predictable problems, **barking up the wrong tree** is the equivalent of mathematician [incorrectly] analyzing the "beautiful tree" with simple properties using BM while **barking up the right tree** is the equivalent of mathematician [correctly] analyzing the "sexy tree" with complex properties using SM. Figuratively speaking, there are many "beautiful trees" choices but only a few "sexy trees" choices. So what actually is this so-called SM for RH and P&TPC? The answer is illustrated using the "mathematical impasse" phenomenon for Completely Predictable problem involving even numbers and their gaps (in Diagram 1) and for Incompletely Predictable problem involving prime numbers and their gaps (in Diagram 2).

Diagram 1. Legend: Even numbers = E, even number gaps = eGap.

E	2		4		6		8		10		12	
eGap		2		2		2		2		2		2

Question: Prove the proposal that even number gaps are always constant and non-varying. Answer: Finite calculations shown in Diagram 1 depict and support even number gaps [= 2] is constant and non-varying but even numbers are infinite in magnitude requiring an infinite number of calculations ("mathematical impasse") in order to show these gaps will always be constant and non-varying.

Obtaining rigorous proof then consist of recognizing this as Completely Predictable problem deriving a **Completely Predictable 'non-varying' equation** for calculating all even numbers which will [intrinsically] contain simple property "all even number gaps = 2". This equation is literally the 'Simple Container' containing all even numbers.

Diagram 2. Legend: Prime numbers = P, prime number gaps = pGap.

P	2		3		5		7		11		13	
pGap		1		2		2		4		2		4

Question: Prove the proposal that apart from first prime number gap [=1] followed by next two consecutive prime number gaps [= 2], prime number gaps are always even numbers and varying. Answer: Finite calculations shown in Diagram 2 depict and support prime number gaps [after the third one] are even numbers and varying but prime numbers are infinite in magnitude requiring an infinite number of calculations ("mathematical impasse") in order to show these gaps will always be even numbers and varying.

Obtaining rigorous proof then consist of recognizing this as Incompletely Predictable problem deriving an **Incompletely Predictable 'varying' equation** for calculating all prime numbers which will [intrinsically] contain complex property "all prime gaps are even numbers and perpetually varying". This equation is literally the 'Complex Container' containing all prime numbers.

Below is another approach to grasp the concepts of "Incompletely Predictable" versus "Completely Predictable" and "Beautiful Mathematics" versus "Sexy Mathematics".

Why read my book? How Riemann hypothesis (RH) is successfully solved is one of the main reason to justify reading this book. Many current mathematical theorems contain proofs that depend on deriving the correct solution for RH. It is often quoted along the line that "successfully solving RH will immediately result in proving 500 theorems or more at once". One big reason why RH, Polignac's and Twin prime conjectures are notoriously difficult to solve despite their rigorous proofs being literally "right under our nose" is simply because of mathematicians failing to crucially recognize and accept that entities or numbers such as nontrivial zeros and prime numbers associated with these open problems are Incompletely Predictable entities! However, after having recognize and accept these entities as "Incompletely Predictable" [as described next]; it still takes the author a mind-boggling effort to rigorously prove these problems with "correctness" and "completeness" in the sense that the total steps of mathematical arguments must be numerically "100% complete" and mathematically "100% correct" in all relevant submitted proofs under peer review.

What is the meaning of "Incompletely Predictable" versus "Completely Predictable"? The set of Natural numbers 1, 2, 3, 4, 5, 6,.... are made up of the two subsets of Even numbers 2, 4, 6, 8, 10,... and Odd numbers 1, 3, 5, 7, 9, 11,... The two subsets of Even and odd numbers are "Independent" and "Completely Predictable". Example: the next even number after 2,984 (which is the 2,984 / 2 = 1,492nd Even number) is [easily] seen to [independently] be 2,984 + 2 = 2,986 (which is the 2,986 / 2 = 1,493rd Even number). The set of Natural numbers 1, 2, 3, 4, 5, 6,.... are made up of the Number '1' (which is neither Prime nor Composite) plus the two subsets of Prime numbers 2, 3, 5, 7, 11,... and Composite numbers 4, 6, 8, 9, 10, 12,... The two subsets of Prime and Composite numbers are "Dependent" and "Incompletely Predictable". Example: the next sixth prime number 13 [after the fifth prime number 11] has to be [not easily and dependently] computed from "scratch" as: 2 is first prime number, 3 is second prime number, 4 is the first composite number, 5 is the third prime number, 6 is the second composite number, 7 is the fourth prime number, 8 is the third composite number, 9 is the fourth composite number, 10 is the fifth composite number, 11 is the fifth prime number, 12 is the sixth composite number, and finally our desired 13 is the sixth prime number.

How were the proofs for Riemann hypothesis, Polignac's and Twin prime conjectures derived? Using colloquially-speaking "Beautiful Mathematics" (for Completely Predictable problems) versus "Sexy Mathematics" (for

207

Incompletely Predictable problems). Recommend the general public to read the rest of this book to understand more.

"Thanks Bernhard! ... Very impressive!" adapted from feedback given by my learned friend Les from Australia in his 2:16 PM June 30, 2019 email reply to my layman explanations on Riemann hypothesis, Polignac's and Twin prime conjectures [as literally given above]. This feedback use, incorporating [first] name identity of person providing it, is approved by Les.

As outlined in (Helpful B., April 12, 2019) containing rigorous proof for RH and explaining Gram points, our 'overall' complex properties consist of three variants of Dirichlet Sigma-Power Laws precisely manifesting the required exact and inexact Dimensional analysis homogeneity. These novel Laws are derived from Dirichlet eta function, the proxy function for Riemann zeta function. As outlined in (Helpful B., April 26, 2019) containing rigorous proofs for P&TPC, our 'overall' complex properties consist of Plus-Minus Gap 2 Composite Number Alternating Law being precisely obeyed by all even number prime gaps apart from first even number prime gap precisely obeying Plus Gap 2 Composite Number Continuous Law. These Laws are derived using novel research method Information-Complexity conservation.

Finally, another useful mental picture on why Incompletely Predictable problems such as Riemann hypothesis, Polignac's and Twin prime conjectures are so difficult to solve is that they require complex mathematical arguments belonging to **'Special-Class-of-Mathematical-Problems with Solitary-Proof-Solution'** whereas Completely Predictable problems such as proving even number gaps = 2 and odd number gaps = 2 only require simple mathematical arguments based on mathematical calculus or geometrical gradient method belonging to '**General-Class-of-Mathematical-Problems with Multiple-Proof-Solutions**'. Thus Completely Predictable problems can easily be "separately" and "independently" solved whereas Incompletely Predictable problems have to be "combined together" and "dependently" solved with difficulty.

Useful Overall Perspective on Incompletely Predictable entities whereby the word "number" [singular noun] or "numbers" [plural noun] could easily be used interchangeably with the word "entity" [singular noun] or "entities" [plural noun]: Mathematics for Incompletely Predictable Problems makes all mathematical arguments valid and complete in my Helpful B., April 12, 2019 paper (based on first key step of converting Riemann zeta function into its continuous format version) and my Helpful B., April 26, 2019 paper (based on second key step of applying Information-Complexity conservation to Sieve of Eratosthenes). Nontrivial zeros and two types of Gram points calculated using this function plus prime and composite numbers computed using this Sieve are defined as Incompletely Predictable entities. Euler product formula alternatively and exactly represents Riemann zeta function but utilizes product over prime numbers (instead of summation over natural numbers). Hence prime numbers are encoded in this function demonstrating deep connection between them. Direct spin-offs from first step consist of proving Riemann hypothesis and explaining manifested properties of both Gram points, and from second step consist of proving Polignac's and Twin prime conjectures. These mentioned open problems are defined as Incompletely Predictable problems.

14 Jelena 'Shining Light'

I have practiced in the specialty field of Anesthesia, Intensive Care, Pain Medicine, Medicinal cannabis, and Addiction Medicine. When I was training in Anesthesia from 2009 to 2013, I was often told that the ability to effectively communicate with my patients is paramount for good patient care. Useful idiom when practicing medicine: Patients 1st, Doctors and Nurses 2nd, Administration and Regulatory Body 3rd.

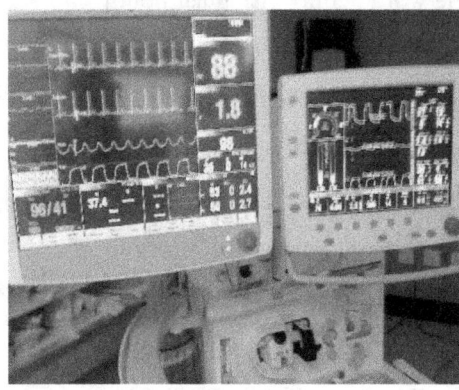

Rebelling against the pressure-cooker life being a doctor, I have always been privileged to be involved in my spare-time hobby of solving open problems in Number theory of Riemann hypothesis, Polignac's and Twin prime conjectures. I estimate it took me about three years from March 2016 to April 2019 to obtain rigorous proofs for these three open problems. To many, mathematical literature is often seen as an impenetrable wall of logic, symbols and formulas. I recommend this book to readers who want to get a meaningful glimpse of what is behind the wall and how the wall can be penetrated. I have endeavored to write this book describing the artistic, creative and human, and spiritual aspect of mathematical enterprise.

With three older sisters and two younger brothers, I am the eldest son of a domineering father. My mother died from a stroke at 75 years of age in 2016. Being raised on a rigmarole of high expectations and little praise, I developed an unwanted psychological Dependent Personality trait with negative consequences but I resolved this issue in Year 2000 by initiating good father-son relationship with lots of help from family and friends.

On Monday May 14, 2012 my youngest daughter Jelena (meaning 'Shinning Light' in Russian) was born 13 weeks premature with a tiny birth weight of 1010 grams (2. 2 pounds). She spent 7 weeks in Neonatal Intensive Care Unit.

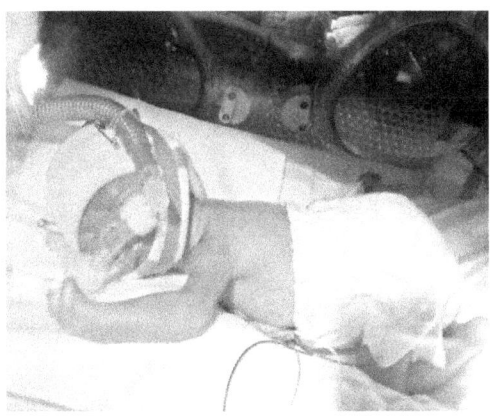

From Olaf Helmer and Nicholas Rescher: On the Epistemiology of the Inexact Sciences, P-1513 (October 13, 1959): "In medicine, exact explanation of causes of diseases, concise diagnosis and absolute predictability of outcome of treatment are difficult, if not impossible!" Thus there is a popular saying **"Medicine is an inexact science"**. Neonates do feel pain and require analgesic relief. Remifentanil is a potent, short-acting synthetic opioid analgesic drug with an effective biological half-life of 3 to 10 minutes. As this drug is esterase metabolized, it is not dependent on immature liver enzymes for metabolism. Therefore its theoretical advantage is that it provide superior analgesia of an opioid without causing prolonged respiratory depression.

For a good cause, my wife Jocelyn and I enrolled Jelena in the **premi-remi study** on May 21, 2012 at 28.2 weeks gestation during her PICC line insertion procedure in left ankle saphenous vein. With primary aim to determine efficacy of remifentanil infusion for alleviating pain in neonates requiring insertion of central venous lines for their medical care, this study is similar to the study "Remifentanil for percutaneous intravenous central catheter placement in preterm infant: a randomized controlled trial" by (Lago P, 2008). Thus Jelena became part of history contributing data as one of the recruited neonatal subjects in this study based on randomized double-blind controlled clinical trial.

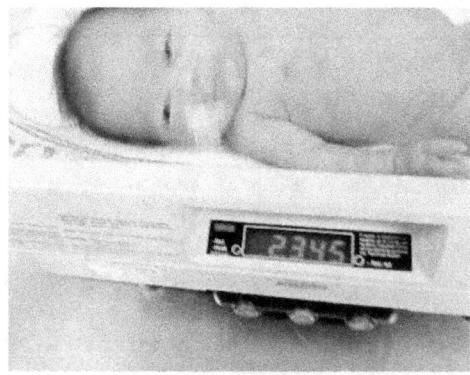

With a feeding tube up her nose, Jelena weight was exactly 2345 grams on July 13, 2012.

{Publication Warning: Timing of events in this chapter are altered with real identities of people, places, businesses, entities, and so on hidden or fictionalized so that they cannot be easily recognized or discovered. Any potentially libelous claims and statements that might invade a person's privacy are altered and minimized.}

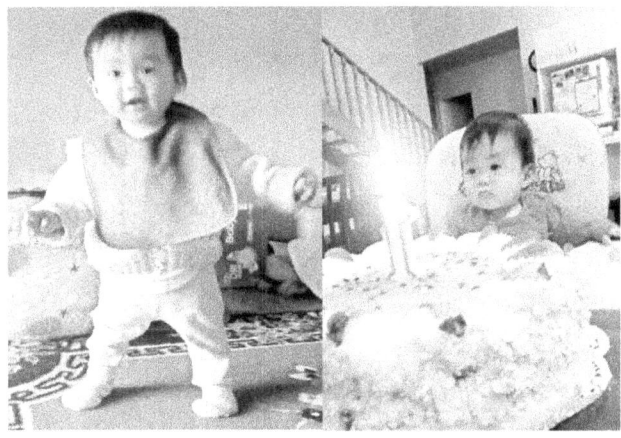

Jelena was born 13 weeks early on May 14, 2012. By her first birthday, she is an inspiration and a fighter developing into a normal healthy child. It is a no-brainer I will always be proud of her. I have to stop my full time Anesthesia career in 2013 to help look after my five children. I continue to do some Anesthesia in rural parts of Australia until 2016. At beginning of 2019, my medical career and livelihood were in jeopardy. I can hardly support my family financially as I am unable to work casually in general practice due to regulatory requirements. These mainly arise from unwarranted allegations levelled against me (now using my Pen Name **Dr. Bernhard Helpful** below) by Big Brother Health Department for practicing Addiction Medicine.

This practice by Dr. Helpful in 'Sleepy Suburb' relates to dutifully combating the rampant and endemic society illness of drug addiction in a chaotic *en masse* manner for up to 123 difficult and vulnerable patients in 2017 and 2018 when their 101 year-old Dr. Elderly who has been practicing Addiction Medicine in his large Dosing Clinic catering for 175 Drug Dependent Persons and for 20 years was graciously retired off on Christmas Day December 25, 2017. Dr. Helpful was also instrumental in providing common-sense and fair opinions to the assessing psychiatrist in late 2017 on Dr. Elderly's fitness to practice medicine such that Dr. Helpful advocate he should continue to carry so-called S4 and S8 license endorsements for dangerous drug but only use these drugs in the general population [and not for Addiction Medicine]. Here is the Medical Conspiracy. It was subsequently discovered in second half of 2019 that Big Brother Health Department has been seeking a scapegoat to blame for a few cases of drug overdoses which may be caused by Dr. Elderly with this perhaps contributing to the death of one young woman in 2017 – see further explanations below. One feels that the cliché phrase "Only the good die young" may apply here. This phrase actually means that highly-regarded people who are morally-upright, kind and compassionate tend to die at a younger age than most people do. It originated in a proverb by Greek historian Herodotus (c. 484 BC – c. 425 BC) in 445 BC. He wrote, "Whom the gods love dies young."

The Diagnostic and Statistical Manual of Mental Disorders (May 2013), Fifth Edition, often called the DSM-V or DSM 5, is the more recent version of American Psychiatric Association's gold-standard text on names, symptoms, and diagnostic features of every recognized mental illness – including addictions. The DSM 5 criteria for substance use disorders are based on decades of research and clinical knowledge. Substance use disorders span a wide variety of problems arising from substance use, and cover 11 different criteria:

1. Taking the substance in larger amounts or for longer than you're meant to.
2. Wanting to cut down or stop using the substance but not managing to.
3. Spending a lot of time getting, using, or recovering from use of the substance.
4. Cravings and urges to use the substance.
5. Not managing to do what you should at work, home, or school because of substance use.
6. Continuing to use, even when it causes problems in relationships.
7. Giving up important social, occupational, or recreational activities because of substance use.
8. Using substances again and again, even when it puts you in danger.
9. Continuing to use, even when you know you have a physical or psychological problem that could have been caused or made worse by the substance.
10. Needing more of the substance to get the effect you want (tolerance).
11. Development of withdrawal symptoms, which can be relieved by taking more of the substance.

For Severity of Substance Use Disorders, DSM 5 allows clinicians to specify how severe or how much of a problem the substance use disorder is, depending on how many symptoms are identified. Two or three symptoms indicate a mild substance use disorder, four or five symptoms indicate a moderate substance use disorder, and six or more symptoms indicate a severe substance use disorder. Clinicians can also add "in early remission," "in sustained remission," "on maintenance therapy," and "in a controlled environment". *Physician, heal thyself* (Greek: Ἰατρέ, θεράπευσον σεαυτόν – Iatre, therapeuson seauton), sometimes quoted in the Latin form, Medice, cura te ipsum, is an ancient proverb appearing in the Bible at Luke 4:23. In harsh judgement, this proverb seems to be applicable to Dr. Elderly. But good Clinical Practice in Modern Medicine boils down to always employing 'Evidence-based Practice' which roughly equates to balancing 'Evidence-based Medicine' with 'Doctor Experience' & 'Patient Expectation'.

Despite the well-known pharmacodynamics mediated phenomenon in development of drug tolerance with prolonged use of certain drugs that belong to the class of Benzodiazepines and Hypnotics, Dr. Elderly's misguided usage of multiple types of these drugs and in large doses for incorrectly perceived 'opioid-sparing' role did perhaps result in a number of drug overdose cases with harmful consequences. Big Brother Health Department was keen to prosecute Dr. Elderly with resulting loss of dignity. Dr. Helpful non-judgmentally rationalize with assessing psychiatrist that Dr. Elderly's unusual practice was caused by poor understanding rather than medical incompetence endangering patient life. Through negotiation, Dr. Helpful was then asked by Big Brother Health Department to initially "cover" and subsequently "take over" the care of Dr. Elderly's patients.

With Opioid Replacement license obtained on May 3, 2017 and extensive pharmacology knowledge gained from qualifications and experiences in Anesthesia, Intensive Care and Pain Medicine from 2009 to 2013; Dr. Helpful commence seeing this large group of patients under "quota of 100 patients" maiden approval for July 1, 2017 to June 30, 2018 period very soon after starting off with the [usual] initial "quota of 5 patients" approval which was unlawfully dated April Fool's Day on April 1, 2017 [whereby this date is before the May 3, 2017 date of license]. This large "quota of 100 patients" was furthermore unlawfully back-dated on October 13, 2017 to that effect by Big Brother Health Department. Note that the overall goal of good patient care is always for Big Brother Health Department to work closely with all doctors in order to prevent & minimize patient harm.

On February 5, 2019 Mr. Auditor from Big Brother Health Department falsify about sending Dr. Helpful a crucial email containing an important attachment; and furthermore on a separate February 9, 2019 email made a false admission that the alleged February 5, 2019 email was actually sent to Dr. Helpful. This despicable act was carried out by Mr. Auditor on February 5, 2019 with intentional, unlawful & impersonating use of [incorrect] email address Dr.Bernhard.Unhelpful@gmail.com instead of [correct] email address Dr.Bernhard.Helpful@gmail.com. In addition, the senior Dr. Boss from Big Brother Health Department made a number of false allegations against Dr. Helpful about some of his confidential patient care not adhering to required regulations. All above primitive blame-game unfair plays have since been elegantly proven by Dr. Helpful. These acts breached the desperately needed shared responsibilities & mutual trust between a regulatory body & doctors when working as a team to effect "harm minimization" or optimize $\frac{Harm}{Benefit}$ Ratio" for patient care. Importantly, these nasty issues were eventually resolved successfully in a mutually acceptable manner. A popular saying goes like this "Time heals all things" will undoubtedly apply to healing of the strained relationship "wounds" that have since existed between Dr. Helpful and Big Brother Health Department.

Although Dr. Helpful has always provide safe treatments for these patients, he did fail to complete some mandatory paper-works due to factors such as time constraint and health practice location change on December 31, 2017 resulting in lack of access to previous medical records for these patients. Being a Christian, Dr. Helpful was grateful to tackle all above mentioned trials and tribulations with faith in God and support from his close friends and family.

Finally, here is Dr. Helpful choice for top three Mega-Achievements by mankind:

(1) American astronaut and aeronautical engineer Neil Armstrong (August 5, 1930 – August 25, 2012) was the first person to walk on the Moon on July 21, 1969 and spoke the now-famous words, "That's one small step for [a] man, one giant leap for mankind."

(2) NASA Voyager 1 and Voyager 2 space probes both launched in 1977 to study the outer planets with computing power available in these probes less than that in a modern mobile phone.

(3) China Economic Miracle. People's Republic of China as economic superpower with market economy currently calculated in 2019 as the

218

world's second largest economy by nominal gross domestic product (GDP) and the world's largest economy by purchasing power parity. Until 2015, China was the world's fastest-growing major economy with growth rates averaging 6% over 30 years. She is the world's largest manufacturing economy and trading nation; and also the world's fastest-growing consumer market and second-largest importer of goods. She is a net importer of services products.

In 1998 and 2003, Dr. Helpful extensively tour China, Hong Kong and Macau. Building of numerous skyscrapers in all major Chinese cities occur rapidly overnight enabling the largest recorded-in-history human mass migration [from rural China to urban China]. Since the August 8 - 24, 2008 Summer Olympic Games in Beijing, transport development such as on China high speed train running from 200km/hr to 350km/hr with reasonable priced tickets occur at a phenomenal pace.

Beijing, being administrative capital of China and rich in ancient Chinese history, is often symbolized by the Forbidden City (Chinese: 故宫 ; pinyin: *Gùgōng*) and the Summer Palace (simplified Chinese: 颐和园 traditional Chinese: 頤和園; pinyin: *Yíhéyuán*). Shanghai is the symbol of China economic superpower success. Shenzhen, the major city in Guangdong Province, is China's Silicon Valley. It was one of the fastest-growing cities in the world in the 1990's and 2000's contributing to China Economic Miracle. The equivalent Silicon Valley in southern San Francisco Bay Area of California, United States of America, is home to many start-up and global technology companies such as Apple, Facebook and Google.

16 Incompletely Predictable numbers, Gram points and A228186

Completely Unpredictable numbers arising from totally chaotic physical processes in nature give rise to countable infinite set (CIS) of measured true random numbers. In this sense, computational pseudorandom number generator employing deterministic logic cannot be a source for true random numbers. There are two types of Predictable number: CIS of Completely, and CIS of Incompletely Predictable numbers with former "contained" in [Simple Container] equation or algorithm obeying Simple Elementary Fundamental Laws, and later "contained" in [Complex Container] complex equation or algorithm obeying Complex Elementary Fundamental Laws.

Prime numbers behave pseudorandomly with a strange mixture of order (structure) and chaos. It is with this pseudorandomly behavior that prime [and composite] numbers are regarded as 'Pseudorandom numbers'. We allude readers here that usage of the term 'Pseudorandom number' is deemed to be synonymous with usage of our devised term 'Incompletely Predictable number'. The concepts behind 'chaos' and 'order' in prime numbers as Pseudorandom or Incompletely Predictable numbers are (respectively) exemplified using the following popular quotes obtained from the May 5, 1975 inaugural lecture 'The First 50 Million Prime Numbers' at Bonn University by mathematician Don Zagier. These two quotes are "...despite their simple definition and role as the building blocks of the natural numbers, the prime numbers belong to the most arbitrary and ornery objects studied

220

by mathematicians: they grow like weeds among the natural numbers, seeming to obey no other law than that of chance, and nobody can predict where the next one will sprout" and "...the prime numbers exhibit stunning regularity, that there are laws governing their behavior, and that they obey these laws with almost military precision".

A Completely, and Incompletely Predictable number is location-defined as a number whose position is independently determined by simple calculations using simple equation/algorithm without, and dependently determined by complex calculations using complex equation/algorithm with needing to know related positions of all preceding numbers in its neighborhood. Both types of Predictable number exist as either rational [integers or fractions of two integers] numbers (Q) or irrational [algebraic or transcendental] numbers ($R - Q$). A well-defined set of $R - Q$ will twice obey relevant location definition in CIS $R - Q$ themselves and in CIS numerical digits after decimal point of each $R - Q$.

97 is an Incompletely Predictable number whose precise position is determined by computing positions of all preceding 24 prime numbers (P) using complex algorithm Sieve of Eratosthenes to conclude that 97 is the 25th P. Calculated using simple algorithm, 97 is also [i = (97+1)/2] 49th odd number (O) which is a Completely Predictable number. 98 & 99 are respectively [i = 98/2] 49th even number (E) & [i = (99+1)/2] 50th O which are Completely Predictable numbers calculated using simple algorithm. Determined indirectly using complex algorithm Sieve of Erastosthenes, 98 & 99 are respectively also the 72nd & 73rd composite numbers (C) which are Incompletely Predictable numbers.

Computing Riemann zeta function (or more specifically its *proxy* Dirichlet eta function) and Sieve of Eratosthenes will, respectively, supply Incompletely Predictable nontrivial zeros, Gram[y=0] & Gram[x=0] points and **P** & **C**. CIS of nontrivial zeros (denoted by imaginary part parameter t) = CIS of transcendental numbers = 14.134725, 21.022040, 25.010858, 30.424876, 32.935062, 37.586178,... [rounded off to six decimal places]. CIS of all **P** = Countable Finite Set (CFS) of all even **P** + CIS of all odd **P** = 2, 3, 5, 7, 11, 13,... whereby **P** '2' when treated as **E** is also regarded as a Completely Predictable number.

The three sets of nontrivial zeros, Gram[y=0] points and Gram[x=0] points, respectively, will dependently constitute three sets of Origin intercepts (or simultaneous x- & y-axes intercepts), x-axis intercepts and y-axis intercepts. Traditional 'Gram points' (or Gram[y=0] points) are relevant x-axis intercepts with choice of index 'n' for 'Gram points' historically chosen such that first 'Gram point' corresponds to t value which is larger than (first) nontrivial zero located at t = 14.134725. By convention, first six Gram[y=0] points will occur with following values [rounded off to six decimal places]: at n = -3, t = 0; at n = -2, t = 3.436218; at n = -1, t = 9.666908; at n = 0, t = 17.845599; at n = 1, t = 23.170282; at n = 2, t = 27.670182.

The two sets of **P** 2, 3, 5, 7, 11, 13,... and **C** 4, 6, 8, 9, 10, 12,... will dependently constitute set of natural numbers (**N**) 1, 2, 3, 4, 5, 6,... minus first **N** '1'. Whole numbers (**W**) = **N** plus '0'. '0' & '1' are special numbers being neither **P** nor **C** as they represent nothingness (zero) and wholeness (one), and the idea of having factors for '0' & '1' is meaningless. Treating '0'

& '1' here as Completely or Incompletely Predictable numbers is also meaningless.

CIS of numbers derived from well-defined simple/complex algorithms or equations are "dual numbers" displayed as Completely & Incompletely Predictable number. Examples of **Q** '2' as **P** (& **E**), '97' as **P** (& **O**), '98' as **C** (& **E**) and '99' as **C** (& **O**) are described above. Examples of **R** -- **Q** are described next. First & only negative Gram[y=0] point (by convention at n = -3) with Completely Predictable y = 0 value is obtained by substituting Completely Predictable t = 0 resulting in $\zeta(\frac{1}{2} + it) = \zeta(\frac{1}{2}) = -1.4603545$, an Incompletely Predictable transcendental number [rounded off to seven decimal places] calculated as a limit similar to limit for Euler-Mascheroni constant or Euler gamma – its precise (1st) position can only be determined by computing positions of all preceding (nil) Gram[y=0] points. With exception of this first Gram[y=0] point, all t values from Gram[x=0] points, Gram[y=0] points, and nontrivial zeros (Gram[x=0,y=0] points) are Incompletely Predictable transcendental numbers – these are respectively associated with Completely Predictable x = 0, y = 0, and simultaneous x = 0 & y = 0 values. First 'Gram point' (by convention at n = 0 & is associated with Completely Predictable x = 0 value from Incompletely Predictable t = 17.845599 substitution) is actually the 4th Gram[y=0] point whose precise (4th) position can only be determined by computing positions of all preceding (three) Gram[y=0] points. First nontrivial zero associated with simultaneous x = 0 & y = 0 value [equating to $\zeta(s) = 0$ whereby $s = \sigma + it = \frac{1}{2} + it$] is Completely Predictable occurring with Incompletely Predictable t = 14.134725 value substitution – its precise (1st) position can only be determined by computing positions of all preceding (nil) nontrivial zeros.

Countable finite set (CFS) of Completely Predictable simple properties intrinsically present in Simple Container simple equations or algorithms help us solve Completely Predictable problems containing countable infinite set (CIS) of Completely Predictable numbers; whereas CFS of Completely Predictable complex properties intrinsically present in Complex Container complex equations or algorithms help us solve Incompletely Predictable problems containing CIS of Incompletely Predictable numbers.

Simple properties are inferred from a phrase like: "...the simple equation or algorithm by itself will intrinsically incorporate actual location [and actual positions] of all Completely Predictable numbers". Solving Completely Predictable problems endowed with simple properties which are amendable to simple treatments using usual mathematical tools such as Calculus will result in their Simple Elementary Fundamental Laws-based solutions. Complex properties are inferred from a phrase like: "...the complex equation or algorithm by itself will intrinsically incorporate actual location [but not actual positions] of all Incompletely Predictable numbers". Solving Incompletely Predictable problems endowed with complex properties which are amendable to complex treatments using unusual mathematical tools such as Information-Complexity conservation and exact & inexact Dimensional analysis homogeneity as well as usual mathematical tools such as Calculus will result in their Complex Elementary Fundamental Laws-based solutions.

Consider x for real number (**R**) values > 1. Let y be Set **R** such that (Simple Container simple equation) $y = 2x$ or $y = 2x - 1$. This Simple

224

Container will "contain" the complete uncountable infinite set (UIS) **R** [straight line of infinite length] commencing from Cartesian point (x=1, y=2) or (x=1, y=1). Computing y = 2x or y = 2x - 1 an infinite number of times – a "mathematical impasse" – will not *per se* result in its Simple Elementary Fundamental Laws-based solution for gradient or slope = 2. This gradient (simple property) is obtained by trigonometrically calculating tangent of y = 2x or y = 2x - 1 straight line which = 2 or mathematically analyzing y = 2x or y = 2x - 1 equation using Differential Calculus viz. $\frac{dy}{dx} = \frac{d(2x)}{dx}$ or $\frac{d(2x-1)}{dx}$ = 2. Note: applying Integral Calculus from Fundamental Theorem of Calculus to y = 2x or y = 2x - 1 for interval [1, +∞], viz. $\int_1^\infty (2x)dx$ or $\int_1^\infty (2x-1)dx = [x^2 + C]_1^\infty$ or $[x^2 - x + C]_1^\infty = (\infty^2 + C) - (1^2 + C)$ or $(\infty^2 - \infty + C) - (1^2 - 1 + C)$ result in Simple Elementary Fundamental Laws-based solution for area (simple property) of ∞ size enclosed by mentioned straight line & x-axis.

By considering x > 1 integer number (**Z**) values for Simple Container simple algorithm y = 2x or y = 2x - 1, we obtain "contained" complete Set **E** or Set **O**. Computing **E** or **O** infinitely often – a "mathematical impasse" – will not *per se* result in its Simple Elementary Fundamental Laws-based solution for gap between any two consecutive **E** (**E** gap) or **O** (**O** gap) will both = 2. This gradient-equivalent **E** gaps or **O** gaps (simple property) is obtained by transforming these algorithms from their "discrete" formats into equivalent "continuous" formats [viz. "discrete" Δx = 1 → "continuous" Δx = 0] resulting in their gradients using either tangent method or Differential Calculus method as per previous paragraph. Then **E** or **O** gaps, both = 2, is numerically identical and mathematically equivalent to relevant gradients, both also = 2. The same method of transforming "discrete" formats into "continuous" formats required to solve Riemann hypothesis involves applying Riemann integral to "discrete-like" simplified Dirichlet eta function (in summation format) to obtain "continuous-like" Dirichlet Sigma-Power Law (in integral format).

Similar to Incompletely Predictable 'varying gaps' [equivalent to 'varying gradients'] between consecutive **P** (**P** gaps) & consecutive **C** (**C** gaps), we have Incompletely Predictable 'varying gaps' [equivalent to 'varying gradients'] between consecutive nontrivial zeros (nontrivial zero gaps), consecutive Gram[y=0] points (Gram[y=0] points gaps) & consecutive Gram[x=0] points (Gram[x=0] points gaps). These 'varying gaps' or 'varying gradients' (complex properties) are geometrically related to different shapes and sizes of spirals as depicted in the figure below on Riemann zeta function. Note from this figure that mathematically defined nontrivial zeros [as t values obtained when setting either $\zeta(s) = 0$ or $\eta(s) = 0$] is exactly equivalent to geometrically defined Gram[x=0,y=0] points [as all the 'Origin' intercepts].

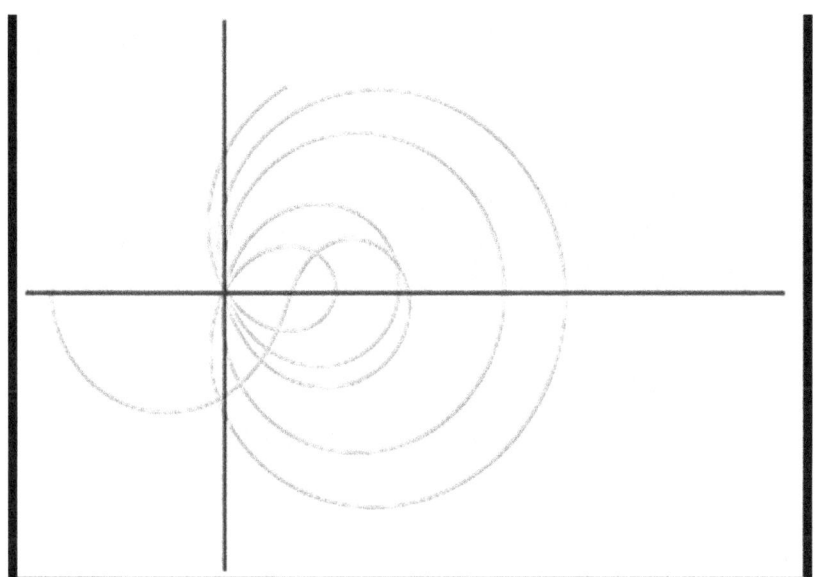

Gram's Law and traditional 'Gram points'

In mathematics, logarithm (log) is the inverse function to exponentiation. Log to base 10 is called the common logarithm and has many applications in science and engineering. Log to base e ($\approx$ 2.718) which is usually denoted by 'ln' [as it is in this book] is called the natural logarithm with widespread use in mathematics and physics because of its simpler derivative. Log to base 2 is called the binary logarithm and is commonly used in computer science.

Named after Danish mathematician Jørgen Pedersen Gram (June 27, 1850 – April 29, 1916), traditional 'Gram points' (or Gram[y=0] points which are **x-axis intercepts** shown in figure above) are other conjugate pairs values on critical line defined by $\text{Im}\{\zeta(\frac{1}{2} \pm \imath t)\} = 0$. They obey Gram's Rule and Rosser's Rule with interesting characteristic properties as outlined by our brief exposition below. Z function is used to study Riemann zeta function on critical line. Defined in terms of Riemann-Siegel theta function & Riemann zeta function by $Z(t) = e^{i\theta(t)}\zeta(\frac{1}{2} + it)$ whereby $\theta(t) = \arg(\Gamma(\frac{(2it+1)}{4}))$ $- \frac{ln\pi}{2}t$; it is also called Riemann-Siegel Z function, Riemann-Siegel zeta function, Hardy function, Hardy Z function, & Hardy zeta function. The algorithm to compute $Z(t)$ is called Riemann-Siegel formula. Riemann zeta function on critical line, $\zeta(\frac{1}{2} + it)$, will be real when $\sin(\theta(t)) = 0$. Positive real values of t where this occurs are called 'Gram points' and can also be described as points where $\frac{\theta(t)}{\pi}$ is an integer. Real part of this function on critical line tends to be positive, while imaginary part alternates more regularly between positive & negative values. That means sign of $Z(t)$ must be opposite to that of sine function most of the time, so one would expect nontrivial zeros of $Z(t)$ to alternate with zeros of sine term, i.e. when θ takes on integer multiples of π. This turns out to hold most of the time and is known as Gram's Rule (Law) – a law which is violated infinitely often though. Thus Gram's Law is statement that nontrivial zeros of $Z(t)$ alternate with 'Gram points'. 'Gram points' which satisfy Gram's Law are called 'good', while those that do not are called 'bad'. A Gram block is an interval such that its very first & last points are good 'Gram points' and all 'Gram points' inside this interval are bad. Counting nontrivial zeros then reduces to counting all 'Gram points' where Gram's Law is satisfied and adding the count of nontrivial zeros inside each Gram block. With this process we do not have to locate nontrivial zeros, and we just have to accurately compute $Z(t)$ to show that it changes sign.

Ratio Study and Inequations

A mathematical equation, containing one or more variables, is a statement that values of two ['left-hand side' (LHS) and 'right-hand side' (RHS)] mathematical expressions is related as equality: LHS = RHS; or as inequalities: LHS < RHS, LHS > RHS, LHS ≤ RHS, or LHS ≥ RHS. A ratio is one mathematical expression divided by another. The term 'unnecessary' Ratio (R) for any given equation is explained by two examples: (1) LHS=RHS and with rearrangement, 'unnecessary' R is given by $\frac{LHS}{RHS}=1$ or $\frac{RHS}{LHS}=1$; and (2) LHS>RHS and with rearrangement, 'unnecessary' R is given by $\frac{LHS}{RHS}>1$ or $\frac{RHS}{LHS}<1$.

Consider exponent y ∈ all **R** values & base x ∈ **R**≥0 values for mathematical expression $\frac{x}{y}$. Equations such as $x^1 = x$, $x^0 = 1$ & $0^y = 0$ are all valid. Simultaneously letting both x & y = 0 is an incorrect mathematical action because xy as function of two-variables is not continuous & is thus undefined at Origin. But if we elect to intentionally carry out this "balanced" action [equally] on x & y, we obtain (simple) inequation $0^0 \neq 1$ with associated perpetual obeyance of '=' equality symbol in x^y for all applicable **R** values except when both x & y = 0. The Number '1' value in this inequation is justified by two arguments: I. Limit of x^y value as both x & y tend to zero (from right) is 1 [thus fully satisfying criterion "x^y is right continuous at the Origin"]; and II. Expression x^y is product of x with itself y times [and thus x^0, the "empty product", should be 1 (no matter what value is given to x)].

Mathematical operator 'summation' must obey the law: We can break up a summation across a sum or difference but not across a product or quotient viz, factoring a sum of quotients into a corresponding quotient of sums is an incorrect mathematical action. But if we elect to carry out this action equally on LHS & RHS products or quotients in a suitable equation, we obtain two (unique) 'necessary' R denoted by R1 for LHS and R2 for RHS whereby R1 ≠ R2 relationship will always hold. We define 'Ratio Study' as intentionally performing this incorrect [but "balanced"] mathematical action on suitable equation [equivalent to one (non-unique) 'unnecessary' R] to obtain its inequation [equivalent to two (unique) 'necessary' R].

Let C denote complex numbers. Set C is a field (but not an ordered field). Thus it is not possible to define a relation between two given (z1 & z2) C as z1 < z2 since inequality operation here is not compatible with addition and multiplication. But performing Ratio Study to obtain inequations involving C does not involve defining a relation between two C.

A228186

The Hybrid method of Integer Sequence classification enables meaningful division of all integer sequences into either Hybrid or non-Hybrid integer sequences. My exotic A228186 (Helpful B., Hybrid integer sequence A228186) integer sequence was published on The On-line Encyclopedia of Integer Sequences website in 2013. It is the first ever [infinite length] Hybrid integer sequence synthesized from Combinatorics Ratio. In 'Position i' notation, let i = 0, 1, 2, 3, 4, 5,..., ∞ be complete set of natural numbers.

A228186 "Greatest k > n such that ratio R < 2 is a maximum rational number with R = $\dfrac{\text{\textit{Combinations With Repetition}}}{\text{\textit{Combinations Without Repetition}}}$ " is equal to [infinite length] non-Hybrid (usual garden-variety) integer sequence A100967 except for finite 21 'exceptional' terms at Positions 0, 11, 13, 19, 21, 28, 30, 37, 39, 45, 50, 51, 52, 55, 57, 62, 66, 70, 73, 77, and 81 with their values given by relevant A100967 terms plus 1.

The first 49 terms [from Position 0 to Position 48] of A100967 (Noe, November 23, 2004) "Least k such that binomial(2k+1, k-n) ≥ binomial(2k, k)" are listed below: 3, 9, 18, 29, 44, 61, 81, 104, 130, 159, 191, 225, 263, 303, 347, 393, 442, 494, 549, 606, 667, 730, 797, 866, 938, 1013, 1091, 1172, 1255, 1342, 1431, 1524, 1619, 1717, 1818, 1922, 2029, 2138, 2251, 2366, 2485, 2606, 2730, 2857, 2987, 3119, 3255, 3394, and 3535. For those 21 'exceptional' terms: at Position 0, A228186 (= 4) is given by A100967 (= 3) + 1; at Position 11, A228186 (= 226) is given by A100967 (= 225) + 1; at Position 13, A228186 (= 304) is given by A100967 (= 303) + 1; at Position 19, A228186 (= 607) is given by A100967 (= 606) + 1; etc.

Here is a useful concept: Commencing from Position 0 onwards "in the limit" that this Position approaches 82, A228186 Hybrid integer sequence becomes (and is identical to) A100967 non-Hybrid integer sequence for all Positions ≥ 82. Discovered in 2013, we will now devotedly explain A228186 in full details below due to its additional role in helping us understand the mathematical concept "in the limit…." employed in this book. A228186 has remarkable properties and whether there are other similar Hybrid Integer Sequences remains to be determined.

A228186: A unique Hybrid Integer Sequence

Hybrid method of Integer Sequence classification		
Main classes of integer sequence (infinite length)	**Non-hybrid integer sequence (Class I)**	**Hybrid integer sequence (Class II)**
Composition of relevant integer sequence	'Usual' terms: made up of only 1 integer sequence: AXXXXXX (f(x)).	'Usual' and 'exceptional' terms (presumptively) derived from only 2 integer sequences in this classification system. 4 main varieties made up of either (i) [AXXXXXX (f(x)) AXXXXXX (f(y))] {near-identical integer sequences} with finite or infinite overlap, and (ii) [AXXXXXX (f(x)) AYYYYYY (f(y))] {non-identical integer sequences} with finite or infinite overlap.
Descriptive positions of 'exceptional' terms	Not applicable	'Exceptional' terms are either isolated, or interspersed with 'usual' terms. In principles all hybrid integer sequences with infinite overlap must be of the interspersed type, and those with finite overlap can be of either interspersed or isolated types.
Theoretical number of all possible subtypes	1	$Ж_6 \rightarrow 4$

232

Legend: AXXXXXX and AYYYYYY represent individual integer sequence; f(x) and f(y) represent individual functional equation involving variable x and y respectively.

Ж In the real mathematical world, the theoretical number of all useful / practical (as opposed to all possible) subtypes of Class II Hybrid integer sequence should be limited to 4 (of the finite overlap variety) for the following reasoning. Notwithstanding the fact that the infinite overlap variety will result in the dilemma or arbitrariness of which near-identical or non-identical individual integer sequences should be designated as belonging to the 'usual' or 'exceptional' terms, their resultant hybrid integer sequence is simply a convenient mathematical combination / marriage of their individual integer sequence components with the (infinite) sum total of all its terms simply being exactly given by adding the (infinite) terms of each individual components. Reverse (and forward) modelling, akin to "engineering", of new hybrid integer sequences can take place at will by simply mathematically combining any two [or more] available infinite integer sequences – this do not result in any information gain from the resultant "totally predictable" hybrid integer sequences of this nature. On the contrary, the hybrid integer sequences of the finite overlap variety are "not totally predictable" with reverse engineering deemed to be extremely difficult if not impossible, and crucially it also simply CANNOT be fully described using only a solitary integer sequence, or using more than one in a simplistic "linear" manner. Also, the actual sum total of terms from this unique variety of integer sequence is less than the simple summation of the terms from the two involved infinite integer sequences if derived as such. In other words using the analogous language of thermodynamics, the increase in orderliness gleaned from the 'exceptional' terms being incorporated onto this hybrid integer sequence variety has resulted in the loss of information resulting from the remaining discarded 'exceptional' terms – this in essence equating to an increase in entropy.

Based on a new paradigm for our understanding of the classification of integer sequence, A228186 is a novel 'Hybrid integer sequence' with its specific denoted subtype. Our goal is to deem the associated mathematical work as ground-breaking research work in this relatively unknown scientific area in order to further the relevant knowledge here. The preliminary effort on our intended new, robust and functional classification method for integer sequence is outlined above. We eagerly await constructive criticisms from the wider scientific community.

Heavily illustrated with the help of Integer Sequence A000001 (Sloane, 1964), 'Number of groups of order n'. OEIS (founded in 1964 by N. J. A. Sloane); a much less satisfying alternative (but nevertheless insightful) classification of integer sequence is given in the following relevant paragraphs. The essential background preliminary arguments required to use the relevant keywords 'sequence' and 'pseudorandom' in "Pseudorandom infinite-length integer sequence" (PIIS) is predominantly based on the exclusion criteria and reproducibility principle – this exercise will justify the use of our coined PIIS and its role as an alternative classification of integer sequence. A sequence of integer numbers can theoretically be characterized by its finite-length [viz. finite-length integer sequence (FIS)] or infinite-length [viz. infinite-length integer sequence (IIS)], and their reproducibility [which loosely equates to predictability] or irreproducibility [which loosely equates to unpredictability].

Although useful as part of overall argument contributions, we will only briefly touch on the FIS type due to its relatively redundant nature. By logical deductions, a (literally) imaginary and non-existent FIS must always consist of a finite arbitrary set of chosen and memorized integers derived from a "perfect" random (or even a "slightly imperfect" pseudorandom) number generator, and that this set of integers can be mathematically proven [likely via various mathematical axioms] to be totally irreproducible, unpredictable & non-deterministic *per se* and also likewise by the same number generator – otherwise, this FIS will have all of its terms to be predictable and reproducible, and thus the total number of terms able to be extended to infinity (and belonging to the IIS class by default). In this setting, we can then in effect discard the "random infinite-length integer sequence" terminology not least due to its virtually useless mathematically descriptive label and revelation capacity. One can then argue that all "proper" integer sequences will always have to be IIS in character and thus also always having to be reproducible and infinite in nature.

Next, we can propose IIS to consist of two mutually exclusive classes viz. Class I: the traditional "non-pseudorandom infinite-length integer sequence" (which is the usual common-garden variety smooth integer sequences typified by many published integer sequences, and Class II: the contemporary "pseudorandom infinite-length integer sequence" – a simplistic & rarity example is currently typified by the solitary A228186 [this one] with A000001 from 'The Online Encyclopedia of Integer Sequences' seemingly typifying a much more difficult example. The true nature of PIIS remains to be elucidated; for instance whether the pseudorandom components [as

235

expounded below] of all existing PIIS can be of varying finite or infinite length and complexity in both their positions and values. Again using sheer logical deductions, finite pseudorandom components of PIIS commencing from its designated initial point will have to terminate at a subsequent designated point in the relevant sequence – otherwise if they fail to terminate, they will obviously have to be (by default) of the infinite pseudorandom components variety.

A typical definition of 'pseudorandomness' here is that, given one value of the sequence, finding the previous (or next) value is hard. Moreover, the task of finding all previous (or next) [infinite] values of the sequence would also take an infinite amount of time. However, our A228186 sequence is quite smooth and it should not be difficult to find terms in either direction [with the proviso] as long as n is not less than 82 – this also with the crucial contrasting implication that finding all the pseudorandom ('exceptional') terms/component contained within n = 0, 1,..., 81 would only take a finite amount of time. So we are also using the word 'pseudorandom' in a different manner in A228186 with the intended definition as follows. A228186 is pseudorandom in the sense that, given a sequence of N consecutive last bits of this sequence [n = 82, 83,..., ∞], it is extremely hard if not impossible to find an index starting a sequence with those final bits. An important corollary observation & caveat, based on the underlying arguments given above, is that when the pseudorandom components were to exist as infinite number of terms, they will still be reproducible and predictable [in the strictest sense of the words] despite hypothetically or conceivably needing an infinite amount of time to elucidate all of their infinite available pseudorandom terms. In addition, all pseudorandom terms in the integer sequences with infinite pseudorandom components will be viewed in its entirety as pseudorandom ('exceptional') terms always interspersed amongst the non-pseudorandom ('usual') terms. Thus by aesthetic reasoning alone, one would be tempted or inclined to suggest that pseudorandom components of the infinite type do not really exist in nature. However & perhaps paradoxically, nature is counter-intuitively and abundantly blessed with integer sequences having just such pseudorandom components of the infinite type – this being typified by A000001, formerly M0098 N0035, given as 0, 1, 1, 1, 2, 1, 2, 1, 5, 2, 2, 1, 5, 1, 2, 1, 14, 1, 5, 1, 5, 2, 2, 1, 15, 2, 2, 5, 4, 1, 4, 1, 51, 1, 2, 1, 14, 1, 2, 2, 14, 1, 6, 1, 4, 2, 2, 1, 52, 2, 5, 1, 5, 1, 15, 2, 13, 2, 2, 1, 13, 1, 2, 4, 267, 1, 4, 1, 5, 1, 4, 1, 50, 1, 2, 3, 4, 1, 6, 1, 52, 15, 2, 1, 15, 1, 2, 1, 12, 1, 10, 1, 4, 2,….

Features of A228186

A228186 is completely defined by the following mathematical statement: Greatest k > n such that ratio R < 2 is a maximum rational number with $R = \dfrac{Combinations\ with\ repetition}{Combinations\ without\ repetition}$. Least k > n such that binomial(2k+1, k-n-1) ≥ binomial(2k, k), or one more than this value if n is in [the 21 'exceptional' terms] {0, 11, 13, 19, 21, 28, 30, 37, 39, 45, 50, 51, 52, 55, 57, 62, 66, 70, 73, 77, 81} also completely define all terms in A228186.

At this point in time, to the best of our knowledge A228186 is the only known Class II Hybrid integer sequence of the specific subtype "[AXXXXXX(f(x)) AXXXXXX(f(y))] with finite overlap of the interspersed type". Each integer sequence made contributions to the 'usual' & 'exceptional' terms in the following manner & proportion:

A100967(+0) occupying all consecutive positions when n is from 82 to ∞; and A100967(+1) & A100967(+0) respectfully occupying the designated positions for the 21 'exceptional' terms & 61 'usual' terms when n is from 0 to 81. [The convention employed here being A100967(+0) & A100967(+1) respectfully symbolizing add 0 and add 1 to every single A100967 term.] Based solely on these unique, finite and totally predictable pseudorandom 'exceptional' terms, we advocate A228186 as a novel true pseudorandom infinite-length integer sequence. The consequent role from this bold proposal on A228186 – correct to the best of our knowledge at the time of composing A228186 – and its highly significant connections with integer sequence A100967, combinatorics, and the mathematical fields of Chaos and Fractals could then be deemed to consist of ground-breaking research work in this relatively unknown scientific area. We will further elaborate all above points in materials below.

The topics of permutations (P) and combinations (C) come under combinatorics. P is an ordered C. There are two types of P and two types of C with their relevant formulae given below, given n = 0, 1, 2,..., ∞ [Symbol ! represent the factorial operation viz. the factorial of a positive integer n denoted by n! is the product of all positive integers less than or equal to n. Thus n! = n X (n-1) X (n-2) X (n-3) X ... X 3 X 2 X 1.]:

P with repetition: $k^{(n+2)}$ Equation (1)

P without repetition: $\dfrac{k!}{(k-n-2)!}$ Equation (2)

238

C with repetition: $\dfrac{(k+n+1)!}{(n+2)!(k-1)!}$ Equation (3)

C without repetition: $\dfrac{k!}{(n+2)!(k-n-2)!}$ Equation (4)

Numerically, (1) > (2) > (3) > (4) always holds true and R = $\dfrac{(3)}{(4)}$.

Least k > n such that binomial(2k+1, k-n-1) ≥ binomial(2k, k), or one more than this value if n is in {0, 11, 13, 19, 21, 28, 30, 37, 39, 45, 50, 51, 52, 55, 57, 62, 66, 70, 73, 77, 81} completely define all terms in A228186 [see Fig 1, Graphs related to A228186 below]. It consists of minority finite 21 'exceptional' terms (denoted by k+1) and majority infinite 'usual' terms (denoted by k). From the formula below for both 'usual' and 'exceptional' terms alike, we can also estimate n as approximately 0.83√(k) - 1 and 0.83√(k+1) - 1 if we know k and k+1 respectively.

"Greatest k > n such that R < 2 is a maximum [= $\dfrac{(k+n+1)!(k-n-2)!}{k!(k-1)!}$ < 2 is a maximum]" results in the complete set of A228186 and essentially represents a study on inherent properties (predominantly on mathematical patterns) for fractional part of the argument for R. These naturally present properties can intrinsically be determined graphically, by calculations, and by mathematical deductions based on the key concept "maxima R values < 2 equates to fract(R) being the maxima occurring at a particular k for 'usual' terms and k+1 for 'exceptional' terms" as further outlined below. This then will also provide sound explanations for the occurrence of actual number and position for all known 'exceptional' terms.

Binomial is Combinations without repetition. "Least k > n such that

binomial(2k+1, k-n-1) ≥ binomial(2k, k) $\left[\;=\dfrac{(2k+1)!}{(k-n-1)!(k+n+2)!} \right.$

$\left. \geq \dfrac{(2k)!}{(k!)^2} \right]$" is the exact replica of A100967 integer sequence which

was equivalently expressed previously as the inequality [in slightly different format here with using the notation: n = 0, 1, 2,..., ∞] "Greatest k > n such that binomial(2k+1, k-n-1) < binomial(2k, k)".

Intra-R analysis: Graphically plot fract(R) on y-axis versus k on x-axis for each n using integers from 0 onwards. Then for each graph with its finite number of peaks and troughs observed, maxima R values < 2 equates to fract(R) being the maxima occur at a particular k for 'usual' terms, and k+1 for 'exceptional' terms respectively. As a real physical entity, this particular R value will computationally equate to the last (and highest) peak on Right Hand Side of x-axis; and its k or k+1 value is contained in A228186 for that particular n. The value k when [incorrectly] used for an 'exceptional' term will result in fract(R) being a minima with relatively much smaller value than this maxima value.

Inter-R analysis: Graphically plot each of the 2 individual data sets derived from fract(R using k) and fract(R using k+1) on y-axis versus n on x-axis for n from 0 onwards. Then for each of the distinct graphical line with its own peculiar finite number of peaks and troughs observed, fract(R using k) > fract(R using k+1) always holds true except at the 21 'exceptional' terms whereby fract(R using k) is a nadir value close to 0 and fract(R using k+1) is a zenith value close to 1. With progressively higher 'exceptional' terms, their fract(R using k+1) values will overall be monotonously rising steadily approaching a value of 1 (boundary condition), which then limit the total number of possible 'exceptional' terms. This intuitive mathematical deduction utilizing the imposed boundary condition supports our preposition that number of 'exceptional' terms is finite.

In effect both the fract(R using k) and its closely-related fract(R using k+1), with a constant value difference of 1 amongst their corresponding terms, could be seen to reflect near-identical A228186 integer sequences. A228186 can also be perceived to consist of an infinite number of 'usual' (pure) terms tainted by a finite number of 'exceptional' (impure) terms, and can be eloquently designated by various nonlinear algorithms and formulae application thus manifesting the typical property of Chaos known as "deterministic pseudorandomness". Resulting relevant graphs employing various calculations based on A228186 will spontaneously manifest the typical property of Fractals known as "self-similarity" with simultaneous depiction of [opportunistic] recurrences in identical sets of the same 21 'exceptional' terms.

The 21 'exceptional' terms can be derived by employing the following steps. Perform an infinite number of iterations on formula $k = \text{Round}(0.3807 + 1.43869(n+1) + 1.44276(n + 1)^2$ from n = 0 onwards while simultaneously substituting each n and its derived k value into R. It then follows that, respectively, the 'usual' [and 'exceptional'] terms will always be from maxima R values < 2 [and ≥ 2]; or equivalently expressed as its reciprocal, $\frac{1}{R}$, being from minima rational numbers with values ≥ 2 [and < 2]. The occurrence of any n and k with its calculated fract(R using k) being a minima close to 0 is the signature that this specific k is an 'exceptional' term with the corrected 'exceptional' term needing to be defined by k+1 instead. Alternatively, add 1 to each k value obtained from iterations on the formula and substitute using this (k+1) value instead. The graphical trend of the R obtained using this k+1 values will largely mirror but always be smaller than their corresponding counterparts using k values. The signature for an 'exceptional' term is then denoted by the calculated fract(R using k+1) having a maxima value close to 1 [but now with all other corrected 'usual' terms to be defined by k instead].

Denoted by the syntax product(f,var,a,b), this function returns the product of f where var is evaluated for all integers from a to b. Importantly, 'R using k' and 'R using k+1' can be further simplified

to $\dfrac{product(n,n,k+1,k+n+1)}{product(n,n,k-n-1,k-n,k-1)}$ and

$\dfrac{product(n,n,k+2,k+n+2)}{product(n,n,k-n,k)}$ respectively, whereby k =

Round(0.3807 + 1.43869(n+1) + 1.44276$(n+1)^2$.

In summary, to the best of our knowledge A228186 represents a novel and freshly discovered Class II Hybrid integer sequence of the extremely specific subtype '[AXXXXXX(f(x)) AXXXXXX(f(y))]' with finite overlap of the interspersed type' & a true 'Pseudorandom infinite-length integer sequence' based solely on its 21 unique pseudorandom 'exceptional' terms. We are strong advocates of this bold statement because of the following observations. Although the positions of the 21 'exceptional' terms are seemingly random in nature, they are also totally predictable. Moreover they only symbolize the tiny finite number of differences (of constant value 1) between two different combinatorics-flavored infinite series contained in A228186 and A100967 with their remaining infinite terms being exactly the same. Like A100967 (with one component from combinatorics, viz. Combinations without repetition), A228186 (with two components from combinatorics, viz. Combinations with and without repetition) is a genuine and real physical entity [albeit with differing number of combinatorics components] able to be vividly depicted graphically in their various formats. It constitutes a pioneer study on fractional part of the argument for ratio R, a rational number simply defined using two selected components from combinatorics, and should encourage future niche research studies into this previously little-known scientific area.

Finally, the pseudorandom nature of A228186 owes much of its explanations (and importantly, deep-seated connections) to certain non-frivolous aspects derived from the ubiquitous and exciting mathematical fields of Chaos and Fractals – in particular and respectively, this refer to "deterministic pseudorandomness" and "self-similarity".

Graph related to A 228186

Figure 1: Scatterplot of A228186(n)

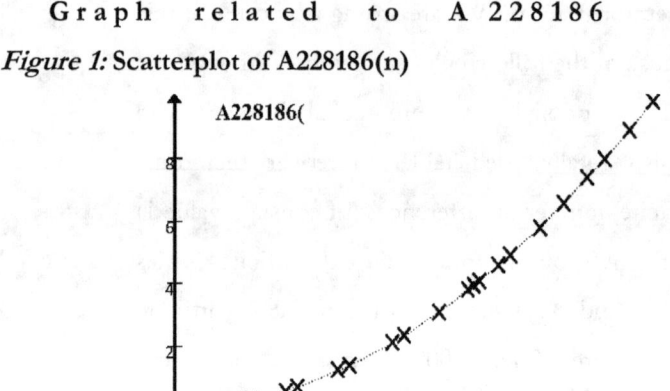

Legend: The 21 'Exceptional' terms denoted by X.

Figure 1 above clearly depict the visual image on our 21 'Exceptional' terms. We respectfully advocate that this graph represents a stunning break-through research result culminating in the very first known Hybrid Integer Sequence A228186.

T o start this chapter, I will provide the following interesting research

ideas in relation to Medicine dubbed Paradoxical-to-Modern Science Law. A *paradox* is an argument that produces an inconsistency, typically within logic or common sense. Although most logical paradoxes are known to be invalid arguments, they can still be valuable in promoting critical thinking. Not surprisingly, some have revealed errors in definitions assumed to be rigorous, and have even caused axioms of mathematics and logic to be re-examined (e.g. Russell's paradox).

Employing lateral thinking in an analogous manner, I devise the rule: In the event that initial seemingly *Paradoxical Practices and Phenomena* (PPP) can rigorously be validated by research with accompanying sound theoretical arguments, they are then able to be reclassified as final proven *Scientific Practices and Phenomena* (SPP). We intuitively provide this rule as below:

Paradoxical-to-Modern Science Law

Paradoxical Science (*PPP*)

$\downarrow$ *Research validation* $\uparrow$ (*Reverse direction*)

Modern Science (*SPP*)

Limbo Basket: Those Practices and Phenomena not able to be easily classified as either PPP or SPP on a temporary or permanent basis.

Legend: (*Reverse direction*) = current disproved or superseded SPP demoted to PPP

At my previous workplace in relevant medical fields, the diverse terms *Practices and Phenomena* could easily be referring to Anaesthesia and Intensive Care Practices and Phenomena. Thus, the relevant *Paradoxical Anaesthesia and Intensive care Practices and Phenomena* in question will become *Modern Anaesthesia and Intensive care Practices and Phenomena* when validated by research. Common sense will tell us here that especially with regards to *Practices*, many of these entities would be prone to be in the Limbo Basket of being neither classified as Paradoxical Practices nor Scientific Practices.

A somewhat related concept is 'Off-label drug use'. This is defined as use of pharmaceutical drugs for an unapproved indication or in an unapproved age group, dosage, or route of administration whereby the ability to prescribe drugs for uses beyond the officially approved indications is commonly carried out to good effect by healthcare providers. Indications for generic drug Aspirin are to reduce fever and relieve mild to moderate pain from conditions such as muscle aches, toothaches, common cold, and headaches, etc. It is also known to "thin the blood". An off-label drug use example would be the commonly accepted practice of giving 300 mg soluble Aspirin to a person suspected of having a heart attack whereby soluble Aspirin is not pharmaceutically registered for this purpose.

In contrast, 100 mg Aspirin commonly marketed under brand-name 'Cartia' (endowed with enteric coating to lessen its Gastrointestinal side effect of causing peptic ulceration) is registered for this purpose under the approved indications for its use in known cardiovascular or cerebrovascular disease, as an antiplatelet agent for prophylaxis against acute myocardial infarction, unstable angina, transient ischaemic attack and cerebrovascular accident (stroke).

246

There are potential beneficial roles of occasional multiple narcotic use in modern pain management:-

Point 1. Opioid analgesic drugs tend to exhibit incomplete cross-tolerance so that even when a patient has developed a high level of tolerance to one drug from the opioid class, they may find that a different opioid drug will still be effective – this is the concept/rationale for employing **Opioid Rotation**.

Point 2. The reasons for incomplete cross-tolerance is likely because of variations in opioid receptor affinity and occupancy levels at equianalgesic doses, as well as additional mechanisms of action possessed by some narcotic drugs such as NMDA antagonist action of Methadone, or SNRI activity of Tramadol or Tapentadol – the additional mechanisms of these drugs led to the potential use of **Multiple Opioids in acute or chronic pain scenario** on acute (short-term), subacute (medium-term), or chronic (long-term) basis. In comparison to Methadone (full agonist mu receptor with receptor affinity comparable to heroin), Buprenorphine has a ceiling effect to analgesia with "safer" less potential of respiratory depression due to partial mu-opioid receptor agonist (with receptor affinity >> Methadone/heroin but approximately = Nalaxone/Naltrexone) and fewer psychotomimetic & dysphoric effect due to kappa-opioid receptor antagonist. Note that Methadone and Buprenorphine are the two common Opioid Replacement drugs acceptably and legally used to control drug addiction in many societies.

Point 3. Drug dependent patients requiring controlled/restricted, for example, dangerous narcotic "S8 controlled drugs" (with or without significant acute or chronic pain component) should only be prescribed

247

under a treating doctor with regulatory approval for that particular Drug X. Another nominated doctor in the same practice during the treating doctor's absence may prescribe Drug X. A doctor may judiciously prescribe Drug X or another "S4 restricted drug" (e.g. the Benzodiazepines) or S8 controlled drug" in an emergency setting for acute pain control whereby individual variation / pharmacokinetic / pharmacodynamic principles involved for the opioid-tolerant [as opposed to opioid-naïve] patient usually dictate the use of larger than normal doses of Drug X due to narcotic tolerance or larger than normal doses of another narcotic drug to competitively displace the original Drug X from mu-opioid receptors.

Point 4. Multi-modal approach to acute and chronic pain management may involve combining physical therapy (e.g. physiotherapy or Spinal Cord Stimulator for low back pain) and multiple drug therapy viz. time-limited narcotic trial / minimizing chronic narcotic use (as persistent non-cancer pain can have poor response to narcotic e.g. from (Farzana Mitra, January 15, 2013) *A feasibility study of transdermal buprenorphine versus transdermal fentanyl in the long-term management of persistent non-cancer pain*. Pain Medicine, 14 (1). pp. 75 – 83. https://doi.org/10.1111/pme.12011; only 11 - 13% have pain relief at 12 months from Fentanyl / Buprenorphine patches and 1/3 cease their use from side effect) and adding pain adjunct medications to narcotic such as tricyclic antidepressant Amitriptylline with Number Needed to Treat (NNT) 2.1, Pregabalin with NNT 4 – 6, Clonidine, Non-steroidal anti-inflammatory drugs (NSAIDS), Paracetamol, topical Lignocaine patch, and Capsaicin cream. NNT is the number of patients you need to treat to prevent one additional bad outcome (death, stroke, etc.). For example, if a drug has an NNT of 5, it means you have to treat 5 people with the drug to

248

prevent one additional bad outcome.

Examples of short-term use are Fast-onset Alfentanil to blunt intubation response + later on add short acting Fentanyl or Morphine during General Anesthesia / Total Intravenous Anesthesia (TIVA) use of potent ultra-short acting Remifentanil + later on add Fentanyl or Morphine; Use non-narcotic drugs and/or a larger dose of non-Methadone or non-Suboxone narcotic drug(s) to compete with Methadone or Suboxone at the receptor level in emergency acute pain management. Use of Tramadol + Methadone is a chronic long term basis case example of managing patient's narcotic addiction + chronic pain from (for example) severe Rheumatoid Arthritis.

In the vast majority of cases, doctors should usually endorse / follow the general recommendation of using either solitary Suboxone or Methadone alone without adding any other narcotic in opioid dependent patients even when they have chronic pain. Adjusted for bio-availability, the Opioid Replacement equi-analgesic doses of Methadone & Buprenorphine as "Subutex" or "Suboxone" [Buprenorphine-Naloxone combination] prescribed for opioid dependent patients' narcotic addiction (Opioid-tolerant patients) are much higher than the doses of the same drugs used for acute (often Opioid-naïve patients) or chronic (often Opioid-tolerant patients) pain management. Example, calculations using the app 'Opioid Calculator': Opioid Replacement 18mg of Buprenorphine sublingually per day = oral Morphine Equivalent Daily Dose (oMEDD) of 720 mg per day. 40 mcg/hr (960 mcg or 0.9 mg/day) topical Buprenorphine (Norspan) patch = oMEDD of 80 mg/day only. Owing to the pharmacokinetic long half-life of

249

Methadone, the time required to achieve dose stability varies from 35 to 325 h (13.5 days). Thus, oMEDD <30 mg/day led to oMEDD:Methadone ratio 2:1, oMEDD 31 to 99 led to oMEDD:Methadone ratio 4:1, oMEDD 100 to 299 led to oMEDD:Methadone ratio 8:1, oMEDD 300 to 499 led to oMEDD:Methadone ratio 12:1, oMEDD 500 to 999 led to oMEDD:Methadone ratio 15:1, oMEDD 1000 to 1200 led to oMEDD:Methadone ratio 20:1 – example, for subacute pain control Mr. BG 80 mg Methadone per day = oMEDD 320 mg/day + 500mg per day Tapentadol = oMEDD 200mg/day + (say) 600 mg/day PRN Tramadol = oMEDD 120mg/day => Total oMEDD 640 mg/day [on weaning pathway for Tapentadol and Tramadol]. Other examples: Patients with chronic Low Back Pain and chronic neuropathic pancreatitis pain on opioid replacement Suboxone + "as required" PRN Tapentadol (Palexia) chronic use with approval from appropriate regulatory health organization.

Point 5. Opioid tolerant patients can get Opioid Withdrawal Syndrome such as increased pain, abdominal cramps, etc. with insufficient narcotics and Opioid Intoxication of respiratory and CNS depression with excessive narcotics.

In the Class of n-Variable Equations with n = 2 [which translate to 2-Variable Equations], when computationally depicted by 2-dimensional graphs with their x- and y-axes relevantly defined; they often have one or more points of intersection on (i) x-axis, and/or (ii) y-axis, and/or (iii) both x- and y-axes [formally known as the 'Origin']. The Origin, often labeled with capital letter 'O', is defined as the point where the vertical y-axis and the horizontal x-axis intersect each other. Not all functions, though, will have intercepts; which are where the graph crosses either the x-axis (viz. the x-axis intercept, often referred to as "zeros or roots of the equation"), or the y-axis (viz. the y-axis intercept), or both the x- and y-axes (viz. the Origin intercept). There are eight possible Categories of Intercepts for 2-Variable Equations:

Category I Intercept: comprising of nil intercept

Category II Intercept: comprising of single x-axis intercept(s) only

Category III Intercept: comprising of single y-axis intercept(s) only

Category IV Intercept: comprising of single Origin intercept(s) only

Category V Intercept: comprising of double x- & y-axes intercept(s)

Category VI Intercept: comprising of double x-axis & Origin intercept(s)

Category VII Intercept: comprising of double y-axis & Origin intercept(s)

Category VIII Intercept: comprising of triple x-, y-axes & Origin intercept(s)

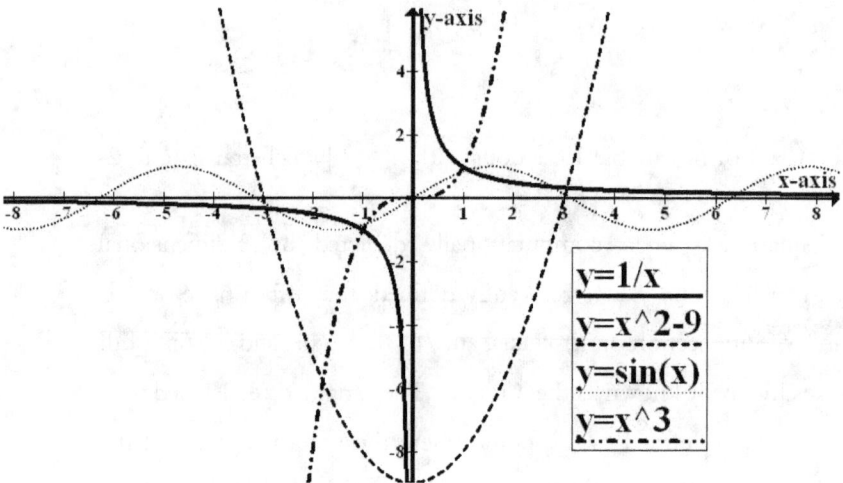

The above figure depicts simple formulae $y = \dfrac{1}{x}$, $y = x^2 - 9$, $y = \sin(x)$

and $y = x^3$ with their axes intercepts.

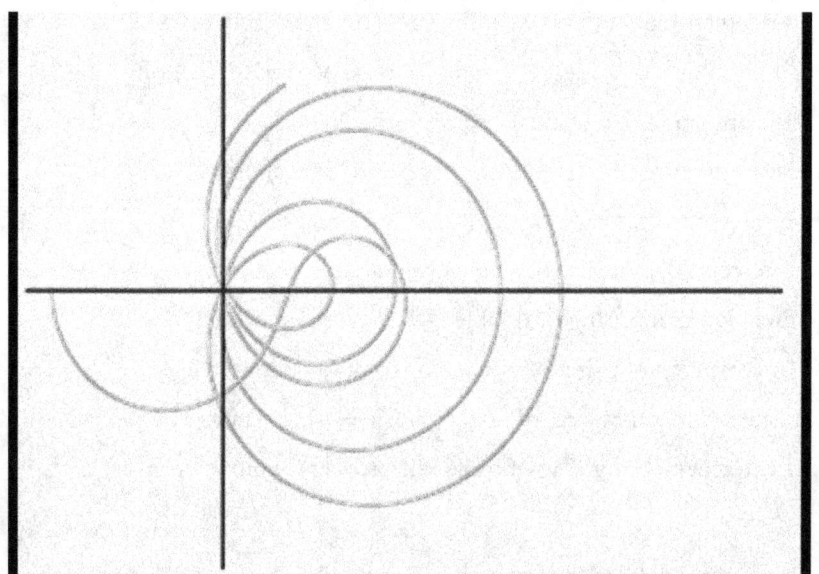

The above graph shows all nontrivial zeros as Origin intercepts when $\sigma = \dfrac{1}{2}$.

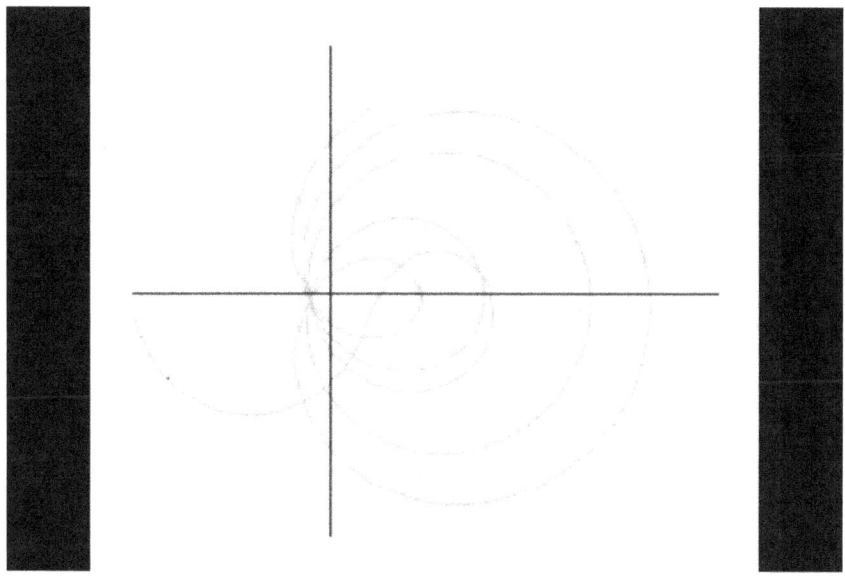

The above graph shows nil nontrivial zeros when $\sigma = \frac{2}{5}$.

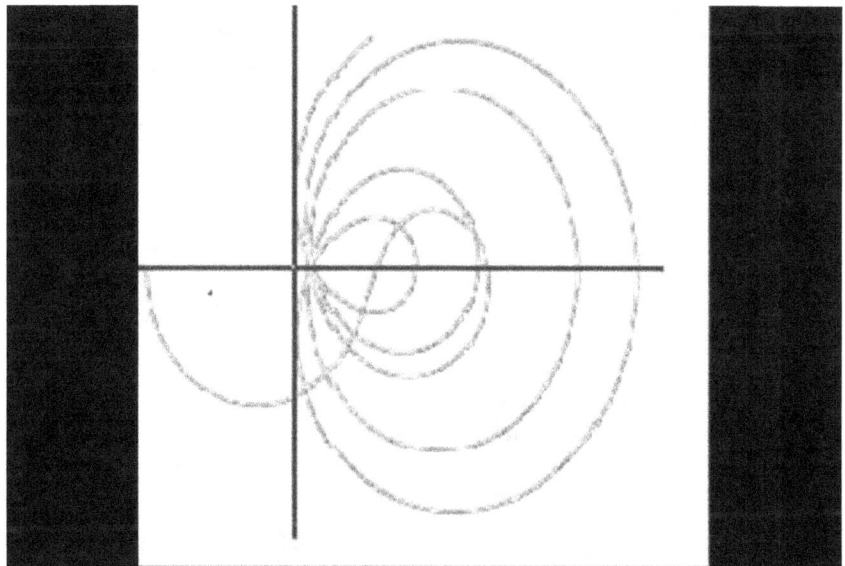

The above graph shows nil nontrivial zeros when $\sigma = \frac{3}{5}$.

We surmise from above three graphs on Riemann zeta function that when σ

253

$= \frac{1}{2}$; Gram[x=0,y=0] conjecture (Riemann hypothesis), Gram[y=0] conjecture, and Gram[x=0] conjecture can be combined as 'grand' Dirichlet-Gram-Riemann conjecture. When $\sigma \neq \frac{1}{2}$; Gram[x=0,y=0] (Riemann hypothesis) conjecture cannot exist and we only have 'virtual' Gram[y=0] conjecture associated with 'virtual' Gram[y=0] points and 'virtual' Gram[x=0] conjecture associated with 'virtual' Gram[x=0] points. 'Virtual' Gram[y=0] points and 'virtual' Gram[x=0] points have totally different values to their corresponding Gram[y=0] points and Gram[x=0] points.

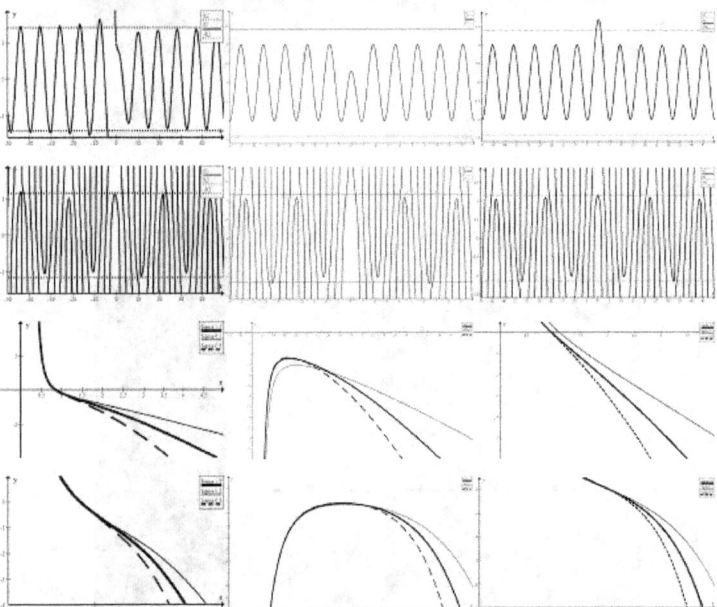

We depict the above computer-generated "combined" miniaturized version graphs from (Helpful B., Rigorous Proof for Riemann Hypothesis Using the Novel Sigma-power Laws and Concepts from the Hybrid Method of Integer Sequence Classification, 2016) and (Helpful B., Key Role of Dimensional Analysis Homogeneity in Proving Riemann Hypothesis and Providing

254

Explanations on the Closely Related Gram Points, 2016) depicted using $\sigma = \frac{1}{5}$, $\sigma = \frac{1}{2}$, and $\sigma = \frac{4}{5}$ thus giving us a snapshot on "physical manifestations" of various Laws – taking into account the 'Errata Notice' previously mentioned in Chapter 11 on mathematical errors present in relevant equations. These graphs are derived out of Gram[x=0,y=0] conjecture (Riemann hypothesis), Gram[y=0] conjecture and Gram[x=0] conjecture. Respectively, the mentioned σ values are from Left to Right seen as three vertical columns of 'grouped' miniature graphs.

Table 1: **Completely Predictable problem of Even-odd number pairing**. Note the non-varying relationship between even and odd numbers.

Set of even number

{Completely Predictable Straight line INTERFACE}

Set of odd number

Table 2: **Incompletely Predictable problem of Prime-composite number pairing**. Note the varying relationship between prime and composite numbers.

Set of prime number

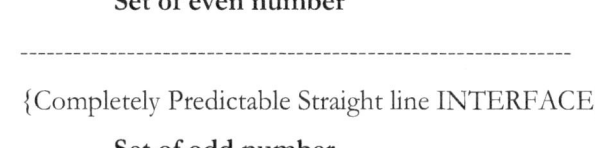

{Completely Predictable Jagged line INTERFACE}

Set of composite number

Table 3: **Completely Predictable problem on intercepts of y = sin x function**. Note the non-varying relationship between [solitary] 'Origin' intercept and x-axis intercepts.

> **Set of x-axis intercepts in y = sin x**
>
> --
>
> {Completely Predictable Straight line INTERFACE}
>
> **Set of 'Origin' intercept in y = sin x**

Table 4: **Incompletely Predictable problem on intercepts of Riemann zeta function (or its *proxy* Dirichlet eta function)**. Note the varying relationship between all three types of Gram points.

> **Set of Gram[y=0] points ('usual' Gram points)**
>
> vvvvvvvvvvvvvvvvvvvvvvvvvvvvvvvvvvvvvvvvvvvvvv
>
> {Completely Predictable Jagged line INTERFACE}
>
> **Set of Gram[x=0,y=0] points (nontrivial zeros)**
>
> vvvvvvvvvvvvvvvvvvvvvvvvvvvvvvvvvvvvvvvvvvvvv
>
> {Completely Predictable Jagged line INTERFACE}
>
> **Set of Gram[x=0] points**

Simple Containers: simple algorithms ("discrete") y = 2x & y = 2x - 1 and simple equations ("continuous") y = 2x, y = 2x - 1 & y = sin x. Complex Containers: complex algorithm ("discrete") Sieve of Eratosthenes and complex equation ("continuous") Riemann zeta function (or its *proxy* Dirichlet eta function). Completely Predictable problems: even-odd number pairing given by "discrete" y = 2x and y = 2x - 1, straight line pairing given by "continuous" y = 2x and y = 2x - 1, and axes intercepts given by "continuous" y = sin x. Incompletely Predictable problems: prime-

256

composite number pairing given by "discrete" Sieve of Eratosthenes and axes intercepts given by "continuous" Riemann zeta function. In 'Numerical relationship interface' and 'Axes intercept relationship interface', CFS of simple & complex properties are present infinitely often at both types of interface for, respectively, Completely & Incompletely Predictable problems.

We can now provide a brief synopsis on Incompletely Predictable entities for Riemann zeta function. Gram points and virtual Gram points are dependently calculated using complex equation Riemann zeta function, ζ (s), or its proxy Dirichlet eta function, η(s), in critical strip (denoted by $0 < \sigma < 1$). Gram[y=0], Gram[x=0] and Gram[x=0,y=0] points respectively refer to x-axis, y-axis and Origin intercepts at critical line (denoted by $\sigma = \frac{1}{2}$). Gram[y=0] and Gram[x=0,y=0] points are respectively synonymous with traditional 'Gram points' and nontrivial zeros. Virtual Gram[y=0] and virtual Gram[x=0] points respectively refer to x-axis and y-axis intercepts at non-critical lines (denoted by $\sigma \neq \frac{1}{2}$). Virtual Gram[x=0,y=0] points do not exist. Activities to prove associated open problem in number theory Riemann hypothesis and explain Gram[y=0] and Gram[x=0] points equate to solving Incompletely Predictable problems.

Solving our three open problems of Riemann hypothesis, Polignac's and Twin prime conjectures will involve obtaining and analyzing these Completely Predictable CFS of complex properties that are (intrinsically) present in complex equation & algorithm instead of hitting a "**mathematical impasse**" by analyzing CIS of Incompletely Predictable entities (extrinsically) generated by these equation & algorithm.

Obviously, a lot of materials in the previous chapter will belong to this chapter [and *vice versa*] due to "overlapping" presence of Completely Predictable and Incompletely Predictable entities in both 'Axes Intercept relationship Interface' and 'Numerical Relationship Interface'.

At 'Numerical relationship interface', uncountable complex numbers (C) include uncountable real numbers (**R**). **R** = countable rational numbers (**Q**) + uncountable irrational numbers (**R − Q**). **R − Q** = countable algebraic numbers + uncountable transcendental numbers. **Q** include countable integers (**Z**) which include countable whole numbers (**W**) which in turn include countable natural numbers (**N**). **N** is constituted by either countable even numbers (**E**) and countable odd numbers (**O**) or countable prime numbers (**P**), countable composite numbers (**C**) and Number '1'.

Then (i) Set **N** = Set **E** + Set **O**, (ii) Set **N** = Set **P** + Set **C** + Number '1', and (iii) Set **N** ⊂ Set **W** ⊂ Set **Z** ⊂ Set **Q** ⊂ Set **R** ⊂ Set **C**. With increasing magnitude, arbitrary Set **X** belongs to countable finite set (CFS), countable infinite set (CIS) or uncountable infinite set (UIS). Cardinality of Set **X**, $|\mathbf{X}|$, measures the "number of elements" in Set **X**. E.g. Set even **P** has CFS of even **P** with $|\text{even } \mathbf{P}| = 1$, Set **N** has CIS of **N** with $|\mathbf{N}| = \aleph_0$, and Set **R** has UIS of **R** with $|\mathbf{R}| = c$ (cardinality of the continuum).

In comparison at the 'Axes intercept relationship interface', uncountable

Generated curves of Riemann zeta function include three types of Gram points viz. countable Gram[x=0,y=0] points, countable Gram[x=0] points & countable Gram[y=0] points. Then (i) Set all Gram points = Set Gram[x=0,y=0] points + Set Gram[x=0] points + Set Gram[y=0] points, and (ii) Set all Gram points ⊂ Set Generated curves. Note that Set negative Gram[y=0] point has CFS of negative Gram[y=0] point with |negative Gram[y=0] point| = 1.

Dependently calculated using complex algorithm Sieve of Eratosthenes, **P** and **C** are Incompletely Predictable numbers. Occurring over 2000 years ago (circa 300 BC), ancient Euclid's proof on infinitude of **P** in totality [viz. $|\mathbf{P}| = \aleph o$ for Set **P**] predominantly by *reductio ad absurdum* (proof by contradiction) is earliest known but not the only proof for this simple problem in number theory.

Since then dozens of proofs have been devised such as three chronologically listed: Goldbach's Proof using Fermat numbers (written in a letter to Swiss mathematician Leonhard Euler, July 1730), Furstenberg's Topological Proof in 1955 (Furstenberg, 1955), and Filip Saidak's Proof in 2006 (Saidak, 2006). The strangest candidate is likely to be Furstenberg's Topological Proof.

In 2013, Yitang Zhang proved a landmark result showing some unknown even number 'N' < 70 million such that there are infinitely many pairs of P that differ by 'N' (Zhang, 2014). By optimizing Zhang's bound, subsequent Polymath Project collaborative efforts using a new refinement of GPY sieve in 2013 lowered 'N' to 246; and assuming Elliott-Halberstam conjecture and its generalized form have further lower 'N' to 12 and 6, respectively. Then 'N' has intuitively more than one valid values such that there are infinitely many pairs of P that differ by each of those 'N' values [thus proving existence of more than one Subset odd Pi with |odd Pi | = ℵo]. We can only theoretically lower 'N' to 2 (in regards to P with 'small gaps') but there are still an infinite number of E prime gaps (in regards to P with 'large gaps') that require "the proof that each will generate its unique set of infinite P".

Two *overall* complex properties are perpetual presence of firstly, critical line location for Set **nontrivial zeros** of Riemann zeta function (indicating Riemann hypothesis to be true) and secondly, |**E** prime gaps| = ℵo and Subsets odd **P** derived from associated **E** prime gaps all having |odd **P**| = ℵo (indicating Polignac's and Twin prime conjectures to be true).

Solving open problems of Riemann hypothesis, Polignac's and Twin prime conjectures involve deriving Complex Elementary Fundamental Laws-based solutions of Sigma-Power Laws for Riemann hypothesis, Plus Gap 2 Composite Number Continuous Law and Plus-Minus Gap 2 Composite Number Alternating Law for Polignac's and Twin prime conjectures obtained by (respectively) undertaking certain non-negotiable mathematical steps as outlined in relevant parts of my research papers for 'Riemann hypothesis mathematical foot-prints' & 'Prime and Twin prime mathematical foot-prints'. We next provide collective statements for our two interfaces.

Collective statements on Axes intercept relationship interface:

For intercepts of y = sin x: Set of all axes intercepts = Set of x-axis intercepts + Single Origin intercept. Set of x-axis intercepts > Set of Origin intercept inequality forms a Completely Predictable "exact non-varying" Axes intercept relationship interface. Example of simple properties due to interactions between two independent sets of x-axis and Origin intercepts is the Completely Predictable timing of both intercepts eternal occurrences whereby they are all given by a simple algorithm: x-axis intercepts = $k\pi$, where k is an **Z** and π = 3.14159 [rounded off to five decimal places] is a transcendental number.

For intercepts of Riemann zeta function: Set of all axes intercepts = Set of x-axis intercepts + Set of y-axis intercepts + Set of Origin intercepts. Set of x-axis intercepts = Set of y-axis intercepts = Set of Origin intercepts equality forms an Incompletely Predictable "exact varying" Axes intercept relationship interface. Examples of complex properties due to interactions between two dependent sets of Gram[y=0] points and nontrivial zeros are Gram's Law & its violation, and Gram block which are Completely Predictable to periodically occur an infinite number of times, albeit in an Incompletely Predictable manner.

Collective statements on Numerical relationship interface:

For **E-O** number pairing: Set **N** = Set **E** + Set **O**. Set **E** = Set **O** equality forms a Completely Predictable "exact non-varying" Numerical relationship interface. Example of simple properties due to interactions between two independent Set **E** and Set **O** are **E** gap = 2 and **O** gap = 2 whose eternal occurrences are Completely Predictable in timing.

For straight line pairing: [Set of two lines in combined length] = [Set

261

of y = 2x line in length] + [Set of 2x - 1 line in length]. [Set of y = 2x line in length] = [Set of y = 2x - 1 line in length] forms a Completely Predictable "exact non-varying" Numerical relationship interface. Examples of simple properties due to interactions between two independent sets of y = 2x line and y = 2x - 1 line is their manifestation as two parallel infinite length straight lines perpetually separated apart by Completely Predictable *horizontal* distance = 0.5, *vertical* distance = 2, and *perpendicular* distance = √(0.25 - 0.0625) = √0.1875 = 0.433 [rounded off to three decimal places] -- this is an algebraic number.

For **P-C** number pairing: **N** = Set **P** + Set **C** + The number '1'. Set **C** > Set **P** inequality forms an Incompletely Predictable "exact varying" Numerical relationship interface. Examples of complex properties due to interactions between two dependent Set **P** and Set **C** relate to Composite Gap 2 Number appearances governed by Plus-Minus Gap 2 Composite Number Alternating Law and Plus Gap 2 Composite Number Continuous Law. These Completely Predictable laws are computationally applicable [albeit with Incompletely Predictable timing] on an eternal basis to designated **P** (generated from all even **P** gaps minus even **P** gap = 2 in the first law, and even **P** gap = 2 in the second law).

With same underlying complex equation, proof of Set Gram[x=0,y=0] points location on critical line should *dependently* confirm Set Gram[y=0] points & Set Gram[x=0] points location on this line. With same underlying complex algorithm, proof of ℵo cardinality value for Subsets odd **P** (from CIS **E** prime gaps = 2, 4, 6, 8, 10,...) constituting Set all **P** minus even **P** '2' should *dependently* confirm this cardinality value for Subset even **C** & Subset odd **C** (from CFS composite gaps = 1 & 2) constituting Set all **C**.

262

Thispter refer to my research paper *Solving Incompletely Predictable*

problem Riemann hypothesis with Dirichlet Sigma-Power Law which is fully

outlined in Appendix 1.

Abstract for this paper:

Riemann hypothesis proposed all nontrivial zeros to be located on critical line of Riemann zeta function. Treated as Incompletely Predictable problem, we obtain Dirichlet Sigma-Power Law as final proof of solving this problem. This Law is derived as equation and inequation from original Dirichlet eta function (proxy function for Riemann zeta function). Performing a parallel procedure help explain closely related Gram points.

Mathematical Footprints as six identifiable steps to prove this hypothesis:

Step 1 Use $\eta(s)$, proxy for $\zeta(s)$, in critical strip.

Step 2 Apply Euler formula to $\eta(s)$.

Step 3 Obtain "simplified" Dirichlet eta function which intrinsically incorporates actual location [but not actual positions] of all nontrivial zeros[#1].

Step 4 Apply Riemann integral to "simplified" Dirichlet eta function in discrete (summation) format.

Step 5 Obtain Dirichlet Sigma-Power Law in continuous (integral) format as equation or inequation.

Step 6 Note exact and inexact DA homogeneity on their fractional exponents.

Footnote #1: Respectively, Gram[y=0] points, Gram[x=0] points & Gram[x=0,y=0] points (nontrivial zeros) are Incompletely Predictable entities with actual positions determined by setting $\sum \text{Im}\{\eta(s)\} = 0$, $\sum \text{Re}\{\eta(s)\} = 0$ & $\sum \text{ReIm}\{\eta(s)\} = 0$ to dependently calculate relevant positions of all preceding entities in neighborhood. Respectively, actual location of Gram[y=0] points, Gram[x=0] points & nontrivial zeros; and virtual Gram[y=0] points, virtual Gram[x=0] points & "absent" nontrivial zeros occur precisely at $\sigma = \frac{1}{2}$; and $\sigma \neq \frac{1}{2}$.

Outline of proof for Riemann hypothesis. To simultaneously satisfy two mutually inclusive conditions:

I. With rigid manifestation of exact DA homogeneity, Set nontrivial zeros with |nontrivial zeros| = $\aleph$o is located on critical line (viz. $\sigma = \frac{1}{2}$) when 2(1 − σ) [or 2(σ + 1)] as $\sum$(all fractional exponents) = whole number '1' [or '3'] in Dirichlet Sigma-Power Law[#2] as equation [or inequation].

II. With rigid manifestation of inexact DA homogeneity, Set nontrivial zeros with |nontrivial zeros| = $\aleph$o is not located on non-critical lines (viz. $\sigma \neq \frac{1}{2}$) when 2(1 − σ) [or 2(σ + 1)] as $\sum$(all fractional exponents) = fractional number '≠1' [or '≠3'] in Dirichlet Sigma-Power Law[#2] as equation [or inequation].

Footnote #2: Derived from original $\eta(s)$ (proxy for $\zeta(s)$) as equation or inequation, this Law symbolizes end-result proof on Riemann hypothesis.

We now provide an outline of exact and inexact Dimensional analysis (DA) homogeneity. Respectively for 'base quantities' such as length, mass and time; their fundamental SI 'units of measurement' meter (m) is defined as distance travelled by light in vacuum for time interval 1/299 792 458 s with speed of light c = 299,792,458 ms^{-1}, kilogram (kg) is defined by taking fixed numerical value Planck constant h to be 6.626 070 15 X 10^{-34} Joules·second (Js) [whereby Js is equal to kgm^2s^{-1}] and second (s) is defined in terms of ΔvCs = Δ(133Cs)h f s = 9,192,631,770 s^{-1}. Derived SI units such as J and ms^{-1} respectively represent 'base quantities' energy and velocity. The word 'dimension' is commonly used to indicate all those mentioned 'units of measurement' in well-defined equations.

DA is an analytic tool with DA homogeneity and non-homogeneity (respectively) denoting valid and invalid equation occurring when 'units of measurements' for 'base quantities' are "balanced" and "unbalanced" across both sides of the equation. E.g. equation 2 m + 3 m = 5 m is valid and equation 2 m + 3 kg = 5 mkg is invalid (respectively) manifesting DA homogeneity and non-homogeneity.

Let (2n) and (2n-1) be 'base quantities' in Dirichlet Sigma-Power Laws formatted in simplest forms as equations and inequations. E.g. DA on exponent $\frac{1}{2}$ in $(2n)^{\frac{1}{2}}$ in simplest form is correct but DA on exponent $\frac{1}{4}$ in equivalent $(2^2n^2)^{\frac{1}{4}}$ *not* in simplest form is incorrect.

Fractional exponents as 'units of measurement' given by $(1 - \sigma)$ for equations and $(\sigma + 1)$ for inequations when $\sigma = \frac{1}{2}$ coincide with exact DA homogeneity[#3]; and $(1 - \sigma)$ for equations and $(\sigma + 1)$ for inequations when

$\sigma \neq \frac{1}{2}$ coincide with inexact DA homogeneity[#4].

Footnotes #3, #4: Exact and inexact DA homogeneity occur in Dirichlet Sigma-Power Laws as equations or inequations for Gram[y=0] points, Gram[x=0] points and Gram[x=0,y=0] points (nontrivial zeros). Law of Continuity is a heuristic principle whatever succeed for the finite, also succeed for the infinite. Then these Laws which inherently manifest themselves on finite and infinite time scale should "succeed for the finite, also succeed for the infinite".

Respectively for equations and inequations, exact DA homogeneity at $\sigma = \frac{1}{2}$ denotes $\sum$(all fractional exponents) as $2(1 - \sigma)$ and $2(\sigma + 1)$ equates to ["exact"] whole number '1' and '3'; and inexact DA homogeneity at $\sigma \neq \frac{1}{2}$ denotes $\sum$(all fractional exponents) as $2(1 - \sigma)$ and $2(\sigma + 1)$ equates to ["inexact"] fractional number '$\neq$1' and '$\neq$3'.

The above preliminary materials should now put readers in a strong position to understand my research paper. The main body of my research paper – readers to refer to this area as provided in Appendix 1 – then outline using correct and complete mathematical arguments how rigorous proof for Riemann hypothesis and explaining the closely related Gram points are derived. These correct and complete mathematical arguments will not currently be elaborated upon in this book.

Rigorous proof for Riemann hypothesis is summarized as Theorem Riemann I – IV below [Note: QED = *quod erat demonstrandum*]:

As preliminary, we supply the following important mathematical arguments.

266

For $0 < \sigma < 1$, then $0 < 2(1-\sigma) < 2$. The only whole number between 0 and 2 is '1' which coincide with $\sigma = \frac{1}{2}$. When $0 < \sigma < \frac{1}{2}$ and $\frac{1}{2} < \sigma < 1$, then $0 < 2(1-\sigma) < 1$ and $1 < 2(1-\sigma) < 2$.

For $0 < \sigma < 1$, $2 < 2(\sigma + 1) < 4$. The only whole number between 2 and 4 is '3' which coincide with $\sigma = \frac{1}{2}$. When $0 < \sigma < \frac{1}{2}$ and $\frac{1}{2} < \sigma < 1$, then $2 < 2(\sigma +1) < 3$ and $3 < 2(\sigma +1) < 4$.

Legend: R = all real numbers. For $0<\sigma<1$, σ consist of $0<R<1$. For $0 < 2(1-\sigma) < 2$ and $2 < 2(\sigma +1) < 4$, $2(1-\sigma)$ and $2(\sigma +1)$ must (respectively) consist of $0 < R < 2$ and $2 < R < 4$.

An important caveat is that previously used phrases such as "fractional exponent σ" and "$\sum$(all fractional exponents) = whole number '1' [or '3'] and fractional number '$\neq$1' [or '$\neq$3']", although not incorrect per se, should respectively be replaced by "real number exponent σ" and "$\sum$(all real number exponents) = whole number '1' [or '3'] and real number '$\neq$1' [or '$\neq$3']#5" for complete accuracy. We apply this caveat to Theorem Riemann I – IV.

Footnote #5: As whole numbers $\subset$ real numbers, one could also depict this phrase as "$\sum$(all real number exponents) = real number '1' [or '3'] and real number '$\neq$1' [or '$\neq$3']".

Theorem Riemann I. Derived from proxy Dirichlet eta function, "simplified" Dirichlet eta function will exclusively contain de novo property for actual location [but not actual positions] of all nontrivial zeros.

Proof. The phrase "actual location [but not actual positions] of all nontrivial zeros" can be validly shortened to "actual location of all nontrivial zeros" as used in Theorem Riemann II, III and IV. The proof for Theorem Riemann I is now complete as it successfully incorporates proof for Lemma 3.1 QED.

Theorem Riemann II. Dirichlet Sigma-Power Law [in continuous (integral) format] as equation and inequation which are both derived from "simplified" Dirichlet eta function [in discrete (summation) format] will exclusively manifest exact DA homogeneity in equation and inequation only when real number exponent $\sigma = \frac{1}{2}$.

Proof. The proof for Theorem Riemann II is now complete as it successfully incorporates proofs from Proposition 3.2 on derivation for equation and inequation of Dirichlet Sigma-Power Law [with both containing de novo property for "actual location of all nontrivial zeros"] and Proposition 3.3 on manifestation of exact DA homogeneity in Dirichlet Sigma-Power Law as equation and inequation when real number exponent $\sigma = \frac{1}{2}$ QED.

Theorem Riemann III. Real number exponent $\sigma = \frac{1}{2}$ in Dirichlet Sigma-Power Law as equation and inequation satisfying exact DA homogeneity is identical to σ variable in Riemann hypothesis which propose σ to also have exclusive value of $\frac{1}{2}$ (representing critical line) for "actual location of all nontrivial zeros", thus fully supporting Riemann hypothesis to be true with further clarification by Theorem Riemann IV.

Proof. Since $s = \sigma \pm \imath t$, complete set of nontrivial zeros which is defined by $\eta(s) = 0$ is exclusively associated with one (and only one) particular $\eta(\sigma \pm \imath t) = 0$ value solution, and by default one (and only one) particular σ [conjecturally] $= \frac{1}{2}$ solution. When performing exact DA homogeneity on Dirichlet Sigma-Power Law as equation and inequation [with both containing de novo property for "actual location of all nontrivial zeros"], the phrase "If real number exponent σ has exclusively $\frac{1}{2}$ value, only then will exact DA homogeneity be satisfied" implies one (and only one) possible mathematical solution. Theorem Riemann III reflect Theorem Riemann II on presence of

268

exact DA homogeneity for $\sigma = \frac{1}{2}$ in Dirichlet Sigma-Power Law as equation and inequation. This Law has identical σ variable as that referred to by Riemann hypothesis [whereby σ here uniquely refer to critical line]. The proof for Theorem Riemann III is now complete as it independently refers to simultaneous association of confirmed (i) solitary $\sigma = \frac{1}{2}$ value in Dirichlet Sigma-Power Law as equation and inequation satisfying exact DA homogeneity and (ii) critical line defined by solitary $\sigma = \frac{1}{2}$ value being the "actual location [but with no request to determine actual positions]" of all nontrivial zeros as proposed in original Riemann hypothesis QED.

Theorem Riemann IV. Condition 1. All $\sigma \neq \frac{1}{2}$ values (non-critical lines), viz. $0 < \sigma < \frac{1}{2}$ and $\frac{1}{2} < \sigma < 1$ values, exclusively does not contain "actual location of all nontrivial zeros" [manifesting de novo inexact DA homogeneity in equation and inequation], together with Condition 2. One (and only one) $\sigma = \frac{1}{2}$ value (critical line) exclusively contains "actual location of all nontrivial zeros" [manifesting de novo exact DA homogeneity in equation and inequation], fully support Riemann hypothesis to be true when these two mutually inclusive conditions are met.

Proof. Condition 2 Theorem Riemann IV simply reflect proof from Theorem Riemann III [incorporating Proposition 3.3] for "actual location of all nontrivial zeros" exclusively on critical line manifesting de novo exact DA homogeneity $\sum$(all real number exponents) = whole number '1' for equation [or '3' for inequation]. The proof for Condition 2 Theorem Riemann IV is now complete QED. Corollary 3.4 confirms de novo inexact DA homogeneity manifested as $\sum$(all real number exponents) = real number '$\neq$1' for equation [or '$\neq$3' for inequation] by all $\sigma \neq \frac{1}{2}$ values (non-critical lines) that are exclusively not associated with "actual location of all nontrivial zeros". Applying inclusion-exclusion principle: Exclusive presence of nontrivial zeros on critical line for Condition 2 Theorem Riemann IV implies exclusive absence of nontrivial zeros on non-critical lines for Condition 1 Theorem Riemann IV. The proof for Condition 1 Theorem Riemann IV is now complete QED.

The outline of conclusion section of my research paper is provided next.
In our Hybrid method of Integer Sequence classification, a formula is either non-Hybrid or Hybrid integer sequence. Inequation with two 'necessary' Ratio (R) or equation with one 'unnecessary' R contains non-Hybrid integer sequence. Equation with one 'necessary' R contains Hybrid integer sequence. "In the limit" Hybrid integer sequence approach unique Position X, it becomes non-Hybrid integer sequence for all Positions ≥ Position X.

Consider kinetic energy (KE) in MJ with m_o = rest mass in kg and v = velocity in ms^{-1}. In classical mechanics concerning low velocity with v << c, Newtonian KE $= \frac{1}{2}m_o v^2$. In relativistic mechanics concerning high velocity with v ≥ 0.01c, Relativistic KE $= \frac{m_o c^2}{\sqrt{(1-(v^2/c^2))}} - m_o c^2$. Obtained from the later by binomial approximation or by taking first two terms of Taylor expansion for reciprocal square root, the former approximates the later well at low speed.

We arbitrarily divide DA homogeneity into inexact DA homogeneity for ["<100% accuracy"] Newtonian KE and exact DA homogeneity for ["100% accuracy"] Relativistic KE. "In the limit" ["<100% accuracy'] Newtonian KE at low speed approach ['100% accuracy'] Relativistic KE at high speed, we achieve perfection.

Analogy: "In the limit" all three version of Dirichlet Sigma-Power Laws for Gram[y=0] points, Gram[x=0] points and nontrivial zeros as '<100% accuracy' inequations approach perfection as '100% accuracy' equations, compliance with inexact DA homogeneity becomes compliance with exact DA homogeneity. We note R1 terms in all inequations contain (2n) and (2n - 1) 'base quantities' but these are not endowed with fractional exponent (σ + 1) as relevant 'unit of measurement'. As Incompletely Predictable problems, we gave relatively elementary proof of Riemann hypothesis and explain closely related Gram points whereby various "meta-properties" such as exact and inexact DA homogeneity occur in (respectively) equations and inequations of relevant Dirichlet Sigma-Power Laws. Harnessed key benefit from successful proof for Riemann hypothesis is often stated as "With this one solution, we have proven five hundred theorems or more at once". This apply to important theorems in number theory that rely on properties of Riemann zeta function or Dirichlet eta function such as location of trivial and nontrivial zeros. E.g., we delineate prime number theorem by prime counting function $\pi(x)$ [which is defined as number of primes $\leq$ x].

271

21 Solving Polignac's and Twin Prime Conjectures

This chapter refer to my research paper *Solving Incompletely Predictable*

problems Polignac's and Twin Prime conjectures using Information-Complexity
conservation which is fully outlined in Appendix 2.

Abstract for this paper:

Prime numbers are Incompletely Predictable numbers calculated using complex algorithm Sieve of Eratosthenes. Involving proposals that prime gaps and associated sets of prime numbers are infinite in magnitude, Twin prime conjecture deals with even prime gap 2 and is a subset of Polignac's conjecture which deals with all even prime gaps 2, 4, 6, 8, 10,.... Treated as Incompletely Predictable problems, we solve these conjectures with research method Information-Complexity conservation to get Plus Gap 2 Composite Number Continuous Law and Plus-Minus Gap 2 Composite Number Alternating Law.

Mathematical Footprints as six identifiable steps to prove these conjectures: Step 1 Considering $x \in N$, obtain Dimensions $(2x - 2)^1$, $(2x - 4)^1$, $(2x - 5)^1$, $(2x - 7)^1$, $(2x - 8)^1$, $(2x - 9)^1$, ..., $(2x - \infty)^1$ with specific groupings to constitute all elements of Set P [culminating in obtaining all prime gaps (= E prime gaps + Solitary O prime gap) with |all prime gaps| = $\aleph$o]. Note Dimension $(2x - 2)^1$ represents x = 1 (Number '1') which is neither P nor C. Step 2 Considering $i \in E$, confirm perpetual recurrences of individual E prime gap = i (associated with its unique odd Pi) occur only when depicted as specific groupings of these Dimensions endowed with exponent '1' for all ranges of x.

Step 3 Perform DA on exponent '1' in these Dimensions.

Step 4 Perform DA on equation Set odd P $= \sum_{i=2}^{\infty} Subset\ odd\ Pi$ to obtain $|$odd P$| = |$odd Pi$| = \aleph 0$ whereby Subset odd Pi is derived from its associated unique E prime gap $=$ i with $|$E prime gaps$| = \aleph 0$.

Step 5 Confirm 'Prime number' variable and 'Prime gap' variable complex algorithm "containing" all P with knowing their overall actual location [but not actual positions][#1].

Step 6 Derive Plus-Minus Gap 2 Composite Number Alternating Law and Plus Gap 2 Composite Number Continuous Law using Information-Complexity conservation.

Footnote #1: This phrase implies all P (and C) are treated as Incompletely Predictable numbers. Actual positions will require using complex algorithm Sieve of Eratosthenes to dependently calculate positions of all preceding P (and C) in the neighborhood.

Outline of proof for Polignac's and Twin prime conjectures. Requires simultaneously satisfying two mutually inclusive conditions:

I. With rigid manifestation of DA homogeneity, quantitive[#2] fulfillment by considering i $\in$ E for each Subset odd Pi generated by E prime gap $=$ i from Set E prime gaps occurs only if solitary cardinality value is present in equation Set odd P $= \sum_{i=2}^{\infty} Subset\ odd\ Pi$ with $|$odd P$| = |$odd Pi$| = |$E prime gaps$| = \aleph 0$.

II. With rigid manifestation of DA non-homogeneity, quantitive[#2] fulfillment by considering i $\in$ E for each Subset odd Pi generated by E prime gap = i from Set E prime gaps does not occur if more than one cardinality values are present in equation Set odd P $> \sum_{i=2}^{\infty} Subset\ odd\ Pi$ with | E prime gaps | = $\aleph_0$ having incorrect | Subset(s) odd P | = N (finite value) and/or Set odd P $> \sum_{i=2}^{N} Subset\ odd\ Pi$ with | odd Pi | = $\aleph_0$ having incorrect | E prime gaps | = N (finite value).

Footnote #2: Qualitative fulfillment of | odd P | = | odd Pi | = | all E prime gaps | = $\aleph_0$ equates to Plus-Minus Gap 2 Composite Number Alternating Law being precisely obeyed by all E prime gaps apart from first E prime gap precisely obeying Plus Gap 2 Composite Number Continuous Law. Derived using Information-Complexity conservation, these Laws symbolize "end-result" proof on Polignac's and Twin prime conjectures. Law of Continuity is a heuristic principle whatever succeed for the finite, also succeed for the infinite. Then these Laws which inherently manifest 'Gap 2 Composite Number' on finite and infinite time scale should in principle "succeed for the finite, also succeed for the infinite".

We now provide an outline on Dimensional analysis (DA) on Cardinality and "Dimensions". For 'base quantities' such as length, mass and time; their fundamental SI 'units of measurement' are [respectively] given by meter (m), kilogram (kg) and second (s). The word 'dimension' is commonly used to denote 'units of measurement' in well-defined equations.

DA is an analytic tool with resulting DA homogeneity and non-homogeneity (respectively) denoting valid and invalid equation when 'units of measurements' are "balanced" and "unbalanced" across both sides of the equation. E.g. 2 m + 3 m = 5 m is a valid equation but 2 m + 3 kg = 5 mkg is an invalid equation.

274

We use "Dimensions" to denote well-defined Incompletely Predictable entities obtained from Information-Complexity conservation. Relevant "Dimensions" dependently represent Number '1', P and C. Then by default any (sub)sets of P and C in well-defined equations can also be represented by their corresponding "Dimensions". We can apply Dimensional analysis to "Dimensions" from Information-Complexity conservation and cardinality of relevant sets in certain well-defined equations.

Let X denote E, O, N [which are classified as Completely Predictable numbers], P and C [which are classified as Incompletely Predictable numbers]. For x = 1, 2, 3, 4, 5,..., ∞; consider all X ≤ x. Then this "all X ≤ x" is definition for X-π(x) [denoting "X counting function"] resulting in following two types of equations coined as (I) 'Exact' equation Nπ(x) = E-π(x) + O-π(x) with "non-varying" relationships E-π(x) = O-π(x) for all x = E and E-π(x) = O-π(x) - 1 for all x = O, and (II) 'Inexact' equation N-π(x) = 1 + P-π(x) + C-π(x) with "varying" relationships P-π(x) > C-π(x) for all x ≤ 8; P-π(x) = C-π(x) for x = 9, 11, and 13; and P-π(x) < C-π(x) for x = 10, 12, and all x ≥ 14.

Let "Dimensions" and different (sub)sets of E, O, N, P and C be 'base quantities'. Then exponent '1' of "Dimensions" and cardinality of these (sub)sets in well-defined equations Polignac's and Twin Prime conjectures are corresponding 'units of measurement'. Performing DA on 'Dimensions" for PC pairing are depicted later on [in my paper]. Performing DA on cardinality are depicted next.

275

For Set N = Set E + Set O, then $|N| = |E| + |O| \Rightarrow \aleph o = \aleph o + \aleph o$ thus conforming with DA homogeneity.

For Set N = Set P + Set C + Number '1', then Set N - Number '1' = Set P + Set C and $|N - \text{Number '1'}| = |P| + |C| \Rightarrow \aleph o = \aleph o + \aleph o$ thus conforming with DA homogeneity.

For Set N - Set even P - Number '1'= Set odd P + Set even C + Set odd C, then $|N- \text{even P} - \text{Number '1'}| = |\text{odd P}| + |\text{even C}| + |\text{odd C}| \Rightarrow \aleph o = \aleph o + \aleph o + \aleph o$ thus conforming with DA homogeneity.

Symbolically represented by all available O prime gap = 1 and E prime gaps = 2, 4, 6, 8, 10,...; O composite gap = 1 and E composite gap = 2; and O natural gap = 1; then $|\text{Gap 1 N} - \text{Gap 1 P} - \text{Number '1'}| = |\text{Gap 2 P}| + |\text{Gap 4 P}| + |\text{Gap 6 P}| + |\text{Gap 8 P}| + |\text{Gap 10 P}| + ... + |\text{Gap 1 C}| + |\text{Gap 2 C}| \Rightarrow \aleph o = \aleph o + \aleph o + \aleph o + \aleph o + \aleph o + ... \aleph o + \aleph o$ thus conforming with DA homogeneity. It is known that $|\text{Gap 1 P}| = |\text{Number '1'}| = 1$ and $|\text{Gap 1 N}| = |\text{Gap 1 C}| = |\text{Gap 2 C}| = \aleph o$. Then solving Polignac's & Twin prime conjectures translate to successfully proving $|\text{Gap 2 P}| = |\text{Gap 4 P}| = |\text{Gap 6 P}| = |\text{Gap 8 P}| = |\text{Gap 10 P}| = ... = \aleph o$ with $|\text{E prime gaps}| = \aleph o$.

The above preliminary materials should now put readers in a strong position to understand my research paper. The main body of my research paper – readers to refer to this area as provided in Appendix 2 – then outline using the correct and complete mathematical arguments on how rigorous proofs for Polignac's and Twin prime conjectures are derived. These mathematical arguments will not be further elaborated upon in this book.

Rigorous proofs for Polignac's and Twin prime conjectures are summarized as Theorem Polignac-Twin prime I – IV below [Note: QED = quod erat demonstrandum]:

Theorem Polignac-Twin prime I. Incompletely Predictable prime numbers Pn = 2, 3, 5, 7, 11, ..., ∞ or composite numbers Cn = 4, 6, 8, 9, 10, ..., ∞ are CIS with overall actual location [but not actual positions] of all prime or composite numbers accurately represented by complex algorithm involving prime gaps GPi viz. $P_{n+1} = 2 + \sum_{i=1}^{n} GPi$ or involving composite gaps GCi viz. $C_{n+1} = 4 + \sum_{i=1}^{n} GCi$ whereby prime & composite numbers are symbolically represented here with aid of 'n' notation instead of usual 'i' notation; and i & n = 1, 2, 3, 4, 5, ..., ∞. Number '2' in first algorithm represents P1, the very first (and only even) P. Number '4' in second algorithm represent C1, the very first (and even) C.

Proof. We treat above algorithms as unique mathematical objects looking for key intrinsic properties and behaviors. Each P or C is assigned a unique prime or composite gap. Absolute number of P or C and (thus) prime or composite gaps are infinite in magnitude. As original formulae containing all P or C by themselves (viz. without supplying prime or composite gaps as "input information" to generate P or C as "output complexity"), these algorithms intrinsically incorporate overall actual location [but not actual positions] of all P or C. The proof is now complete for Theorem Polignac-Twin prime I QED.

Theorem Polignac-Twin prime II. Set of prime gaps GPi = 2, 4, 6, 8, 10, ..., ∞ is infinite in magnitude whereby these prime gaps accurately and completely represented by Dimensions $(2x - 7)^1$, $(2x - 8)^1$, $(2x - 9)^1$, ..., $(2x - \infty)^1$ must satisfy Information-Complexity conservation in a consistent manner.

Proof. Part I of Proposition 4.2 proved all P are represented by Dimension $(2x - N)^1$ with N ≥ 7 for any given x value (except for x = 2 & 3 values). Note that although x = 1 is neither P nor C, it is validly represented by Dimension $(2x - 2)^1$. If each P is endowed with a specific prime gap value, then each such prime gap must [via logical mathematical deduction] be represented by Dimension $(2x - N)^1$. We advocate this nominated method of prime gap representation using Dimensions be [purportedly] the only way to achieve Information-Complexity conservation. The preceding mathematical statements are correct as there is a unique prime gap value associated with each P. Proposition 5.1 below based on principles from Set theory provides further supporting materials that prime gaps are infinite in magnitude. The proof is now complete for Theorem Polignac-Twin prime II QED.

Theorem Polignac-Twin prime III. To maintain Dimensional analysis (DA) homogeneity, those Dimensions $(2x - N)^1$ from Theorem Polignac-Twin prime II must contain eternal repetitions of well-ordered sets constituted by Dimensions $(2x - 7)^1$, $(2x - 8)^1$, $(2x - 9)^1$, $(2x - 10)^1$, $(2x - 11)^1$, ..., $(2x - \infty)^1$.

Proof. This Theorem is stated in greater details as "To maintain DA homogeneity, those aforementioned [endowed with exponent 1] Dimensions $(2x - N)^1$ from Theorem Polignac-Twin prime II must repeat themselves indefinitely in following specific combinations – (i) Dimension $(2x - 7)^1$

278

only appearing as twin [two-times-in-a-row] and quadruplet [fourtimes-in-a-row] sequences, and (ii) Dimensions $(2x - 8)^1$, $(2x - 9)^1$, $(2x - 10)^1$, $(2x - 11)^1$, ..., $(2x - \infty)^1$ appearing as progressive groupings of E 2, 4, 6, 8, 10,..., ∞." To accommodate the only even P '2', exceptions to this DA homogeneity compliance will expectedly occur right at beginning of P sequence – (i) one-off appearance of Dimensions $(2x - 2)^1$, $(2x - 4)^1$ and $(2x - 5)^1$ and (ii) one-off appearance of Dimension $(2x - 7)^1$ as a quintuplet [five-times-in-a-row] sequence which is equivalent to (eternal) non-appearance of Dimension $(2x - 6)^1$ at x = 4. [We again note Dimension $(2x - 2)^1$ validly represent Number '1' which is neither P nor C.] These sequentially arranged sets are CFS whereby from x = 11 onwards, each set always commence initially as 'baseline' Dimension $(2x - 7)^1$ at x = O values and always end with its last Dimension at x = E values. Each set also have varying cardinality with values derived from all E; and correctly combined sets always intrinsically generate two infinite sets of P and, by default, C in an integrated manner. Our Theorem Polignac-Twin prime III simply represent a mathematical summary derived from Section 3 & 4 of all expressed characteristics of Dimension $(2x - N)^1$ when used to represent P with intrinsic display of DA homogeneity. See Proposition 5.2 below for further details on DA aspect. The proof is now complete for Theorem Polignac-Twin prime III QED.

Theorem Polignac-Twin prime IV. Aspect 1. The "quantitive" aspect to existence of both prime gaps and their associated prime numbers as sets of infinite magnitude will be shown to be correct by utilizing principles from Set theory. Aspect 2. The "qualitative" aspect to existence of both

279

prime gaps and their associated prime numbers as sets of infinite magnitude will be shown to be correct by 'Plus-Minus Gap 2 Composite Number Alternating Law' and 'Plus Gap 2 Composite Number Continuous Law'.

Proof. Required concepts from Set theory involve cardinality of a set with its 'well ordering principle' application. Supporting materials for these concepts based on 'pigeonhole principle' in relation to Aspect 1 are outlined in Proposition 5.1 below. 'Plus-Minus Gap 2 Composite Number Alternating Law' is applicable to all E prime gaps [apart from first E prime gap = 2 for twin primes]. The prime gap = 2 situation will obey 'Plus Gap 2 Composite Number Continuous Law'. These Laws are essentially Laws of Continuity inferring underlying intrinsic driving mechanisms that enables infinity magnitude association for both prime gaps & prime numbers to co-exist. By the same token, these Laws have the important implication that they must be applicable to those relevant prime gaps on a perpetual time scale. Supporting materials in relation to Aspect 2 are found in Proposition 4.2 above. The proof is now complete for Theorem Polignac-Twin prime IV QED.

We note two mutually inclusive conditions: Condition 1. Presence of all Dimensions that repeat themselves on an indefinite basis and with exponent of '1' will give rise to complete sets of P & C ["DA-wise one & only one mathematical possibility argument" associated with inevitable de novo DA homogeneity], and Condition 2. Presence of any Dimension(s) that do not repeat itself (themselves) on an indefinite basis or with exponent other than '1' will give rise to incomplete set of P & C or incorrect set of non-P & non-C ["DA-wise mathematical impossibility argument" associated with inevitable de novo DA non-homogeneity].

When met, these two conditions will fully support the point that CFS Dimensions representations of P & C [with respective prime & composite gaps] are totally accurate. Condition 1 reflect proof from Theorem Polignac-Twin prime III above as all P & C are associated with DA homogeneity when their Dimensions are endowed with exponent of '1'. Condition 2 invoke corollary on inevitable appearance of incomplete P or C or non-P or non-C [associated with DA non-homogeneity] being tightly incorporated into this mathematical framework. See Propositions 5.1 and 5.2, and Corollary 5.3 below for supporting materials on DA homogeneity & non-homogeneity.

We analyze P (& C) in terms of (i) measurements based on cardinality of CIS and (ii) pigeonhole principle which states that if n items are put into m containers, with n>m, then at least one container must contain more than one item. We note that ordinality of all infinite P (& C) is "fixed" implying that each one of the infinite well-ordered Dimension sets conforming to CFS type as constituted by Dimensions $(2x - 7)^1$, $(2x - 8)^1$, $(2x - 9)^1$, $(2x - 10)^1$, $(2x - 11)^1$, ..., $(2x - \infty)^1 1$ on respective gaps for P (& C) must also be "fixed".

The outline of conclusion section of my research paper is provided next.

The harnessed property CIS of [Completely Predictable] natural numbers 1, 2, 3, 4, 5, 6, 7,... having CIS of [Completely Predictable] natural gaps 1, 1, 1, 1, 1, 1,... are constituted by three dependent sets of numbers:

(i)   CIS of [Incompletely Predictable] odd prime numbers 3, 5, 7, 11, 13, 17,... having CIS of [Incompletely Predictable] prime gaps 2, 2, 4, 2, 4,... plus CFS of solitary [Incompletely Predictable] even prime number 2 having CFS of [Incompletely

Predictable] prime gap 1,

(ii)   CIS of [Incompletely Predictable] even and odd composite numbers 4, 6, 8, 9, 10, 12,... having CIS of [Incompletely Predictable] composite gaps 2, 2, 1, 1, 2, 2,.... and

(iii)   CFS of solitary odd number '1' [neither prime nor composite]. Treated as Incompletely Predictable problems endowed with "meta-properties", we gave relatively elementary proofs on Polignac's and Twin prime conjectures based on this harnessed property by performing Dimensional analysis on (sub)sets and "Dimensions" of prime and composite numbers, and obtaining 'Plus-Minus Gap 2 Composite Number Alternating Law' and 'Plus Gap 2 Composite Number Continuous Law'.

Prime number theorem describes asymptotic distribution of prime numbers among positive integers by formalizing intuitive idea that prime numbers become less common as they become larger through precisely quantifying rate at which this occurs using probability. Nontrivial zeros [from 'Axes intercept relationship interface' relevant to Riemann hypothesis] and prime numbers [from 'Numerical relationship interface' relevant to prime number theorem] are Incompletely Predictable entities and numbers.

Deep-seated connections exist between Riemann hypothesis and prime number theorem (which is fully delineated by prime counting function [denoted here with $\pi(x)$]). Solving Incompletely Predictable problem Riemann hypothesis is instrumental in proving efficacy of techniques that estimate $\pi(x)$ efficiently. This should now confirm "best possible" bound for error ("smallest possible" error) of prime number theorem.

In mathematics, logarithmic integral function or integral logarithm li(x) is a special function. Relevant to problems of physics and with number theoretic significance, it occurs in prime number theorem as an estimate of π(x) whereby the form of this special function is defined so that li(2) = 0; viz. li(x) $\equiv \int_2^x \frac{du}{\ln u}$ = li(x) - li(2). There are less accurate ways of estimating π(x) such as conjectured by Gauss and Legendre at end of 18th century. This is approximately x/ln x in the sense $\lim\limits_{x \to \infty} \left(\frac{\pi(x)}{nx/\ln x} \right) = 1$.

Skewes' number is any of several extremely large numbers used by South African mathematician Stanley Skewes as upper bounds for smallest natural number x for which li(x)<π(x). These bounds have since been improved by others: there is a crossing near $e^{727.95133}$ but it is not known whether this is the smallest. John Edensor Littlewood, who was Skewes' research supervisor, proved in 1914 that there is such a [first] number; and found that sign of difference π(x) - li(x) changes infinitely often. This refute all prior numerical evidence that seem to suggest li(x) was always more than π(x).

The key point is [100% accurate] π(x) mathematical tool being "wrapped around" by [less-than-100% accurate] approximate mathematical tool li(x) infinitely often via this 'sign of difference' changes meant that li(x) is the most efficient approximate mathematical tool. Contrast this with "crude" x/ln x approximate mathematical tool where values obtained diverge away from π(x) at increasingly greater rate when larger range of prime numbers are studied.

By an L-function, we generally refer to a Dirichlet series with a functional equation and an Euler product. Contextually, the simplest example of an L-function is Riemann zeta function on which the 1859 Riemann hypothesis is based upon. L-functions are ubiquitous in number theory and hence have applications to mathematical physics and cryptography. They arise from and encode information about a number of mathematical objects and it is necessary to exhibit these objects along with the L-functions themselves since typically we need these objects to compute L-functions.

For examples, L-functions can come from modular forms, elliptic curves, number fields, and Dirichlet characters, as well as more generally from automorphic forms, algebraic varieties, and Artin representations. Broadly based on these examples, the mammoth 'L-functions and Modular Forms Database' (LMFDB) creation was conducted with massive team-effort collaboration from an international group of more than 80 researchers from 12 countries which included prominent mathematicians such as from American Institute of Mathematics in United States, University of Bristol in United Kingdom, and Dartmouth College in United States.

The LMFDB idea was first conceived at an American Institute of Mathematics workshop in 2007. Six years after commencing the LMFDB project [website address http://www.lmfdb.org], its launching was celebrated on May 10, 2016. In effect, LMFDB can be considered an uncharted mathematical terrain providing a detailed atlas of mathematical objects that highlights deep relationships and serves as a guide to latest research happening in physics, computer science and mathematics. Elliptic curves arise naturally in many parts of mathematics and can be described by a simple cubic equation. They also form the basis of cryptographic protocols used by most of the major internet companies including Google, Facebook and Amazon. Modular forms are more mysterious objects constituted by complex functions with an almost unbelievable degree of symmetry.

The two mathematical worlds of elliptic curves and modular forms are remarkably connected via their L-functions. It is this deep connection that was in essence required in the late 20th century by famous British number theorist Andrew John Wiles to successfully achieve his proof of Fermat's Last Theorem. To put into perspective the importance of LMFDB in relation to active research areas such as involving Monstrous moonshine (Moonshine theory), Mathieu moonshine, and Umbral moonshine with their conjectured roles in Quantum gravity and String theory; we think that most physicists would have a positive opinion or consensus on the potential role of these research areas in successfully merging gravity with Grand Unified Theory (GUT) – consisting of the unification of electromagnetism, weak nuclear force, and strong nuclear force – thus giving rise to the holy grail Theory of Everything (TOE).

We briefly divert here to mention that the name 'Standard Model of particle physics', commenced in the 1970s, denotes the theory describing three of the four known fundamental forces in the universe (viz. the electromagnetic, weak, and strong interactions of GUT), as well as classifying all known elementary particles. Despite all its predictive power, it is not

285

'perfect" in that it can't explain gravity, dark matter or dark energy.

String theories assume that fundamental building blocks of the universe are strings instead of point particles. String duality is a class of symmetries in physics that link different String theories, with K3 surfaces appearing almost ubiquitously in string duality. A K3 surface is a complex or algebraic smooth minimal complete surface that is regular and has trivial canonical bundle. Not least because of this difficulty of multiple String theories (and hence multiple possibilities), an alternative view is that all four fundamental forces of nature will always exist as the current *status quo* with gravity obeying laws [perhaps endowed with certain "continuous" Completely Predictable properties] derived from Einstein's Theory of General Relativity and the three forces of GUT obeying laws [perhaps endowed with certain "discrete" Incompletely Predictable properties] based on Quantum mechanics. Alternatively stated, nature will intrinsically never allow the mathematical merging together of those two totally incompatible situations; namely, the "continuous" property on the one hand and "discrete" property on the other hand. Despite this issue, LMFDB with one of its crucial features acting as "intricate catalog of mathematical objects" will, metaphorically speaking, be the source supplying the required mathematical objects in those mentioned research areas.

In the grand scheme of things, this paper manifests the classically encountered phenomenon that pure and applied mathematics during, and resulting from, derivation of many mathematical proofs are largely inseparable. Some of the less conventional aspects of resulting applied mathematics in regards to the following (depicted from biologist-to-physicist point of view with highest-to-lowest decreasing hierarchical order) are

286

intuitively useful:

I. Living Things obeying Complex Emergent Fundamental Laws
II. Living Things obeying Simple Emergent Fundamental Laws
III. Nonliving Things obeying Complex Elementary Fundamental Laws
IV. Nonliving Things obeying Simple Elementary Fundamental Laws

In this context, our Incompletely Predictable problems of Riemann hypothesis, Polignac's and Twin prime conjectures are Nonliving Things obeying Complex Elementary Fundamental Laws. People have often strived to obtain pivotal scientific answers on whether Living Things arise from Nonliving Things via the Evolution process [as per atheist belief] or Living Things arise from Nonliving Things via the Creation process [as per religious belief]. We speculatively hope and selfishly dream that the applied mathematics pathway resulting from solving Riemann hypothesis, Polignac's and Twin prime conjectures could at least one day lead to answering the following question: Could the concocted expression "Living Things seem to exist at the edge of Chaos and Fractals" be mathematically equivalent to the following statement "Living Things must be made up of a combination of Completely Predictable entities, Incompletely Predictable entities, and Completely Unpredictable entities"?

Without going into finer details using Number theory, the irrationality measure (or irrationality exponent or approximation exponent or Liouville-Roth constant) of any real number is a measure of how "closely" it can be approximated by rationals. For a rational number, the irrationality measure is 1. The Thue-Siegel-Roth theorem states that for an algebraic irrational number, viz. real but not rational number, then the irrationality measure is 2. Transcendental irrational numbers have irrationality measure 2 or greater; for instance, the transcendental Euler's number e (= 2.718281828459...) has irrationality measure equal to 2. The [seemingly] simplistic-looking Liouville numbers is typified by Liouville's constant, sometimes also called Liouville's number, a real number defined by $L \equiv \sum_{n=1}^{\infty} 10^{-n!}$ = 0.110001000000000000000001... with '!' denoting factorial. These numbers are irrational numbers of [the relatively more "complex"] transcendental types instead of [the relatively less "complex"] algebraic types; and their numerical make-up consist of just '0' and '1' digits. Despite this apparently simple-looking numerical make-up of Liouville numbers (as opposed to more complicated-looking numerical make-up of e), they are precisely those numbers [paradoxically] having infinite irrationality measure. For the above, we would assign all [Completely Predictable] rational numbers to obeying Simple Elementary Fundamental Laws, and all [Incompletely Predictable] irrational numbers to obeying Complex Elementary Fundamental Laws.

We now endeavor to compare, contrast and reconcile the two entities Living Things and Nonliving Things. Rigorous mathematical proofs must obviously be associated with 100% certainty. This can only apply to Simple and Complex Elementary Fundamental Laws on Nonliving Things. Diverging onto proofs for Simple and Complex Emergent Fundamental Laws on Living Things, one observe that they can never be associated with perfect 100% certainty simply because we are dealing with "ALIVE" Living Things with dynamic spatial and temporal properties that could not be totally predictable. In this setting, the proofs for the Simple cases [e.g.

physiologically modeling Cardiac Output (CO) equals to Heart Rate (HR) multiplied by Stroke Volume (SV) in the Cardiovascular System (CVS)] will comparatively be less challenging to derive than the Complex cases [e.g. physiologically modeling complex Human Brain functions using Neural Networks in the Central Nervous System (CNS)].

Note that the terms Elementary and Emergent are used here in the preceding and subsequent paragraphs to, respectively, denote Nonliving Things and Living Things. In real life situation for Living Things, there will always be the perpetual presence of infinitesimally tiny and unpredictable "Chaos and Fractals physiological variability", for instance, in the Simple Emergent Fundamental Law CO = HR X SR. This variability phenomenon will inevitably occur even in the most relaxed state of a person in deep sleep whereby dynamic processes such as intrinsic neuro-endocrine continuous signal input to the heart must occur on a permanent basis thus giving rise to this variability.

For the medically oriented readers, we finish off this topic by touching on Evidence based Medicine (EBM) and Evidence based Practice (EBP). Both could comply with either Simple or Complex Emergent Fundamental Laws on Living Things (namely, Human Beings in this scenario). EBM is typically depicted pictorially as a 'Pyramidal hierarchy of Literature Review' classifying available medical research materials into [the most powerful] Systematic Reviews down to [the least powerful] Expert Opinion.

Then EBP = Clinician Experience + Patient Expectation + Best Practice; with Best Practice being roughly equated with EBM. For doctors and medical researchers confronted daily with responsibly abiding to and improving up-to-date EBP and EBM, they must be familiar with most statistical tools employed in medical research. The classic example is research hypothesis expressed as a null hypothesis [the "devil's advocate" position] and alternative hypothesis. The level of statistical significance for hypothesis testing is often expressed as the so-called p-value. Whilst there is relatively little justification why a [cut-off] significance level of 0.05 is widely used in academic research [rather than 0.01 or 0.10]; we could be particularly more confident in our results by setting a more stringent level of (say) 0.01 [a 1% chance or less; 1 in 100 chance or less]. Despite this experimental/research tactic, we could strive to, but never, achieve perfect or 100% confidence in our results by setting ever more stringent levels.

We point out that there are overlapping pure and applied mathematics in our rigorous proof for Riemann hypothesis which was proposed more than 150 years in 1859. Conforming to logical arguments above, one could postulate that the lengthy delay in solving this hypothesis is simply because Riemann zeta function contains an infinite number of Incompletely Predictable intercepts demonstrating **Supraminimal Simplicity**, or alternatively stated, contains none of the Completely Predictable infinite intercepts demonstrating **Supramaximal Simplicity** [whereby Supramaximal Simplicity does allow multiple type solutions to prove a particular conjecture]. This will then require a proviso that there is only one [solitary] way using the "Complex Container Research Method" to solve this 'Incompletely Predictable problem' which belongs to the 'Special-Class-of-Mathematical-Problems with Solitary-Proof-Solution'.

Brief discussions on Statistics

With written permission, I gratefully base the following brief discussions on statistics materials present in HyperStat Online Statistics Textbook http://davidmlane.com/hyperstat/index.html, 1993 – 2013 by Professor David M. Lane. Measurement is assignment of numbers to objects or events in a systematic fashion. Four levels of measurement scales are commonly distinguished: nominal, ordinal, interval, and ratio.

There is a relationship between level of measurement and appropriateness of various statistical procedures. For instance, it is foolish to compute mean of nominal measurements. Also, appropriateness of statistical analyses involving means for ordinal level data is controversial. One position is that data must be measured on an interval or a ratio scale for computation of means and other statistics to be valid. Thus, if data are measured on an ordinal scale, median but not mean can serve as measure of central tendency.

The arguments on both sides of this issue is examined in context of a hypothetical experiment designed to determine whether people prefer to work with color or with black and white computer displays. Twenty subjects viewed black and white displays and 20 subjects viewed color displays.

Displays were rated on a 7 point scale where a 1 was the lowest rating and a 7 was the highest rating. This rating scale is only an ordinal scale since there is no assurance that difference between a rating of 1 and a rating of 2 represents same degree of difference in preference as difference between a rating of 5 and a rating of 6.

The mean rating of color display was 5.5 and the mean rating of black and white display was 3.9. The first question experimenter would ask is how likely is it that this big a difference between means could have occurred just because of chance factors such as which subjects saw black and white display and which subjects saw color display. Standard methods of statistical inference can answer this question. Assume these methods led to conclusion that the difference was not due to chance but represented a "real" difference in means. Does the fact that rating scale was ordinal instead of interval have any implications for validity of the statistical conclusion that difference between means was not due to chance?

The answer is unequivocally 'no' with no room for argument here. What can be questioned is whether it is worth knowing that mean rating of color displays is higher than mean rating for black and white displays.

The argument that it is not worth knowing assumes that means of ordinal data are meaningless. Supporting notion that means of ordinal data are meaningless is the fact that examples can be made showing that a difference between means on an ordinal scale can be in opposite direction of what they would have been if "true" measurement scale had been used.

If means of ordinal data are meaningless, why should we care whether difference between two meaningless quantities (the two means) is due to chance or not. We answer this by challenging proposition that means of ordinal data are meaningless. Counter arguments or counter examples are often used to disprove a mathematical proposition, conjecture or hypothesis. There are two counter arguments to the example showing that using an ordinal scale can reverse the direction of difference between means.

The first is philosophical and challenges validity of the notion that there is some unseen "true" measurement scale that is only being approximated by rating scale. The second counter argument accepts the notion of an underlying scale but considers examples to be very contrived and unlikely to occur in real data. Measurement scales used in behavioral research are invariably somewhere between ordinal and interval scales. In preference experiment, it may not be the case that difference between ratings one and two is exactly the same as difference between five and six, but it is unlikely to be many times larger either. The scale is roughly interval and it is exceedingly unlikely that means on this scale would favor color displays while means on the "true" scale would favor black and white displays.

There are case examples where one can validly argue that use of an ordinal instead of a ratio scale seriously distorts conclusions. Consider an experiment designed to determine whether 5-year old children are more distractible than 10-year old children.

Measurement Scales

	No Distraction	Distraction
5-yr	6	3
10-yr	12	8

It appears as though 10-year olds are more distractible since distraction cost them 4 points (12 minus 8) but only cost 5-year olds 3 points (6 minus 3). Thus, it might be that a change from 3 to 6 represents a larger difference than a change from 8 to 12. We must consider the performance of 5-year olds dropped 50% (3/6 X 100%) from distraction but the performance of 10-year olds dropped only 33% (4/12 X 100%).

Which age group is "really" more distractible? Unfortunately, there is no right or wrong answer here. If proportional change is considered, then 5-year olds are more distractible; if amount of change is considered then 10-year olds are more distractible. We must keep in mind that statistical conclusions are not affected by choice of measurement scale even though important interpretation of these conclusions can be.

In above example, a statistical test could validly rule out chance as an explanation of finding that 10-year olds lost more points from distraction than did 5-year olds. However, statistical test will not reveal whether a greater drop necessarily means 10-year olds are more distractible. So, the conclusion that distraction costs 10-year olds more points than it costs 5-year olds is valid. Thus the interpretation depends on measurement issues.

Broadly speaking, statistical analyses provide conclusions about the numbers entered into them. Relating these conclusions to the substantive research issues depends on measurement operations.

Example for Measurement Scales – Assume there were a "true" measurement scale for job satisfaction and that it maps onto a 7-point rating scale as follows:

"True scale" 7-point scale

1 - 5	1
6 - 40	2
41 - 42	3
43 - 75	4
76 - 90	5
91 - 94	6
95 - 100	7

Then if someone's "true" job satisfaction were 55 he or she would have a rated score of 4. Now consider following two sets of job satisfaction scores:

	Group A		Group B	
	True Scale	Rating	True Scale	Rating
	6	2	5	1
	6	2	40	2
	43	4	74	4
	91	6	90	5
	95	7	100	7
Mean	48.2	4.2	61.8	3.8

On "true" scale mean for Group B is 61.8 which is much higher than mean for Group A which is 48.2. However on the 7-point rating scale, mean for B is only 3.8 which is lower than mean for A of 4.2.

As easily seen, I was involved in conducting medical research during my Anesthesia training with results published in the paper "Supramaximal elevation in B-type natriuretic peptide and its N-terminal fragment levels in anephric patients with heart failure: a case series" Journal of medical case reports 2012, Primary author: Helpful, B. and Secondary author: Pussell, B. With explicit written permission obtained from Primary author, the composed conclusion and subsequent discussion below are based upon materials from this paper.

This study achieved the primary endpoint of demonstrating (sustained) supramaximal elevations of BNP and NT-proBNP in only one (obtainable) anephric patient inflicted with congestive heart failure (CHF) which suggested the need for dramatically higher BNP and NT-proBNP cut-off values for anephric patients in CHF with respective magnitudes of the order of 50-fold to 100-fold higher than the usual figures quoted to 'rule in' CHF.

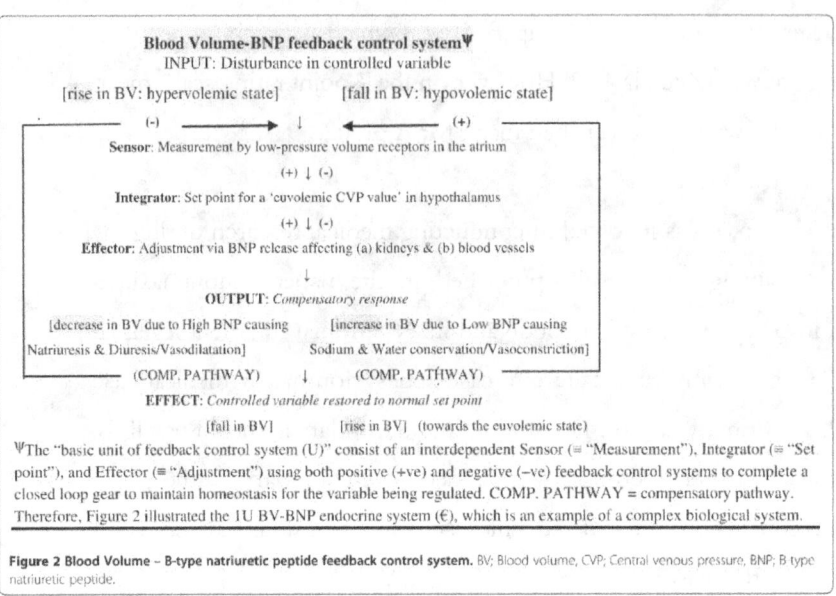

Figure 2 Blood Volume – B-type natriuretic peptide feedback control system. BV; Blood volume, CVP; Central venous pressure, BNP; B type natriuretic peptide.

Introduction: CHF and renal failure (RF) act synergistically to increase the levels of B-type natriuretic peptide (BNP) and its co-secreted biologically inactive N-terminal fragment (NT-proBNP). These two cardiac neuro-hormones are mainly secreted from the ventricles and, to a lesser extent, the atria. They have an established role as useful diagnostic tests for CHF in both the pediatric and adult population, including RF patients.

RF and CHF represent two merging pathologies with a varying spectrum for speed of onset and severity. The intersection of cardiac and renal insufficiency is referred to as cardiorenal syndrome (Type 1 to 5), which is CHF as a result of RF or vice versa. Cardiac dysfunction in end-stage kidney disease (ESKD) patients, whether acute or chronic, is often due to disorders of perfusion (ischemic heart disease) or to disorders of structure and function. The disorders of structure and function and CHF are often collectively termed 'uremic cardiomyopathy' which is commonly associated with left ventricle (LV) hypertrophy secondary to volume overload and hypertension (HT). Cardiac disease accounts for >50% of deaths in patients with End-Stage Kidney Disease (ESKD).

In theory, the truly anephric state represents a unique position for research purposes because any 'distorting interferences' from a failing kidney are eliminated from consideration. These interferences refer mainly to the kidney (an integral component of the feedback control system) acting as: (1) an &/or the end target-organ for the relevant hormones; and/or (2) an &/or the organ contributing to metabolic clearance rate (MCR) for the relevant hormones. This 'significant elimination' holds true for the first interference (with the associated loss of the kidney compensatory pathway) but less so for the second interference because: (i) anephric patients are dependent on, usually, intermittent hemodialysis (HD) to keep them alive and thus providing them with an artificial means to intermittently and variably clear relevant hormones (namely, BNP and NT-proBNP); and (ii) there may be non-renal pathways (via other organs or tissues) variably contributing to MCR for these relevant hormones as well.

Predominantly based on concepts behind the feedback control systems, we devise the interesting 'Blood Volume (BV) – BNP feedback control system' (Figure 2) [reproduced with permission], with further discourse given below to help provide plausible explanations for our findings.

Figure 2 which depicted a typical 1 U Blood Volume– BNP feedback control system, whereby the symbol 'U' stands for the 'basic unit of feedback control system'. The control of a physiological 'state' or 'variable' such as BV is via a complex web utilizing adaptive and integrative mechanisms. The immediate control of the total body water (TBW) endocrine system (€TBW) is predominantly mediated by the 'thirst-antidiuretic hormone (ADH) mechanism'. Thirst affects the input of water and ADH affects the output of water. The delayed control of TBW (a minor control) is mediated by the endocrine system (€s) such as the renin-angiotensin-aldosterone system (RAAS, which also mediates 'control of Blood Pressure (BP)' to some degree). The unit of this €TBW consists of: (i) sensors = osmoreceptors (monitor tonicity), low-pressure volume receptors (monitor BV), and high-pressure baroreceptors (monitor BP); (ii) integrator = hypothalamus; and (iii) effectors = thirst and ADH. Therefore the 'control of TBW' would also directly or indirectly lead to 'control of BP' (and vice versa) because both variables are controlled by RAAS to various degrees.

Similarly, the 'control of BV' variable (mediated by natriuretic peptides such as BNP in conjunction with RAAS) is affected to a varying degree by the 'control of TBW' and 'control of BP' variables. The '↑/↓ BV → ↑/↓ myocardial stretch or tension → ↑/↓ BNP release' homeostatic mechanism leads to the end-target organ effects of: (a) ↓/↑ BV (via (natriuresis and diuresis) / (sodium and water conservation) by kidneys) and (b) ↓/↑ peripheral vascular resistance (via (vasodilatation) / (vasoconstriction) on blood vessels). This signifies that our proposed homeostasis (with integrator = hypothalamus likely connected via autonomic neural pathway) acting through its compensatory pathways is invoked to help restore the disturbance in BV ('hypervolemic' / 'hypovolemic') to its 'euvolemic' set point resulting in improved diastolic relaxation (lusitropy) and decreased myocardial fibrosis. Measurements on BV status are largely carried out by the low-pressure volume receptors in the atrium (which essentially equates to central venous pressure clinical measurements). Likewise, detailed analysis for a 2 U hypothalamic–pituitary axis for prolactin hormone or a 3 U hypothalamic–pituitary–thyroid axis for thyroid hormones could also be carried out.

For a given €, the magnitude of its size and complexity would increase exponentially if a linear increase in the number of U were to occur due to the associated power-law increase in the number of 'controlled variables' along with their 'mini-components' (input, output, effect and compensatory pathway). The total number of possibilities arising from n mini-components when considering the '(↑ or +)' or '(↓ or –)' state (i.e. r = 2) for each mini-component is given by the permutations with repetition formula: nr = n2 from combinatorics. Let the symbol Σ denote 'the sum of'; and xi and yi denote 'n individual factors or causes of endocrine disorder' for i = 1, 2, 3,…, n that tend to have elevating or lowering properties, respectively, on the relevant hormone.

The overall magnitudes of rises or falls in the particular hormonal output (O) (which is a controlled variable), such as BNP, NT-proBNP and prolactin hormonal concentrations, are governed by the net difference between the resultant effect from Σ (effects from xi that tend to increase O) and Σ (effects from yi that tend to decrease O). This overall resultant effect stemming from the absolute difference between the n value for xi (nx) and the n value for yi (ny), namely |nx – ny|, that tends to increase or decrease O respectively would be some nonlinear function of this absolute difference of a synergistic nature. Then this overall resultant effect will be of ever greater cumulative rises or falls (of an exponential nature) in O when $|nx - ny|$ is numerically >1 and constitutes an ever larger integer number.

The '↑/↓ BV causing ↑/↓ myocardial stretch or tension, resulting in ↑/↓ BNP release' is the main mechanism for BNP (and NT-proBNP) pulsatile co-secretion. Other mechanisms such as heart muscle cell damage from myocardial infarct will also lead to BNP and NT-proBNP release.

There are two major cardiac and non-cardiac causes of BNP and NT-proBNP elevations as follows:

First, moderate increases in BNP (100–500ng/L) or NT-proBNP (250–1000ng/L): ventricular dysfunction, ischemic heart disease, pulmonary HT, acute pulmonary embolism, cor pulmonale, septic shock, renal insufficiency, liver cirrhosis, subarachnoid hemorrhage and hyperthyroidism. Second, severe increases in BNP (>500ng/L) or NT-proBNP (>1000ng/L): decompensated heart failure (HF), pulmonary Hypertension (HT), acute pulmonary embolism and septic shock.

In addition to glomerular filtration, BNP is eliminated from plasma mainly through natriuretic peptide receptors and degraded by neutral endopeptidases. By contrast, it is possible that NTproBNP is largely eliminated by glomerular filtration. Levels of both BNP and NT-proBNP are: elevated with ageing, higher in women than in men, higher in RF and CHF of greater severity, and higher in LV systolic HF than LV diastolic HF. Stage of HF (early versus late) and genetic polymorphisms may result in inter-individual variation of BNP and NT-proBNP. Obesity with and without

302

CHF is associated with lower levels of both molecules; obesity with and without CHF is presumably attributed to non-hemodynamic factors such as BMI-related defect in natriuretic peptide secretion (from either ↓myocardial hormone release or ↓synthesis), and ↑BNP metabolism in adipose tissue either via peptide degradation or regulation of clearance receptors.

The end target-organs for BNP are the kidneys and blood vessels. These are associated with the 'kidney compensatory pathway' and 'blood vessel compensatory pathway' respectively. The SIA state corresponds to the loss of the kidney as (a) an end target-organ and (b) a compensatory pathway, although the contribution of the collective blood vessels as an end target-organ and compensatory pathway is still intact. The primary endpoint of our study was to demonstrate the supramaximal elevation of BNP and NT-proBNP in Patient 1. Computing from Figure 2 (together with the 'major cardiac and non-cardiac causes of BNP and NT-proBNP elevations'), this can be seen to be due to multiple x_i (with no identifiable y_i); namely: (i) CHF itself, (ii) decreased MCR for BNP and NT-proBNP, and (iii) total loss of kidney tissue acting as an end-target organ (but intact collective blood vessels acting as an end-target organ) with total disruption of 'kidney compensatory pathway' loop.

The secondary endpoint of our study was to demonstrate the supramaximal elevation of prolactin in Patient 1 as suggested by the persistently high prolactin values obtained between Event X (development of acute CHF) and Event Y (death of the patient). This was due to multiple x_i (with no identifiable y_i); namely, the systemic disorders of: (i) chronic RF, (ii) emotional stress, (iii) epileptic seizures, (iv) pharmacologic factors (anti-

303

HT dopamine synthesis inhibitors methyldopa), and (v) decreased MCR for prolactin. Both the mildly elevated prolactin levels in Patients 2 and 3 mainly reflect the decreased MCR of prolactin due to CKD Stage 3 and anephric status (needing intermittent HD) for each respective patient.

The $n_x = 3$ in Patient 1 for Σ (effects from x_i that tend to increase the 'outputs' of BNP and NTproBNP) with resultant massive and persistent elevation of these two natriuretic peptides. Applying the $n_x = 3$ minus $n_y = 0$ calculation giving a 'relatively large' $|n_x - n_y|$ value of 3 predicts the overall magnitude of rises to be consistent with the supramaximal elevation of these hormones as seen in our study. Similar calculation of a 'relatively large' $|n_x - n_y|$ value of 5 for prolactin x_i and y_i in Patient 1 also showed that they act in concert to greatly increase and maintain the high prolactin 'output' in a synergistic manner to explain its supramaximal elevation.

'Functional' anephric states should occur in adult ESKD patients when their in-situ remnant kidney tissues have totally lost all their functions or have atrophied completely. One could extrapolate that these patients should behave physiologically in a similar manner to surgically-induced anephric (SIA) patients. A corollary to this argument would result in the hypothesis that when ESKD patients develop CHF with supramaximal elevations of BNP and NT-proBNP, they are likely to be functionally anephric. The full significances of this hypothesis in adults are yet to be fully realized.

Supramaximal hormonal elevations when observed in neonates, infants and children will undoubtedly be due to mechanisms similar to that of their adult counterpart; and with the full impact of this hypothesis lying in uncharted territories. These are exciting areas for future medical research. Let us mathematically analyze the following statement in a logical manner: The defined parameters n_x, n_y, and $|n_x - n_y|$ are applicable to both 'anatomical' and 'functional' anephric patients. Because one can safely assume that all supporting criteria for the statement to hold true are present in both sets of patients, then this 'common denominator' statement per se can provide intuitive non-contradictory explanations for the supramaximal elevation phenomenon in all anephric patients inflicted with CHF. This 'common denominator' statement thus lends support to our proposed hypothesis that 'anatomical' and 'functional' anephric patients inflicted with CHF should have similar natriuretic response behavior.

Footnote on 'An infant in temporary anephric and congestive heart failure state manifesting supramaximal elevations of natriuretic peptides'

In October 2010, we encountered the case of a 5-month-old male baby (weight 7kg) with out-of-hospital cardiac arrest (due to commotio cordis) requiring 30 minutes of cardiopulmonary resuscitation before return of spontaneous circulation. He developed multiple organ dysfunction syndrome (MODS) requiring full Intensive Care Unit (ICU) supportive care. The ICU supportive care included therapeutic hypothermia between 33°C and 34°C for the first 48 hours, full invasive ventilation for 12 days for acute CHF with fractional shortening (FS) 31% on echocardiogram (normal >30%) while on multiple inotropic and vasopressor agents, and Continuous Veno-Venous Hemodiafiltration (CVVHDF) for 5 days from Day 3 to 7 for (anuric) acute kidney injury with peak creatinine 93μmol/L (20–50) on Day 8: this probably equates to the baby being a temporary 'functional' anephric patient. Blood tests on Day 7 showed: Hb 88g/L, creatinine 50μmol/L, and supramaximal elevation of NT-proBNP at 173,392ng/L (20,460pmol/L). By Day 10, the patient had not required dialysis for 3 days with blood tests showing: Hb 84g/L, creatinine mildly elevated at 58μmol/L (with good urine output), and lesser magnitude of supramaximal elevation of NT-proBNP at 98,476ng/L (11,620pmol/L). He continued to steadily improve before being extubated onto continuous positive airway pressure on Day 12 with eventual discharge from the pediatric ICU on Day 27. Follow-up of the patient at 12 months post-cardiac arrest revealed remarkable recovery from his MODS with possibly very mild and subtle residual cognitive dysfunction from the hypoxic ischemic encephalopathy, normal renal function (creatinine 23μmol/L at 8 weeks post-arrest), and normal cardiac function (Fractional Shortening 42% on echocardiogram at 7 months post-arrest).

General Chemistry: Stoichiometry

An important field of chemistry is stoichiometry, which is the quantitative relationship between chemical substances in a reaction. The weight of a molecule is the sum of weights of atoms of which it is made. The unit of weight is **dalton (Da)**, one-twelfth the weight of an atom of ^{12}C, and 1000 Da = 1 kilodalton (kDa). The molecular weight of a substance is the ratio of mass of one molecule of substance to mass of one-twelfth the mass of an atom of ^{12}C – being a ratio, it is dimensionless. A **mole** is the gram-molecular weight of a substance, that is, the quantity of a substance whose weight in grams in equal to molecular weight of the substance. There are approximately 6×10^{23} molecules in a mole. (This is Avogadro's Number with best experimental value of $6.02214199 \times 10^{23}$ atoms per mol.)

One very important property of solutions that must be addressed is concentration. Concentration generally refers to the amount of solute contained in a certain amount of solution. To deal with concentration you must keep in mind the distinctions between solute, solvent and solution. There are five units of concentration that are particularly useful to chemists. The first three: **molarity (with its closely related osmolarity)**, **molality (with its closely related osmolality)** and **normality** are dependant upon the mole unit. The last two: **percent by volume** and **percent by weight** have nothing to do with mole, only weight or volume of the solute or substance to be diluted, versus the weight or volume of the solvent or substance in which the solute is diluted. (Note that Percentages can also be determined for solids within solids.)

Molarity is number of moles of solute dissolved in one liter of solution. The units are moles per liter of solution (**M**). A **molality** is number of moles of solute dissolved in one kilogram of solvent. The units are moles per kilogram of solvent (**m**). Note that to prepare a given molality solution, the solvent must be weighed unless it is water. (One liter of water has a specific gravity of 1.0 and weighs one kilogram; so one can simply measure out one liter of water and add the solute to it. Most other solvents have a specific gravity greater than or less than one; thus one must weigh the solvent.)

The diffusion of solvent molecules into a region in which there is a higher concentration of a solute to which the membrane is impermeable is called osmosis. The pressure necessary to prevent solvent migration is the osmotic pressure of the solution. Like vapor pressure lowering, freezing-point depression, and boiling-point elevation, osmotic pressure depends upon the number rather than the type of particles in a solution – in other words, it is a fundamental colligative property of solutions. In an ideal solution, osmotic pressure (P) is related to (the absolute) temperature (T) and volume (V) in the same as the pressure of a gas: $P = nRT/V$, with n the number of particles and R the gas constant. The concentration of osmotically active particles is expressed in **osmoles**, with one osmole (osm) equals the gram-molecular weight of a substance divided by the number of freely moving particles that each molecule liberates in solution. (One milliosmole (mosm) is $1/1000$ of one osm.) Our body fluids is not an ideal solution and although the dissociation of the solute made of an ionizing compound (such as strong electrolytes) is complete, the number of particles free to exert an osmotic effect is reduced owing to interactions between the ions – thus, it is the effective concentration (*activity*) in the body fluids rather than the number of equivalents of an electrolyte in solution that determines its osmotic effect, and this deviation from an ideal solution is greater with the more concentrated the soultion. **Osmolarity** defined as the number of osmoles per liter of solution (osm/L). If the solute is a nonionizing compound, the osmolarity is the same as molarity – otherwise (as per reasoning above) in the solute made of an ionizing compound, this is not true with the osmolarity greater than the molarity. The term tonicity is used to describe osmolality of a solution relative to plasma such that solutions that have same osmolality as plasma are said to be isotonic; those with greater osmolality are hypertonic; and those with lesser osmolality are hypotonic. With respect to argument for whether a solute is made up of an ionizing or a nonionizing compound, a similar line of thinking can be applied for **osmolality**, which is defined as number of osmoles per kilogram of solvent (osm/kg) – thus, osmolality is only the same as molality in the non-ionizing compound case and they are different in the ionizing compound case. *Note that osmolarity is affected by the volume of various solutes in the solution and the temperature, while osmolality is not.*

The concentration of a solution can be stated by the amount of solute in equivalents rather than moles. This is called **normality** (**N**), which is the number of equivalents of solute per liter of solution (eq/L). Electrical equivalence is not necessarily the same as chemical equivalence. One (electrical) equivalent (eq) is 1 mol of an ionized substance divided by its valence. One (chemical) gram equivalent is the weight of a substance that is chemically equivalent to 8.000 g of oxygen. Therefore, the normality (**N**) of a solution can also be expressed as the number of gram equivalents in 1 liter (g/L). For example, one mole of HCl dissociates into 1 eq of H^+ and 1 eq of Cl^-. One equivalent of H^+ = 1 g/L = 1 g and 1 eq of Cl^- = 35.5 g/l = 35.5 g. Thus, a 1 M (mol/L) solution of hydrochloric acid contains 1 mol/L x 2 eq/mol = 2 N (eq/L) = 1 + 35.5 g/l = 36.5 g/L. There is a close relationship between normality and molarity. Normality can only be calculated when we deal with reactions, because normality is a function of equivalents. Normality = molarity x n, where n (eq/mol) = the number of protons (or hydrogen ions) exchanged in a reaction. In other words, the normality of a solution is simply a multiple of the molarity of the solution. Generally, the normality of a solution is just one, two or three times the molarity. In rare cases it can be four, five, six or even seven times as much. Applying the same above line of thinking all over again, we can easily conclude that osmolarity only approximates normality in the case of the ideal solution.

Percentages are easy to calculate because they do not require information about the chemical nature of the substance. Percentages can be determined as percent by weight or percent by volume. Percentages are used more in the technological fields of chemistry (such as environmental technologies) than they are in pure chemistry.

Percent by weight: To make up a solution based on percentage by weight, one would simply determine what percentage was desired (for example, a 20% by weight aqueous solution of sodium chloride) and the total quantity to be prepared. If the total quantity needed is 1 kg, then it would simply be a matter of calculating 20% of 1 kg which, of course is: 0.20 NaCl * 1000 g/kg = 200 g NaCl/kg. In order to bring the total quantity to 1 kg, it would be necessary to add 800g water.

Percent by volume: Solutions based on percent by volume are calculated the same as for percent by weight, except that calculations are based on volume. Thus one would simply determine what percentage was desired (for example, a 20% by volume aqueous solution of sodium chloride) and the total quantity to be prepared. If the total quantity needed is 1 liter, then it would simply be a matter of calculating 20% of 1 liter which, of course is: 0.20 NaCl * 1000 ml/l = 200 ml NaCl/l.

The Stewart hypothesis of acid-base balance focus on definitions of acid and base. In order to understand the differences between the classic ("Siggaard-Anderson") approach to acid-base analysis and the new paradigm Stewart approach, it's helpful to define a few points from which to work. With appropriate written consent, information content in this section are gratefully based upon materials composed by Dr. Pete Watkinson in AnaesthesiaUK website on Acid-base balance https://www.frca.co.uk/article.aspx?articleid=100924 (Created: 23/5/2007, Updated: 18/2/2009).

The first important point is how to define an acid. This is key because from the definition comes the term "acidotic". Most people would think of a patient as being acidotic when the pH is below 7.35. As pH is purely a function of the theoretical hydrogen ion concentration of a solution $-log_{10}$[H+], an acid is being defined as a substance that can donate a proton. This simple definition is in line with modern theory.

There are, however, various other ways of defining an acid. An acid can also be defined as any substance that produces an increased concentration of hydrogen ions when dissolved in water (the 'Arrhenius definition'). The Stewart approach approximates this definition. Not surprisingly, attachment of different meanings to the same terminology has led to considerable confusion throughout the history of acid base physiology.

The second important point to note is that both the Siggard-Anderson and Stewart approaches to acid base analysis are mathematical models. Both approaches do not explain the mechanism by which a person has become biochemically acidotic/alkalotic, nor do they claim to do this, but both models merely attempt to illustrate the area the acid-base disturbance lies.

We now focus on the Siggaard-Anderson approach whereby in the Siggaard-Anderson model, an acid-base disturbance is looked at using a combination of 3 factors:
    # The Henderson-Hasselbach equation
    # The base excess
    # The anion gap.

The Henderson-Hasselbach equation comes from the dissociation equation for carbonic acid, and its use is based on the premise that in normal plasma, bicarbonate is the most important buffer.

From this equation, the only two factors affecting pH are:
 # bicarbonate concentration
 # pCO2

The Henderson-Hasselbach equation provides an approximate relationship between respiratory variable (pCO2), metabolic variable [HCO3-], and resultant pH. A flaw with this approach is that other important buffers exist and play an important role in acid base physiology (e.g. hemoglobin and albumin). HCO3- and pCO2 are not therefore independent. In the simplified dissociation equation below, a rise in pCO2 will cause dissociation to shift to the right as a result of the law of mass action:

$$CO_2 + H_2O \leftrightarrow H^+ + HCO_3^-$$

Protons will be buffered by hemoglobin and albumin, and the bicarbonate levels will rise. So a rise in pCO2 has resulted in a rise in [HCO3-]. The rise in [HCO3-] could easily be mistaken as a metabolic alkalosis, when in fact the true cause was a respiratory acidosis. The base excess concept was evolved to address this problem. It is a method of measuring the metabolic component. The base excess concept works by resetting the sample to a normal pCO2 (5.33kPa) by equilibration, and then titrating it to pH 7.4 using molar acid (now calculated from normograms). The number of mmol /L

312

required equals the base excess, and is therefore a measure of how acidotic or alkalotic the sample is without any contribution of carbon dioxide.

Finally, calculation of the anion gap (see Equation box below) allows classification of a metabolic acidosis into those with a normal or increased anion gap. The anion gap is a measure of the concentration of unmeasured anions (e.g. plasma proteins) and is based on the theory of electrical neutrality (the sum of the positive ions must equal the sum of the negative ions). An increased anion gap suggests the presence of unmeasured organic acid, whereas a normal anion gap implies that the decrease in bicarbonate has been counteracted by an increase in chloride concentration (see following Table).

Table: Causes of metabolic acidosis - Siggard-Anderson approach

Increased anion gap (usually decreased [Cl-]	Normal anion gap (usually increased [Cl-])
Ketoacidosis	Diarrhoea
Alcoholic	Parenteral nutrition
Diabetic	Carbonic anhydrase inhibitors
Starvation	Dilutional acidosis
Hyperosmolar nonketotic coma	Ingestion of HCl or other acid
Lactic acidosis	Renal tubular acidosis
Uraemic acidosis	Ileostomy
Methanol	Ureterosigmoidostomy
Ethylene glycol	Pancreatic fistula
Salicylate	
Paraldehyde	

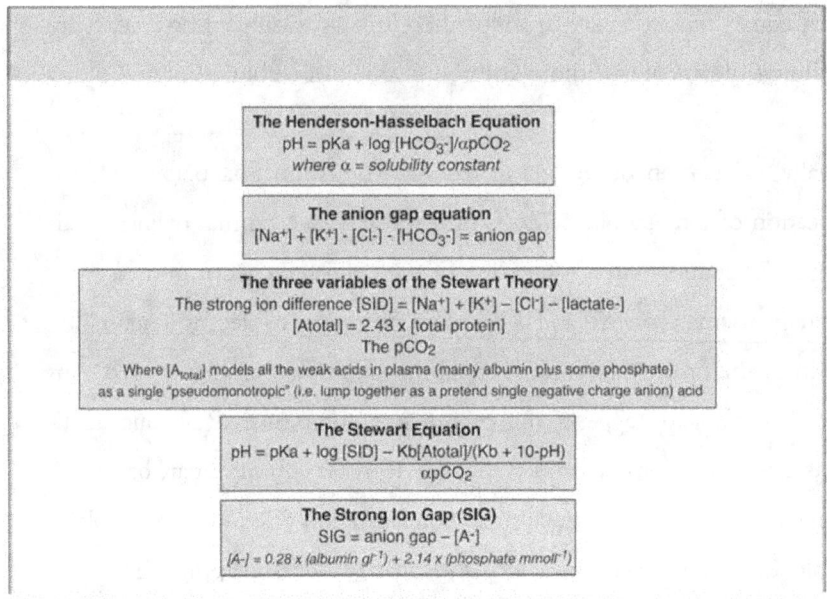

The Henderson-Hasselbach Equation
$pH = pKa + \log [HCO_3^-]/\alpha pCO_2$
where α = solubility constant

The anion gap equation
$[Na^+] + [K^+] - [Cl-] - [HCO_3^-]$ = anion gap

The three variables of the Stewart Theory
The strong ion difference $[SID] = [Na^+] + [K^+] - [Cl^-] - [lactate-]$
$[Atotal] = 2.43$ x [total protein]
The pCO_2
Where $[A_{total}]$ models all the weak acids in plasma (mainly albumin plus some phosphate)
as a single "pseudomonotropic" (i.e. lump together as a pretend single negative charge anion) acid

The Stewart Equation
$pH = pKa + \log \dfrac{[SID] - Kb[Atotal]/(Kb + 10\text{-}pH)}{\alpha pCO_2}$

The Strong Ion Gap (SIG)
SIG = anion gap $- [A-]$
$[A-] = 0.28$ x (albumin gl^{-1}) + 2.14 x (phosphate mmoll^{-1})

Equation Box for Stewart approach.

Stewart took the same system, but looked at from a slightly different angle. He concluded that one might model acid-base disturbances, based on three conceptual contributors described in Equation box above:

1. The Strong Ion difference (SID)

2. Weak Acids in Plasma (A total)

3. PCO2

The law of electrical neutrality means that:

$[Na+] + [K+] + [H+] = [Cl-] + [lactate-] + [HCO3-] + [A-] + [CO32-]$

Ignoring the minimal contribution of [H+], [HCO3-] and [CO32-], and substituting the strong ion difference shows:

$[SID] = [HCO3-] + [A-]$

314

Stewart puts the three variables together in the Stewart Equation described in the equation box. It is interesting to note that if you ignore the contribution of albumin in this equation, it simplifies to the Henderson-Hasselbalch equation. Thus, albumin is the major variable that Stewart has added in, left out by Siggaard-Andersson for reasons of simplicity.

From this comes a practical utilization of Stewart theory in that it can be used to define the Strong Ion Gap (see equation box) which allows metabolic acidosis to be classified in Table below.

Table: Causes of metabolic acidosis – Stewart approach

Low SID and high SIG	Low SID and low/normal SIG
Ketoacidosis	Diarrhoea
Alcoholic	Parenteral nutrition
Diabetic	Carbonic anhydrase inhibitors
Starvation	Saline/dilutional
Lactic acidosis	Renal tubular acidosis
Methanol	Ileostomy
Ethylene glycol	Ureterosigmoidostomy
Salicylate	Pancreatic fistula

When comparing the above two tables with first table using Sigaard Anderson approach and second table using Stewart approach, they illustrate that the two major classifications are broadly similar.

Let us focus on a real life example of a metabolic acidosis. The Sigaard Anderson and Stewart models of acid base disturbance can be illustrated by the following case, commonly encountered in ICU. The table below shows the blood gas analysis of an elderly patient admitted to ICU from operating theatre after a prolonged laparotomy for which he was given 6 litres of saline during the operation:

Table: Commonly encountered post-operative Arterial Blood Gas

ABG	Admission to ICU	Discharge from ICU
Saturation%	95	97
pH	7.28	7.35
pO2	9.7	11.2
pCO2	5.7	5.2
Base excess	-6	-1
Sodium mmol/L	145	138
Potassium mmol/L	4.3	3.9
HCO3- mmol/L	19	24
Chloride mmol/L	115	105
Lactate mmol/L	1.3	1.2
Glucose mmol/L	7.1	6.4

In Siggaard-Anderson's model, the patient has a metabolic acidosis with a normal anion gap. Given the history, a diagnosis of "dilutional" due to saline is the likely explanation.

In the Stewart model, calculation of the SID shows it to be low (normal ~40mmol/L), and the SIG (normally ~0) is approximately normal, (assuming an albumin of 45g/L and a phosphate of 1 mmol/L):

[A-] = 0.28 X (albumin) + 2.14 X (phosphate) = 12.6 + 2.14 = 14.74

[SID] = [HCO3-] + [A-] = 19 + 14.74 = 33.74

Anion gap = [Na+] + [K+] – [Cl-] – [HCO3-] = 145 + 4.3 – 115 – 19 = 15.3

SIG = AG – [A-] = 15.3 – 14.74

Again, given the history, we draw the same conclusion. Neither of these models provide the mechanism by which saline has caused a significant acidosis. So in this case, the patient is in the group with either a decreased strong ion difference, or normal anion gap, depending on whose model we are using. In either case the patient is hyperchloraemic, as serum chloride falls outside the normal range (115 mmol/L). Neither model explains why or even if hyperchloridaemia is causatory in acidosis.

We now focus on the cause of hyperchloraemic acidosis which is thought to be due to a combination of factors. Firstly, normal saline (pH 5-6) possesses little buffering capacity and is being used to replace blood (pH of 7.4) characterized by extensive buffering capacity. The acidic pH of normal saline is due to a combination of dissolved CO_2 and an effect known as Grotthus mechanism (also known as proton jumping) whereby this is the process by which an 'excess' proton or proton defect diffuses through the hydrogen bond network of water molecules or other hydrogen-bonded liquids through the formation and concomitant cleavage of covalent bonds involving neighboring molecules. Secondly, volume expansion causes plasma bicarbonate dilution, and renal bicarbonate wasting.

In summary:

The two main approaches to acid base analysis are the Siggard-Anderson approach (the 'classic') and the Stewart hypothesis.

The three main determinants of acid base status according to Stewart hypothesis are the Strong Ion Difference (SID), bicarbonate (HCO3-) and albumin (A-).

Albumin is the major additional variable of acid base analysis that Stewart has considered.

Stewart's consideration of plasma protein levels acid base analysis is important for ICU patients as albumin levels are often disturbed. Avoidance of such large quantities of saline-containing fluids may help prevent the incidence of hyperchloraemic acidosis on the ICU patients.

The above discussion on 'modern' Stewart hypothesis during my Intensive Care Medicine and Anesthesia training from 2009 to 2013 was what I predominantly learn during that period. Acid-base analysis on human beings (Living Things) represent problems that involve Simple Emergent Fundamental Laws.

We have already mentioned prime counting function in substantial details in previous chapters. In relation to this function, other functions more convenient to work with can also be utilized and they open up a whole new world of marvelous mathematical relationships. An example is Riemann prime counting function (aka prime power counting function), commonly denoted by $J(x)$. This non-infinite series function has jumps of $1/n$ for prime powers p^n, and with it taking a value halfway between the two sides at discontinuities. Amazingly, the prime counting function $\pi(x)$ is related to by $J(x)$ by the Mobius transform. More amazingly still, $J(x)$ is related to Riemann zeta function through the Mellin transform (which is an integral transform).

The word "Apocalypse" comes from Greek, and actually means "to uncover, to reveal". But in modern times it has come to mean the final destruction that is revealed in the Holy Bible from the Book of Revelation (the last book of the New Testament). In the early vision of the Apocalypse, the final blow is a massive earthquake. "Then the spirits brought the kings together in the place that in Hebrew is called Armageddon" (Revelation 16:16 from the Good News Bible, 1988 Australian edition). In biblical terms, Armageddon is the scene of the last battle to be fought between the forces of good and evil, prophesied to happen at the end of time. But Armageddon is also a real place. It is the Greek name for an ancient city in Israel, Megiddo. In Hebrew, "Mount Megiddo" is "Harmegiddo", and "Armageddon" is simply the Greek transliteration of that name. There are thought to be five common scenarios on how our world could end:-

(1) Nuclear holocaust. (2) Doomsday virus. (3) Killer Asteroid. (4) Threat of global warming. (5) Doomsday volcano. Should we add "The Great Earthquake" as the sixth Doomsday machine? Perhaps we should be more optimistic and less pessimistic.

Can we find a book with hidden codes that is supposed to reveal prophesies and predictions for the future? If so, is such a profound allegation scientifically sound? Michael Drosnin in his book 'The Bible Code' (1997, Publisher: Weidenfeld & Nicholson, London) is claiming that decoding the Bible [more specifically, the first five books of the Hebrew bible] allegedly leads to the discovery of prophecies and profound truths of a secular nature, not all of which are related to the Jews. He proposes that the Bible is the only text in which these encoded phrases are found in a statistically significant pattern, and that the chance of this being a random phenomenon is unlikely. Using the equidistant letter sequences (ELS) method, he claims that the assassinations of Yitzhak Rabin, Anwar Sadat and the Kennedy brothers were foretold in the Bible.

The Bible (or Torah) Code is a code alleged to have been intentionally embedded in the Bible. The code is revealed by searching for equidistant letter sequences (ELS). Doron Witztum, Eliyahu Rips and Yoav Rosenberg published their findings in the journal Statistical Science (1994, Vol. 9, No. 3, 429 - 438) under the title of "Equidistant Letter Sequences in the Book of Genesis." When the authors used a randomization test to see how rarely the patterns they found might arise by chance alone they obtained a highly

320

significant result, with the probability p = 0.000016. That is, the probability of getting the results they did was 16 out of one million or 1 out of 62,500. The authors commented that the randomization analysis shows that the effect is significant at the level of 0.00002 and the proximity of ELS's with related meanings in the Book of Genesis is not due to chance. The mathematics was felt "to be solid" and "the numbers held up".

Here is a simplified version on how the Bible Code is supposed to work. The idea is to start at a letter 'x' (which we can vary) and then to pick every nth letter while abiding by the rule (where n is also a number which we can vary). A computer program might read as follows: The starting point is letter 'x'. Input 'x' as the first letter, then 'x plus n' as the second letter, then 'x plus 2n' as the third letter, then 'x plus 3n' as the fourth letter, and so on. Using different values for "N" and "D", one can generate many strings of letters. Imagine wrapping the string of letters around a cylinder in such a way that all the letters can be displayed. Flatten the cylinder to reveal several rows with columns of equal length, except perhaps the last column which might be shorter than all the rest. Now search for meaningful names in proximity to dates. Search horizontally, vertically, diagonally, any which way. A group of Israeli mathematicians did just this and claimed that when they searched for names in close proximity to birth or death dates (as published in the Encyclopedia of Great Men in Israel) they found many matches. Analyze the eventual output result as a long string of letters and see if meaningful words or names might pop out.

A group of Israeli mathematicians did just this and claimed that when they searched for names in close proximity to birth or death dates (as published in Encyclopedia of Great Men in Israel) they found many matches. However, the consensus is that the Bible Code is basically a method of employing some hundreds, if not thousands, of millions of ways of "fishing out words at random, by starting anywhere you like and taking any steps you like".

With such a profound allegation made by Michael Drosnin, there are bound to be many critics. Therefore, in response to his critics, the following challenge was made by Drosnin: "When my critics find a message about the assassination of a prime minister encrypted in Moby Dick, I'll believe them." (Newsweek, 9 June 1997)

The following is the critics' reply. This segment begins with the answer to the question posed above. So, can this challenge thrown down by Drosnin be met? At [http://users.cecs.anu.edu.au/~bdm/dilugim/moby.html] where the webpage entitled "Assassinations Foretold in Moby Dick!" is located, Brendan McKay promptly produced (in 1997) an ELS analysis of Moby Dick predicting not only Indira Ghandi's assassination, but the assassinations of Martin Luther King, John F. Kennedy, Abraham Lincoln, and Yitzhak Rabin, as well as the death of Diana, Princess of Wales. This is despite the fact that English with the vowels included is far less flexible than Hebrew when it comes to making letters into words. (Hebrew doesn't have vowels.) In other words, it is much easier when you think with Hebrew, because they use only consonants – thus, they are not equal in probability.

A note to the credulous: Some of you may have taken this as claiming that the Bible and Moby Dick really predicted the assassinations of famous people. None of these patterns happened by other than pure random chance.

322

No laws of probability are violated here, or even stretched a little. The whole point is that once you learn Drosnin's rules (none) and the method (a bit of messy programming) you can find things like this anywhere. The reason it looks amazing is that the number of possible things to look for, and the number of places to look, is much greater than you imagine. There is an old saying in statistics: "torture your data long and hard enough, and it will tell you anything!" So it is a sleight of hand involving enormous numbers of essentially random characters that are selected by a program that keeps searching and searching, until it finds what you have asked it to.

Let us illustrate another similar kind of statistical "folly" below. In reference to the statistical "folly", the citation for the paper in question is Mantegna, R. N., Buldyrev, S., Goldberger, A., Havlin, S., Peng, C.-K., Simmons, M. and Stanley, H. "Linguistic features of non-coding DNA Sequences", Physical Review Letters 73, 23 (Dec 1994), 3169-3172. This paper observed that introns (base sequences which do not code for any protein) have Zipf power law histograms. Various deep-sounding linguistic conclusions were then drawn.

Unfortunately, Mantegna et. al. were unaware that words of either fixed or variable lengths whose letters are drawn entirely at random from a fixed alphabet with fixed unequal letter probabilities, must exhibit such a power law. Zipf's law has been understood for more than thirty years. In fact, the explanation was the first significant mathematical contribution of Benoit Mandelbrot. Zipf's law has to do with the statistics of a very large class of stochastic processes, and is unrelated to the characteristic structural or grammatical properties of language (whether the genetic code or natural

323

English). Therefore, any power law distribution for the frequency with which various combinations of "letters" appear in a sequence is due simply to a very general statistical phenomenom, and certainly does not indicate some deep underlying process or language.

Here is optimistic dreaming: The year is 2020. There were only a handful of developing countries left on earth, all on the verge of becoming developed nations. Humans are depending less on fossil fuel, and more on alternative and more efficient solar and nuclear energy to generate electricity. Harnessing the limitless (controlled, not cold!) nuclear fusion energy, once thought impossible to achieve, is now possible. Conventional nuclear fission power plants with hazardous radioactive waste products are being replaced with their "cleaner" nuclear fusion counterparts. Electric powered cars and other vehicles, and well-planned public transport systems all use electricity, derived directly from both efficient solar panels and rechargeable batteries, without any pollution. Petrol stations are now replaced by battery recharging stations. Time travel machines have enabled energy mining from black holes and efficient space travel. Human colonies deep in our oceans, far away on giant orbiting space stations, the moon and planet Mars are already well established. Advanced laser beam weapons and missiles using Chaotic control technology easily blasted away any dangerous decent sized asteroids from outer space before they hit earth (and, possibly, wipe out mankind).

With advanced genetic engineering applied to areas such as animal husbandry and agricultural food crops, starvation in Africa and many other poorer nations on earth were wiped out. With the effort from the World Health Organization (WHO), there is harmony and peace on earth and religious freedom everywhere. Thank God inhumane practices, such as the mass scale ethnic cleansing of Albanians by the Serbs in the war-torn province of Kosovo in Yugloslavia, in the early months of 1999, is never to be repeated again. NATO airforce and, eventually, ground troops intervention was required to restore peace. Towards the end of March 1999, NATO commander reported that for the first time in history, war was waged on the Internet, as Yugloslav computer hackers bombarded NATO's communication network with hundreds of unwanted e-mails, thus almost crippling the communication system.

Birth control is widely practiced, people are living longer due to medical advances. Human life expectancy is now close to 150 years. HIV and AIDS is finally declared to be eradicated due to successful vaccine and drug cure. There are prevention and cure for most cancers and tumors. The Internet system, based on quantum computers, is allowing people everywhere to be more energy efficient as they can do or obtain many things without leaving their homes. Greenhouse effect causing adverse global climatic changes is well under control with the cooperation of all nations to abolish greenhouse gas emission. Developmental biologists have mastered the art of inducing complete nuclear reprogramming to achieve the high survival rate in human cloning. The majority of the people on earth voted in the affirmative to clone Albert Einstein using the genetic material obtained from his preserved brain and the whole world waited excitedly for the birth (from a surrogate mother) and growing up of the Genius.

Bold predictions for the year 2020? Correct and time will tell with what we suspect that most of those predictions will not come true. In closing, I leave readers with the following quotation:

"God grant me the serenity to accept things I cannot change, the courage to change things I can, and the wisdom to know the difference."
Serenity Prayer attributed to Reinhold Niebuhr (June 21, 1892 – June 1, 1971).

And with my favorite Bible verses down given below:

1   The Lord is my shepherd;
    I shall not want.

2   He makes me to lie down in green pastures;
    He leads me beside the still waters.

3   He restores my soul;
    He leads me in paths of righteousness for His name's sake.

4   Yea, though I walk through the valley of the shadow of death,
    I will fear no evil; for you are with me;
    Your rod and Your staff, they comfort me.

5   You prepare a table before me in the presence of my enemies;
    You anoint my head with oil;
    My cup runs over.

6   Surely goodness and mercy shall follow me all the days of my life;
    and I will dwell in the house of the Lord forever.

Psalm 23, The Holy Bible (New King James Version).

The commonly used public key encryption system RSA (Rivest-Shamir-Adleman) was first described in 1977 by Ron Rivest, Adi Shamir and Leonard Adleman from Massachusetts Institute of Technology. It gets its security from the difficulty of factoring large integers that are the product of two large prime numbers. Internet transactions in e-commerce depends on the integrity of humongous [non-prime] numbers to be anonymously constituted from its basic prime numbers such as that utilized by RSA. It is often thought that breaching this integrity by being able to easily identify prime numbers constituents of relevant humongous numbers after successfully solving Riemann hypothesis would have massive implication in that it will now brought the whole of e-commerce to its knees overnight.

"Prime numbers can be described as atoms. What mathematicians have been missing is a kind of mathematical number spectrometer. Chemists have an atomic spectrometer machine that, if we give it a molecule, will tell us the atoms that it is built from. Mathematicians have never invented a mathematical version of this. The proof of Riemann hypothesis in 2019 has given us perfect understanding on how prime numbers work, and translating this into essential knowledge allowing construction of this prime number spectrometer. Suddenly all cryptic codes are breakable. No internet transaction would be safe as the whole of e-commerce depends on the integrity of humongous [non-prime] numbers (molecules) to be anonymously or secretly constituted from its basic prime numbers (atoms). In other words, breaching this integrity by identifying the prime numbers constituents of relevant humongous numbers using prime number spectrometer would have massive implication in that it has now brought the whole of e-commerce to its knees overnight."

The truthfulness of the preceding narrative paragraph can now be beautifully refuted in a mathematical manner. Solving Riemann hypothesis,

Polignac's and Twin prime conjectures is simply irrelevant because the CIS of nontrivial zeros and prime numbers must be treated as Incompletely Predictable entities abiding by Complex Elementary Fundamental Laws that are "Incompletely Predictable Laws" [and not Simple Elementary Fundamental Laws that are "Completely Predictable Laws"]. Thus, in principle, we have dispelled the doom-and-gloom prophecy that financial disaster might follow when successful proof of Riemann hypothesis occur.

However, in practice, there may be a twist to this sentiment. Building ever more powerful supercomputers, which are classical computers based on classical physics, could more easily crack crptic codes but this issue can progressively be negated by employing ever larger prime numbers in cryptic codes. The world's smallest transitor made from a single atom was created in 2012. This should hypothetically assist in the future building of the most powerful supercomputers. But the infinitely more powerful quantum computer, based on quantum mechanics phenomena such as superposition and entanglement, could solve problems in minutes which would otherwise take thousands of years due to the theoretical ability of quantum computers to do a huge range of calculations simultaneously rather than sequentially as in classical computers. This will revolutionize research into areas like artificial intelligence, self-driving cars and drug design in a positive manner but will likely impact the desired role of many cryptic codes in a negative manner by easily cracking them ("cryptocalypse").

Quantum computers could easily crack many crytic codes such as that used by RSA in polynomial time by using Shor's algorithm to find the prime number factors of large integers. To circumvent this problem, the use of

328

alternative crytic codes such as lattice-based cryptosystems which are known not to be broken by quantum computers are desirable. Finally, 'quantum cryptography' (as opposed to 'classical cryptography') could potentially fulfill some of the functions of public key cryptography – see clarification below.

Indirect spin-offs from solving Riemann hypothesis are often stated as "With this one solution, we have proven five hundred theorems or more at once". This apply to many important theorems in Number theory (mostly on prime numbers) that rely on properties of Riemann zeta function such as where trivial and nontrivial zeros are / are not located. A classical example is resulting absolute and full delineation of prime number theorem, which relates to prime counting function. This function, usually denoted by $\pi(x)$, is defined as the number of prime numbers $\leq x$. Public-key cryptography that is widely required for financial security in E-Commerce traditionally depend on solving the difficult problem of factoring prime numbers for astronomically large numbers. The intrinsic "Incompletely Predictable" property present in prime numbers, composite numbers, nontrivial zeros and two types of Gram points can never be altered to "Completely Predictable" property. For this stated reason, it is a mathematical impossibility that providing rigorous proofs such as for Riemann hypothesis will ever result in crypto-apocalypse. However, fast supercomputers and the far-more-powerful quantum computers that theoretically allow solving difficult factorization problem in quick time will result in less secure encryption and decryption. Then using quantum cryptography that relies on principles of quantum mechanics to encrypt data and transmit it in a way that cannot be hacked will combat this issue.

25 Pathology Tests

In this section, I will concentrate mainly on the more novel aspects associated with pathology tests that are usually not available in standard textbooks or, for that matter, have never been formulated before. In order to understand them, we will have to touch base with some basic familiar concepts that are usually known to most people.

Guidelines are often drawn up to help clinician manage a clinical condition. For instance, in the "Management of unstable angina guidelines", Medical Journal of Australia, Vol 173, S65-S88, 16 October 2000; the evidence in the guidelines is graded according to the level-of-evidence classifications endorsed by the National Health and Medical Research Council (NHMRC) in 1995. These are:

E1 Level I: Evidence obtained from a systematic review of all relevant randomized controlled trials.

E2 Level II: Evidence obtained from at least one properly designed randomized controlled trial.

E3 Level III: Evidence obtained from all well-designed controlled trials without randomization, well-designed cohort or case-control analytic studies, preferably from more than one center or research group, or from multiple time series with or without the intervention. Dramatic results in uncontrolled experiments (such as the results of the introduction of penicillin treatment in the 1940s) could also be regarded as this type of evidence.

E4 Level IV: Opinions of respected authorities, based on clinical experience, descriptive studies, or reports of expert committees.

Working clinicians often attempt to keep abreast of development by selectively reading those journal articles (among the exponentially expanding literature) that are relevant, valid and applicable. Research articles can be on a diagnostic test, the clinical course & prognosis of a disorder, the etiology or causation, and the usefulness of a therapy. The title, authors, summary and site of a research article will often quickly decide whether the article is worthwhile reading. For instance, in targeting an article on a diagnostic test, once we decided that the article passes these four brief tests, only then do we proceed to read the "Patients and Methods" section.

Then in so doing, one of the most important elements of the proper clinical evaluation of a diagnostic test is "Was there an independent, 'blind' comparison with a 'gold standard' of diagnosis?" In answering this question, we will have come across important parameters (see below for definitions) from "stable properties" and from "frequency-dependent properties".

Gold standard:

		Positive	Negative
Test	Positive	**True +ve [a]**	**False +ve [b]**
Result:	Negative	**False –ve [c]**	**True –ve [d]**

Stable properties:

Sensitivity (Sen) = a/(a+c)

Specificity (Spec) = d/(b+d)

Frequency-dependent properties:

Positive predictive value (+ve Pred value) = a/(a+b)

Negative predictive value (-ve Pred value) = d/(c+d)

Accuracy (Accu) = (a+d)/(a+b+c+d)

Prevalence (Prev) = (a+c)/(a+b+c+d)

Using Bayes' theorem, +ve Pred values can also be calculated as

[(Prev) (Sen)]

[(Prev) (Sen) + (1 - Prev) (1 - Spec)].

The gold standard refers to a definitive diagnosis obtained by biopsy, surgery, autopsy, long-term follow-up or another acknowledged standard. Within reason, the gold standard should have a Sen, Spec, +ve Pred values, -ve Pred values, and Accu of 100%. The ability of a test to discriminate between normal and abnormal individuals is described by its Sen and Spec. Sen and Spec are inversely related to each other, and can be altered by changing the reference interval or the normal range.

In other words, one can only be improved at the expense of the other. When a test has a Sen of 95% (5% false −ve) and a Spec of 95% (5% false +ve), for a disease with a 1% Prev, its +ve Pred value is only 16% but its −ve Pred value is 99%.

The relationship between Prev and +ve Pred value with a Sen of 95% are shown below:

Disease Prev (%)	+ve Pred value (%)
0	0
0.1	2
1	16
5	50
10	68
20	83
50	95
75	98
99	99.9
100	100

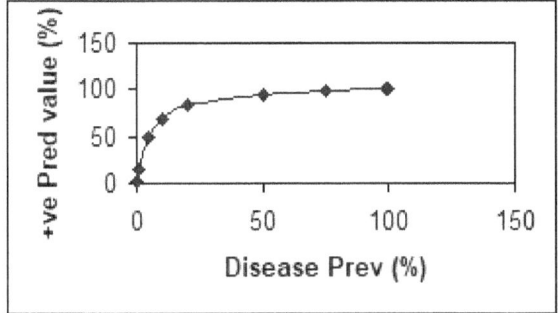

There are different definitions of "normal" such as Gaussian (Normal), Percentile, Risk factor, Culturally desirable, Diagnostic, and Therapeutic – none of them are perfect and each has different properties and consequences. An inevitable consequence arises as follows: when using the Percentile, if the normal range includes the lower 95% of diagnostic test results, then the likelihood that a given patient will be called normal when subjected to this test is 95%. If this same patient undergoes n independent diagnostic tests (independent in the sense that they are probing totally different organs or functions), the likelihood that the patient will be called normal is now 0.95 raised to the power of n. As an example, for n = 35 tests, the patient has only 0.95^{35} or 16.6% of being called normal. The reference interval (RI) for a quantitative test with a Gaussian distribution is defined as the Mean +/- 2 SD, namely 2.5% to 97.5% which encompasses approximately 95% of the results, where SD is the Standard Deviation. (Note that 2 SD is more accurately equal to 95.5%.) Inevitably 5% of entirely normal people will have test results outside the reference interval. Unfortunately, most analytes do not have a Gaussian distribution, generally being skewed towards the higher values – this often causes the yield of a negative value for the lower limit of the RI. Another example is that of Thyroid Stimulating Hormone (TSH) where its reference values are portrayed as logarithmically distributed. Depending on the method used, the RI is often quoted as 0.4 – 5.0 mIU/L, so that 'expected mean' is (0.4+5)/2 or 2.7 mIU/L. However, the actual mean and median values are about 1 mIU/L. Some methods used to alleviate this problem include using logarithm, normal or log-normal probability graph and non-parametric method.

Before a clinician can draw conclusions on a patient's test result he or she usually compares it with the reference interval provided, and often also with the result previously obtained from the same patient. In so doing, the clinician needs to be aware of the effect on the result of some or all the following: analytical variation, biological variation, drugs & disease processes, and prevalence of the disease considered. The reference interval given for a quantitative test result must be appropriate for the patient e.g. physiological factors such as gender, race, age, posture (erect/supine), activity, pregnancy, etc must be taken into consideration if applicable.

Conclusions on test results can be used in one or more ways, for example, to assist a diagnosis, assess the progress of a disease process, determine the extent of dysfunction, assess the effect of treatment, monitor the stability of function, estimate a risk factor, as screening tests for "health", case finding, for research purposes, and so on.

When we talk about the validity and reliability of test results, we need to talk about factors affecting test results. These can be divided into: -

Pre-analytical variables – includes physiological factors, stress, diet, biological variation, drugs, medical history, patient preparation, specimen collection and transport.

Analytical variables – includes precision and accuracy of test method and factors which may interfere with a particular assay. These give rise to the 4 characteristics of analytical methods namely specificity, interference, bias & imprecision which clinical users should be aware.

Post-analytical variables – includes data entry & calculations by laboratory staff, result validation, interpretation of the result, data transfer and the method used to report the results.

Quality control and Quality assurance are the two processes used to control the impact of these variables.

Biological variation and Reference interval

It is crucial to understand the link between biological variation and reference interval. All physiological parameters have dynamic biological variations which are of a continuous nature. However, the monitoring of our many bodily physiological parameters (with different diverse properties) such as a substance concentration or activity, heart rate, heart or brain electrical activities, etc can be on an intermittent (discrete) or continuous basis.

In practice, the observed total (actual) variation of a performed 'Test Y' on 'Patient X' reflects the sum effects of biological (physiological) variation, and pre-analytical, analytical & post-analytical variables. Stated in another way: Total variation = Biological variation + Analytical variation, where Analytical (or Test) variation refers to pre-analytical, analytical & post-analytical variables. (Note that here we are dealing only with the within-person biological variation.) Let us talk about one of the characteristics from the analytical variables, namely analytical imprecision. Imprecision stem from either the sum effects of small changes in the performance of the instrumentation (within-run imprecision), or the sum effects of small changes due to different operators, different batches of reagents, instrument maintenance, etc (between-run or day-to-day imprecision). Test Y is to be performed twice on each of the blood samples collected from Patient X on different days. The difference between the two results for each day is calculated. Then, the within-run imprecision of the test assay (SD_A, standard deviation) is as follows:

$$SD_A = \sqrt{\left(\frac{\text{sum of } (differences)^2}{2 \text{ X number of pairs}} \right)}.$$

Analytical or test imprecision is quoted either as the value for 1 SD_A, or as the coefficient of variation [CV; where CV = 100 (1 SD_A / Mean) % at the mean value which the SD_A was determined]. A good test will have a small CV relative to the absolute value, whilst a poor test will have much wider limits. In real life, a blood sample from Patient X will only have Test Y performed once (and not reported as a mean of several estimations) - therefore, a result for Test Y means that there is a 95% probability that this result lies in the range of +/-2 SD_A. Next, when Test Y is performed on Patient X on two blood samples obtained from two different occasions, the standard error for the two results is $\sqrt{(2\ SD_A^2)}$ = 1.4 SD_A. For 95% probability that two results on Patient X are different, they must differ by more than 2.8 SD_A. Furthermore if the known effect of biological variation SD_B is also taken into account, then total variation $SD_T = \sqrt{(SD_A^2 + SD_B^2)}$. Similarly, for 95% probability that two results on Patient X are different, they must differ by more than 2.8 SD_T.

We have already referred to variation (SD) and variance (SD^2) above. Then $SD_B^2 = SD_T^2 - SD_A^2$ or $SD_B = \sqrt{(SD_T^2 - SD_A^2)}$. However, for most of the purpose of our discussion, we can safely ignore the effects from the pre-analytical, analytical & post-analytical variables as they can be theoretically assumed either to be totally absent or to be of a constant magnitude. In other words, SD_B is approximately taken to be the SD_T, and for 95% probability that two results on Patient X are different, they must differ by more than 2.8 SD_B. Tests can be roughly divided into two groups in terms of their "performance characteristics". This is depicted as below:

Test T1: Large intra-individual variation relative to the inter-individual variation.

The intra-individual variation is about the same as the inter-individual variation. If Test T1 is performed on certain analytes (e.g. plasma sodium or potassium), and it follows that the set-point and the within-person biological variation is about the same for everyone; then the population reference interval is about the same as the biological variation (plus analytical variation) for the individual.

Test T2: Small intra-individual variation relative to the inter-individual variation.
The intra-individual variation is much smaller than the inter-individual variation. On the other hand, for Test T2 performed on certain other analytes (e.g. plasma alkaline phosphatase activity), and it follows that the within-person set points are quite different between individuals as are the variations around the set-points; then the population reference interval is much wider than the biological variation (plus analytical variation) for the individual.

The mean value of the daily test results is termed the set-point for Patient X. As depicted above, the within-person (intra-individual) biological variation for a test may be either of a similar magnitude, or be much smaller, to the population (inter-individual) biological variation (or reference interval). A pathological change in Test T1 analytes will obviously be more readily detected than in Test T2 analytes.

The two main types of analytes' biological variations characteristics

338

also roughly correlate with that of the two types of test performance characteristics.

A simple layman's explanation of the terms Science, Religion and Human body may run as follows:

(1) Science is the study of structure and behavior of physical and natural world and society, especially through observation and experiment. Particular areas of science include computer science, medical science, the science of engineering, etc. The physical sciences include chemistry, physics, etc. The social sciences include psychology, politics, etc. The natural sciences include botany, marine science, zoology, etc. The applied sciences include engineering, computing, etc. Science fiction is a type of writing based on imagined scientific discoveries of the future and often dealing with space travel, life on other planets, and so on.

(2) The human body is best analyzed through study on human physiology and anatomy. Physiology can be defined as science which treats the functions (and the physical and chemical processes involved) of living organism and its parts, and of a species or class of organism and their parts. Its goal is to explain those physical and chemical factors that are responsible for the origin, development, and progression of life. Therefore, physiological studies can be performed on acellular (without enveloping plasma membrane) organisms such as viruses and virus-like agents such as viroids, plasmids, prions, retrons, satellite nucleic acids, satellite viruses; unicellular (single-celled) organisms such as bacteria and amoebae; and multicellular (many-celled) organisms such as plants and animals (including human beings). Small wonders that

studies on physiology can be conducted for all living organisms, plants and animals in this world – thus, we have viral physiology, bacterial physiology, cellular physiology, plant physiology, human physiology, and many more subdivisions. Comparative physiology is a study of organ functions in various types of animals, vertebrate and invertebrate, in an effort to find fundamental relations in the physiology of members of the entire animal kingdom.

Human physiology is the study of the functions of our body (or how our body works) through its mechanisms of action explained in terms of cause-and-effect sequences of physical and chemical processes. With the mechanistic approach, which is employed by physiologists, the mechanisms of action (the "how" of events that occur in the body) are emphasized. With the teleological approach, phenomena that occur in the body are explained in terms of their particular purpose in fulfilling a bodily need (the "why" of body processes), without considering how this outcome is accomplished. Physiology is closely interrelated with anatomy, the study of the structure of our body. The structure and function of the human body are inseparable; and to tell the story of how the body works, we must also provide sufficient anatomical background on the function of the body part being discussed. Medicine is the art and science of the prevention and cure of disease. A person taking medicines is usually taken to imply that he or she is taking one or more drugs to prevent, control or cure a disease or illness. Health is the condition of the body or the mind. Health also refers to the state of being well and free from illness. Pathology is the science of diseases. Obviously, then the study of diseases will involve the study of pathophysiology.

(3) Religion involves the belief in the existence of a supernatural ruling

340

power, the creator and controller of the universe, who has given to man and woman a spiritual nature which continues to exist after death of the body. The great religions of this world with various systems of faith and worship based on such belief include Christianity, Islam, Buddhism, and Hinduism. One of my frequent worries in life is about being falsely influenced by, or wasting precious time in, reading scientific materials (be it journal articles, literature reviews, magazines, books, etc.) from any fields that may be "inaccurate" or "imperfect" due to its underlying mathematical language, research and theories being not quite up-to-date or being superseded by more advanced and correct theories. But our current understanding of science in the 21st Century, based on the work and visions of our scientific forebear and current scientists, has provided the framework to enable us to unravel the important basics or fabrics of science – the knowledge of which will help us in gauging the "soundness" of scientific materials; and perhaps, more significantly, in allowing a person to read any scientific literature and watch or listen to any scientific materials with less apprehension and greater comprehension on what was on offer.

During our journey through this book, we have touched base with many aspects of basic (and sometimes more advanced) mathematics and science, and we have also occasionally explore uncharted territories and sometimes expect the unexpected! However, this exercise would have been a futile one had we not link them together with "common threads of network" (such as Complexity, Chaos and Fractals) underlying the "fabrics of science".

Using scientific aspect and non-scientific aspect criteria, this popular science and medicine book can be roughly separated into two parts. The scientific aspect of this book is about "Basics of Science" and "Human body". The Basics of Science is mainly based on relatively new 'Science of Complexity', and along the way, requiring all-important **Alphabet and Language of Science** to be exposed. These important aspects are centered on two innovative models: Spherical Model of Numbers and Spherical Model of Science. The brief non-scientific aspect of this book is about Basics of Religion based mainly on nature and role of religion in our lives, and along the way, some of the controversies, conflicts and parallels of religion with science exposed. The terminology of "non-scientific aspect criteria" with respect to religion refers to more traditional view that there is nothing scientific about study or practice of various religions – whether true or false, many people will, of course, beg to disagree with this traditional assessment.

The study of religions is mainly based on historical, or origins, science. This type of science deals with the past, is limited because we cannot do experiments directly on past events and as history cannot be repeated, inferences require a deal of guesswork. In contrast to process, or operational, science; inferences and conclusions are closely related to experiments with little room for speculation. It involve doing experiments in the present, making inferences from these results and then carrying out further experiments to test those ideas or hypothesis. This second type of science has given us many valuable advances in knowledge that have given us many wonderful things and has benefited mankind: landing men on the moon, modern medicine, cheap food, electricity, computers, and so on.

EPILOGUE

The author is a self-taught mathematician. His mathematical career pathway is now well established after writing up his two landmark research papers in 2019 on successfully solving Incompletely Predictable problems of Riemann hypothesis, Polignac's and Twin prime conjectures. He looks forward to the general public and wider scientific community future opinion on this area. In any event, he is now content with the legacy of his life achievements in mathematics. Does the coined jargon Sexy Mathematics in this book *cheekily* mean that he has bragging rights to be regarded as a Sexy Mathematician? Probably not but he can always dream this to be true!

The author has deep Chinese heritage and embrace Christianity. His current or future books will contain some Asian cultural flavors and religious inputs. Trust readers have enjoy reading this book and gain something out of it. The author also hopes to have more time in the near future writing planned books concentrating on depicting many more "colorful" benefits arising out of solving Riemann hypothesis, Polignac's and Twin prime conjectures using the "Mathematics for Incompletely Predictable Problems" with again incorporating materials from Medicine, Physiology and Religion.

Bibliography

Farzana Mitra, Shahead Chowdhury, Mike Shelley and Gary Williams (January 15, 2013). A Feasibility Study of Transdermal Buprenorphine Versus Transdermal Fentanyl in the Long-Term Management of Persistent Non-Cancer Pain. Townsville, Australia: Pain Medicine, Volume 14, Issue 1, Pages 75–83 https://doi.org/10.1111/pme.12011.

Furstenberg, H. (1955). On the infinitude of primes. http://dx.doi.org/10.2307/2307043. USA: Amer. Math. Monthly, 62, (5) 353.

Lago P, T. C. (2008). Lago P, TiozRemifentanil for percutaneous intravenous central catheter placement in preterm infant: a randomized controlled trial. . Italy: Ped Anesth 2008; 18: 736 – 744 https://onlinelibrary.wiley.com/doi/abs/10.1111/j.1460-9592.2008.02636.x.

Noe, T. (November 23, 2004). A100967 https://oeis.org/A100967. USA: The On-Line Encyclopedia of Integer Sequences.

Saidak, F. (2006). A New Proof of Euclid's theorem. http://dx.doi.org/10.2307/27642094. USA: Amer. Math. Monthly, 113, (10) 937.

Sloane, N. J. (1964). A000001 https://oeis.org/A000001 (formerly published as M0098 N0035. Number of groups of order n.). USA: The On-Line Encyclopedia of Integer Sequences.

Helpful, B. (2016). Key Role of Dimensional Analysis Homogeneity in Proving Riemann Hypothesis and Providing Explanations on the Closely Related Gram Points. Brisbane: Journal of Mathematics Research.

Helpful, B. (2016). Rigorous Proof for Riemann Hypothesis Using the Novel Sigma-power Laws and Concepts from the Hybrid Method of Integer Sequence Classification. Brisbane: Journal of Mathematics Research.

Helpful, B. (April 12, 2019). Solving Incompletely Predictable problem Riemann hypothesis with Dirichlet Sigma-Power Law: viXra.

Helpful, B. (April 26, 2019). Solving Incompletely Predictable problems Polignac's and Twin Prime conjectures using Information-Complexity conservation: viXra.

Helpful, B. (August 15, 2013). Hybrid integer sequence A228186. Brisbane: The On-Line Encyclopedia of Integer Sequences.

Helpful, B. (2012) and Pussell, B. Supramaximal elevation in B-type natriuretic peptide and its N-terminal fragment levels in anephric patients with heart failure: a case series. Sydney: Journal of medical case reports.

Zhang, Y. (2014). Bounded gaps between primes. http://dx.doi.org/10.4007/annals.2014.179.3.7. USA: Ann. Math. 179(3) (2014) 1121 − 1174.

Solving Incompletely Predictable problem Riemann hypothesis with Dirichlet Sigma-Power Law

Dr. Bernhard Helpful

Published in viXra in 2019

Abstract Riemann hypothesis proposed all nontrivial zeros to be located on critical line of Riemann zeta function. Treated as Incompletely Predictable problem, we obtain Dirichlet Sigma-Power Law as final proof of solving this problem. This Law is derived as equation and inequation from original Dirichlet eta function (proxy function for Riemann zeta function). Performing a parallel procedure help explain closely related Gram points.

Keywords Dimensional analysis, Dirichlet Sigma-Power Law, Gram points, Incompletely Predictable problems, Inequation, Nontrivial zeros, Riemann hypothesis

2010 Mathematics Subject Classification. 11A41, 11M26

1 Introduction

As Incompletely Predictable entities, Gram and virtual Gram points are dependently calculated using *complex* equation Riemann zeta function, $\zeta(s)$, or its proxy Dirichlet eta function, $\eta(s)$, in critical strip (denoted by $0 < \sigma < 1$). Gram[y=0], Gram[x=0] and Gram[x=0,y=0] points respectively refer to x-axis, y-axis and Origin intercepts at critical line (denoted by $\sigma = \frac{1}{2}$). Gram[y=0] and Gram[x=0,y=0] points are respectively synonymous with traditional *'Gram points'* and *nontrivial zeros* with the former further discussed in Segment A2, Appendix A. Virtual Gram[y=0] and virtual Gram[x=0] points respectively refer to x-axis and y-axis intercepts at non-critical lines (denoted by $\sigma \neq \frac{1}{2}$). Virtual Gram[x=0,y=0] points do not exist. Activities to prove associated open problem in number theory Riemann hypothesis and explain Gram[y=0] and Gram[x=0] points equate to solving Incompletely Predictable problems. Claims from these activities are only meaningful when provided with definitions for relevant terms in Segment A1, Appendix A. Dependently calculated using *complex algorithm* Sieve of Eratosthenes, prime and composite numbers as Incompletely Predictable numbers are also depicted there.

In increasing size, arbitrary Set **X** can be countable finite set (CFS), countable infinite set (CIS) or uncountable infinite set (UIS). Cardinality of Set **X**, $|\mathbf{X}|$, measures the "number of elements" in Set **X**. E.g. Set **negative Gram[y=0] point** has CFS of negative Gram[y=0] point with $|\textbf{negative Gram[y=0] point}| = 1$, Set **N** has CIS of natural numbers with $|\mathbf{N}| = \aleph_0$, and Set **R** has UIS of real numbers with $|\mathbf{R}| = c$ (cardinality of the continuum).

$$\zeta(s) = \frac{e^{\left(\ln(2\pi)-1-\frac{\gamma}{2}\right)}}{2(s-1)\Gamma\left(1+\frac{s}{2}\right)}\Pi_\rho\left(1-\frac{s}{\rho}\right)e^{\frac{s}{\rho}} = \pi^{\frac{s}{2}}\frac{\Pi_\rho\left(1-\frac{s}{\rho}\right)}{2(s-1)\Gamma\left(1+\frac{s}{2}\right)}$$

Proposed in 1859 by German mathematician Bernhard Riemann (September 17, 1826 – July 20, 1866), Riemann hypothesis is mathematical statement on $\zeta(s)$ that critical line denoted by $\sigma = \frac{1}{2}$

347

contains complete Set **nontrivial zeros** with $|$**nontrivial zeros**$| = \aleph_0$. Alternatively, this hypothesis is geometrical statement on $\zeta(s)$ that generated curves when $\sigma = \frac{1}{2}$ contain complete Set **Origin intercepts** with $|$**Origin intercepts**$| = \aleph_0$. Depicted in full and abbreviated version, Hadamard product is infinite product expansion of $\zeta(s)$ based on Weierstrass's factorization theorem displaying a simple pole at s = 1. It contains both trivial and nontrivial zeros indicating their common origin from $\zeta(s)$. Set **trivial zeros** occurs at σ = -2, -4, -6, -8, -10,..., ∞ with $|$**trivial zeros**$| = \aleph_0$ due to Γ function term in denominator. Nontrivial zeros occur at s = ϱ with γ denoting Euler-Mascheroni constant.

Remark 1.1. Computationally checking for first 10,000,000,000,000 nontrivial zeros location on critical line implies but does not prove Riemann hypothesis to be true.

Locations of first 10,000,000,000,000 nontrivial zeros on critical line are previously confirmed to be correct. Hardy in 1914[1] and Hardy and Littlewood in 1921[2] showed infinite nontrivial zeros on critical line by considering moments of certain functions related to $\zeta(s)$. This discovery cannot constitute rigorous proof for Riemann hypothesis because they have not exclude theoretical existence of nontrivial zeros located away from this line.

Remark 1.2. We can apply useful concepts from exact and inexact Dimensional analysis homogeneity to well-defined equations and inequations.

Respectively for 'base quantities' such as *length*, *mass* and *time*; their fundamental SI 'units of measurement' meter (m) is defined as distance

348

travelled by light in vacuum for time interval 1/299 792 458 s with speed of light c = 299,792,458 ms^{-1}, kilogram (kg) is defined by taking fixed numerical value Planck constant h to be 6.626 070 15 X 10^{-34} Joules·second (Js) [whereby Js is equal to kgm^{2}s^{-1}] and second (s) is defined in terms of ΔvCs = Δ(^{133}Cs)$_{hfs}$ = 9,192,631,770 s^{-1}. Derived SI units such as J and ms^{-1} respectively represent 'base quantities' *energy* and *velocity*. The word 'dimension' is commonly used to indicate all those mentioned 'units of measurement' in well-defined equations.

Dimensional analysis (DA) is an analytic tool with DA homogeneity and non-homogeneity (respectively) denoting valid and invalid equation occurring when 'units of measurements' for 'base quantities' are "balanced" and "unbalanced" across both sides of the equation. E.g. equation 2 m + 3 m = 5 m is valid and equation 2 m + 3 kg = 5 mkg is invalid (respectively) manifesting DA homogeneity and non-homogeneity.

Let (2n) and (2n-1) be 'base quantities' in Dirichlet Sigma-Power Laws formatted in simplest forms as equations and inequations. E.g. DA on exponent $\frac{1}{2}$ in $(2n)^{\frac{1}{2}}$ in simplest form is correct but DA on exponent $\frac{1}{4}$ in equivalent $(2^2 n^2)^{\frac{1}{4}}$ *not* in simplest form is incorrect. Fractional exponents as 'units of measurement' given by (1 −σ) for equations and (σ + 1) for inequations when σ = $\frac{1}{2}$ coincide with exact DA homogeneity[1]; and (1 −σ) for equations and (σ +1) for inequations when σ ≠ $\frac{1}{2}$ coincide with inexact DA homogeneity[2].

Respectively for equations and inequations, exact DA homogeneity at σ = $\frac{1}{2}$ denotes $\sum$(all fractional exponents) as 2(1−σ) and 2(σ +1)

349

equates to ["exact"] whole number '1' and '3'; and inexact DA homogeneity at $\sigma \neq \frac{1}{2}$ denotes $\sum$(all fractional exponents) as $2(1-\sigma)$ and $2(\sigma +1)$ equates to ["inexact"] fractional number '$\neq 1$' and '$\neq 3$'.

Footnote 1, 2: Exact and inexact DA homogeneity occur in Dirichlet Sigma-Power Laws as equations or inequations for Gram[y=0] points, Gram[x=0] points and nontrivial zeros. *Law of Continuity* is a heuristic principle *whatever succeed for the finite, also succeed for the infinite*. Then these Laws which inherently manifest themselves on finite and infinite time scale should "succeed for the finite, also succeed for the infinite".

Outline of proof for Riemann hypothesis. To simultaneously satisfy two mutually inclusive conditions: I. *With rigid manifestation of exact DA homogeneity*, Set nontrivial zeros with |nontrivial zeros| = $\aleph_0$ is located on critical line (viz. $\sigma = \frac{1}{2}$) when $2(1 - \sigma)$ [or $2(\sigma + 1)$] as $\sum$(all fractional exponents) = whole number '1' [or '3'] in Dirichlet Sigma-Power Law[3] as equation [or inequation]. II. *With rigid manifestation of inexact DA homogeneity*, Set nontrivial zeros with |nontrivial zeros| = $\aleph_0$ is not located on non-critical lines (viz. $\sigma \neq \frac{1}{2}$) when $2(1-\sigma)$ [or $2(\sigma +1)$] as $\sum$(all fractional exponents) = fractional number '$\neq 1$' [or '$\neq 3$'] in Dirichlet Sigma-Power Law[3] as equation [or inequation].

Footnote 3: Derived from original $\eta(s)$ (*proxy* for $\zeta(s)$) as equation or inequation, this Law symbolizes end-result proof on Riemann hypothesis.

Riemann hypothesis mathematical foot-prints. Six identifiable steps to prove Riemann hypothesis: *Step 1* Use $\eta(s)$, *proxy* for $\zeta(s)$, in critical strip. *Step 2* Apply Euler formula to $\eta(s)$. *Step 3* Obtain "simplified"

Dirichlet eta function which intrinsically incorporates *actual location [but not actual positions]* of all nontrivial zeros[4]. *Step 4* Apply Riemann integral to "simplified" Dirichlet eta function in discrete (summation) format. *Step 5* Obtain Dirichlet Sigma-Power Law in continuous (integral) format as equation or inequation. *Step 6* Note exact and inexact DA homogeneity on their fractional exponents.

Footnote 4: Respectively Gram[y=0] points, Gram[x=0] points and nontrivial zeros are Incompletely Predictable entities with actual positions determined by setting $\sum \text{Im}\{\eta(s)\} = 0$, $\sum \text{Re}\{\eta(s)\} = 0$ and $\sum \text{ReIm}\{\eta(s)\} = 0$ to *dependently* calculate relevant positions of all preceding entities in neighborhood. Respectively actual location of Gram[y=0] points, Gram[x=0] points and nontrivial zeros; and virtual Gram[y=0] points, virtual Gram[x=0] points and "absent" nontrivial zeros occur precisely at $\sigma = \frac{1}{2}$ and $\sigma \neq \frac{1}{2}$.

2 Riemann zeta and Dirichlet eta functions

L-functions form an integral part of 'L-functions and Modular Forms Database' (LMFDB) with far-reaching implications. In perspective, $\zeta(s)$ is simplest example of an L-function. $\zeta(s)$ is a function of complex variable s $(= \sigma \pm \iota t)$ that analytically continues sum of infinite series $\zeta(s) = \sum_{n=1}^{\infty} \frac{1}{n^s} = \frac{1}{1^s} + \frac{1}{2^s} + \frac{1}{3^s} + \cdots$. The common convention is to write s as $\sigma + \iota t$ with $i = \sqrt{-1}$, and σ and t real. Valid for $\sigma > 0$, we write $\zeta(s)$ as $\text{Re}\{\zeta(s)\} + \iota \cdot \text{Im}\{\zeta(s)\}$ and note that $\zeta(\sigma + \iota t)$ when $0 < t < +\infty$ is the complex conjugate of $\zeta(\sigma - \iota t)$ when $-\infty < t < 0$.

Also known as alternating zeta function, $\eta(s)$ must act as *proxy* for $\zeta(s)$ in critical strip (viz. $0 < \sigma < 1$) containing critical line (viz. $\sigma = \frac{1}{2}$)

because $\zeta(s)$ only converges when $\sigma > 1$. This implies $\zeta(s)$ is undefined to left of this region in critical strip which then requires $\eta(s)$ representation instead. They are related to each other as $\zeta(s) = \gamma \cdot \eta(s)$ with proportionality factor $\gamma = \frac{1}{(1-2^{1-s})}$ and $\eta(s) = \sum_{n=1}^{\infty} \frac{(-1)^{n+1}}{n^s} = \frac{1}{1^s} - \frac{1}{2^s} + \frac{1}{3^s} + \cdots$.

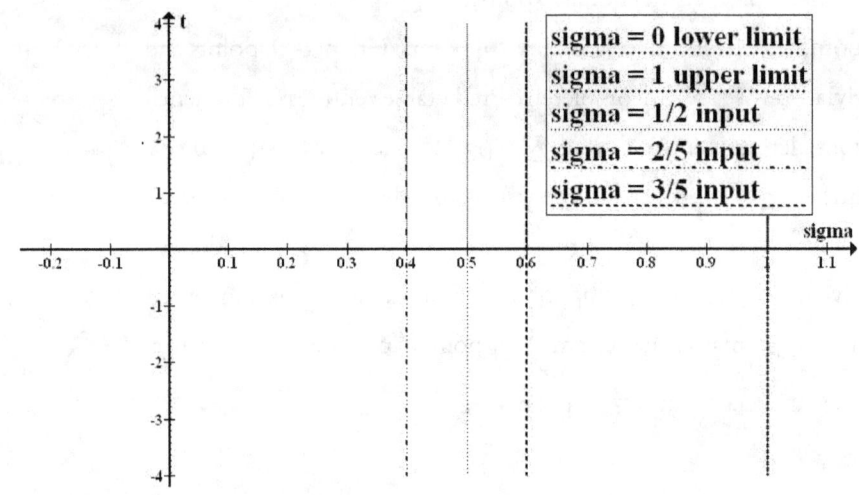

Fig. 1 INPUT for $\sigma = \frac{1}{2}, \frac{2}{5}, and \frac{3}{5}$. $\zeta(s)$ has CIS of Completely Predictable trivial zeros at $\sigma =$ all negative even numbers and CIS of Incompletely Predictable nontrivial zeros at $\sigma = \frac{1}{2}$ for various t values.

352

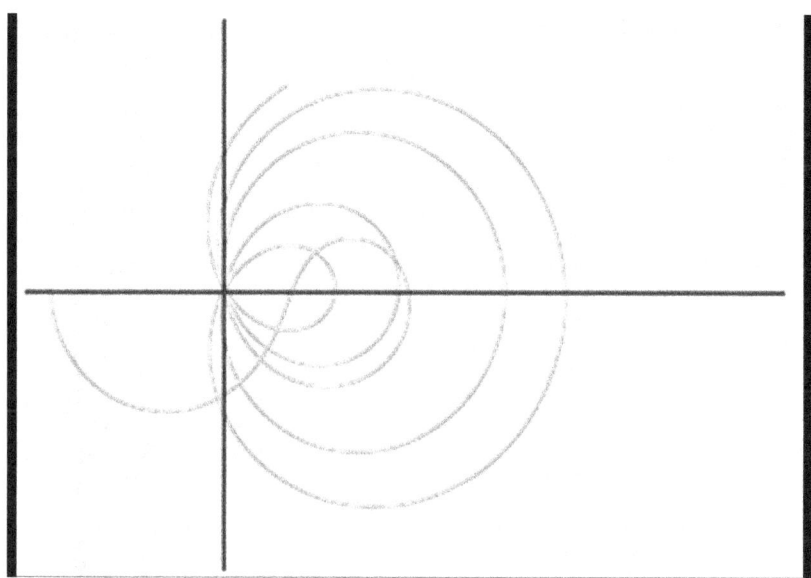

Fig. 2 OUTPUT for $\sigma = \frac{1}{2}$. Schematically depicted polar graph of

$\zeta(\frac{1}{2}+\imath t)$ plotted along critical line for real values of t running from 0 to

34, horizontal axis: as $Re\{\zeta(\frac{1}{2}+\imath t)\}$, and vertical axis: $Im\{\zeta(\frac{1}{2}+\imath t)\}$. Note

presence of Origin intercepts which are totally absent in Figures 3 and

4 [with identical axes definitions].

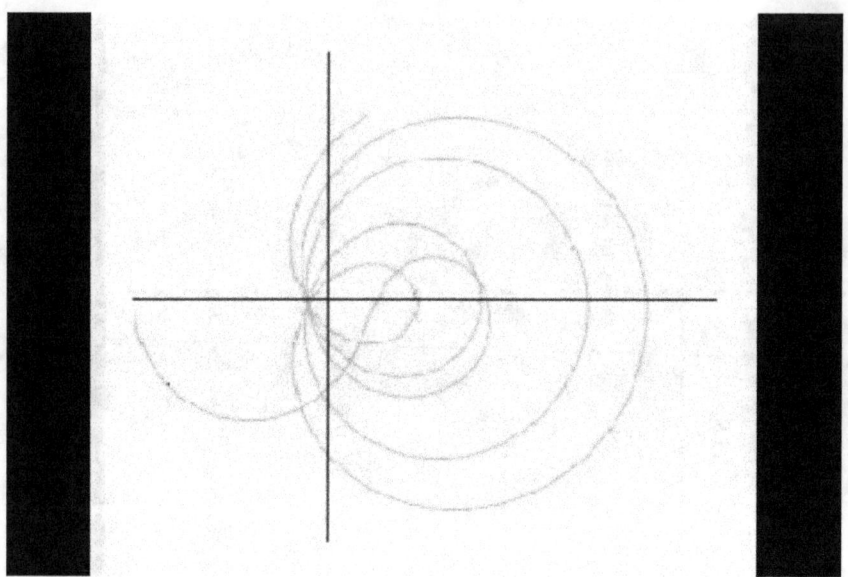

Fig. 3 OUTPUT for $\sigma = \frac{2}{5}$.

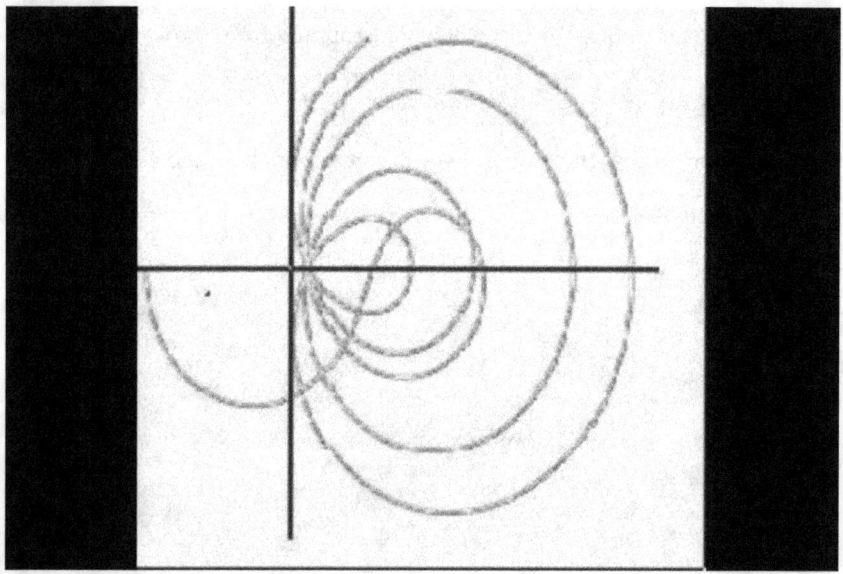

Fig. 4 OUTPUT for $\sigma = \frac{3}{5}$.

354

$$\zeta(s) = \sum_{n=1}^{\infty} \frac{1}{n^s} \qquad\qquad (1)$$

$$= \frac{1}{1^s} + \frac{1}{2^s} + \frac{1}{3^s} + \cdots$$

$$= \prod p \; prime \; \frac{1}{(1-p^{-s})}$$

$$= \frac{1}{(1-2^{-s})} \cdot \frac{1}{(1-3^{-s})} \cdot \frac{1}{(1-5^{-s})} \cdot \frac{1}{(1-7^{-s})} \cdot \frac{1}{(1-11^{-s})} \cdots \frac{1}{(1-p^{-s})} \cdots$$

Eq. (1) is defined for only $1 < \sigma < \infty$ region where $\zeta(s)$ is absolutely convergent. There are no zeros located here. In Eq. (1), equivalent Euler product formula with product over prime numbers [instead of summation over natural numbers] can also represent $\zeta(s)$.

$$\zeta(s) = 2^s \pi^{s-1} \sin\left(\frac{\pi s}{2}\right) . \Gamma(1-s) . \zeta(1-s) \qquad\qquad (2)$$

With $\sigma = \frac{1}{2}$ as symmetry line of reflection, Eq. (2) is Riemann's functional equation valid for $-\infty < \sigma < \infty$. It can be used to find all trivial zeros on horizontal line at $t = 0$ occurring πs when σ = -2, -4, -6, -8, -10,..., ∞ whereby $\zeta(s) = 0$ because factor $\sin\left(\frac{\pi s}{2}\right)$ vanishes. Γ is gamma function, an extension of factorial function [a product function denoted by ! notation whereby n! = n(n−1)(n−2)...(n−(n−1))] with its argument shifted down by 1, to real and complex numbers. That is, if n is a positive integer, $\Gamma(n) = (n-1)!$

$$\zeta(s) = \frac{1}{(1-2^{1-s})} \sum_{n=1}^{\infty} \frac{(-1)^{n+1}}{n^s}$$

$$= \frac{1}{(1-2^{1-s})} \left(\frac{1}{1^s} - \frac{1}{2^s} + \frac{1}{3^s} + \cdots\right) \qquad\qquad (3)$$

Eq. (3) is defined for all $\sigma > 0$ values except for simple pole at $\sigma = 1$. As alluded to above, $\zeta(s)$ without $\frac{1}{(1-2^{1-s})}$ viz. $\sum_{n=1}^{\infty} \frac{(-1)^{n+1}}{n^s}$ is $\eta(s)$. It is a holomorphic function of s defined by analytic continuation and is mathematically defined at $\sigma = 1$ whereby analogous trivial zeros with presence only for $\eta(s)$ [but not for $\zeta(s)$] on vertical straight line $\sigma = 1$ are found at $s = 1 \pm i.\frac{2\pi k}{\ln(2)}$ where $k = 1, 2, 3, 4, 5, ..., \infty$.

Figure 1 depict complex variable s ($= \sigma \pm it$) as INPUT with x-axis denoting real part Re$\{s\}$ equating to σ; and y-axis denoting imaginary part Im$\{s\}$ equating to t. Figures 2, 3 and 4 respectively depict $\zeta(s)$ as OUTPUT for real values of t running from 0 to 34 at for $\sigma = \frac{1}{2}$ (critical line), for $\sigma = \frac{2}{5}$ (non-critical line), and for $\sigma = \frac{3}{5}$ (non-critical line) with x-axis denoting real part Re$\{\zeta(s)\}$ and y-axis denoting imaginary part Im$\{\zeta(s)\}$. There are infinite types-of-spirals possibilities associated with each σ value arising from all infinite σ values in critical strip. Mathematically proving all nontrivial zeros location on critical line as denoted by solitary for $\sigma = \frac{1}{2}$ value equates to geometrically proving all Origin intercepts occurrence at solitary $\sigma = \frac{1}{2}$ value. Both result in rigorous proof for Riemann hypothesis.

3 Prerequisite lemma, corollary and propositions for Riemann hypothesis

Original equation $\eta(s)$, *proxy* for $\zeta(s)$, is treated as unique mathematical object with key properties and behaviors. Containing all x-axis, y-axis and Origin intercepts, it will intrinsically incorporate *actual location [but not actual positions]* of all Gram[y=0] points, Gram[x=0] points and nontrivial zeros. Proofs on lemma, corollary and propositions on nontrivial zeros depict exact and inexact DA homogeneity in both derived equation and inequation. Parallel

356

procedure on Gram[y=0] and Gram[x=0] points in Appendix B depict exact and inexact DA homogeneity in similarly derived equations and inequations.

Lemma 3.1. "Simplified" Dirichlet eta function is derived directly from Dirichlet eta function with Euler formula application and it will intrinsically incorporate actual location [but not actual positions] of all nontrivial zeros.

Proof. Denote complex number (C) as $z = x + i \cdot y$. Then $z = \mathrm{Re}(z) + i \cdot \mathrm{Im}(z)$ with $\mathrm{Re}(z) = x$ and $\mathrm{Im}(z) = y$; modulus of z, $|z| = \sqrt{(\mathrm{Re}(z)^2 + \mathrm{Im}(z)^2)} = \sqrt{(x^2 + y^2)}$; and $|z|^2 = x^2 + y^2$.

Euler formula is commonly stated as $e^{ix} = \cos x + i \cdot \sin x$. Euler identity (where $x = \pi$) is $e^{i\pi} = \cos \pi + i \cdot \sin \pi = -1 + 0$ [or stated as $e^{i\pi} + 1 = 0$]. The n^s of $\zeta(s)$ is expanded to $n^s = n^{(\sigma + it)} = n^\sigma e^{ln(n) \cdot i}$ since $n^t = e^{t \, ln(n)}$. Apply Euler formula to n^s result in $n^s = n^\sigma (\cos(t \, ln(n)) + i \cdot \sin(t \, ln(n))$. This is written in trigonometric form [designated by shorthand notation n^s(Euler)] whereby n^σ is modulus and $t \, ln(n)$ is polar angle (argument). Apply n^s (Euler) to Eq. (1). Then $\zeta(s) = \mathrm{Re}\{\zeta(s)\} + i \cdot \mathrm{Im}\{\zeta(s)\}$ with $\mathrm{Re}\{\zeta(s)\} = \sum_{n=1}^{\infty} n^{-\sigma} \cos(t \, ln(n))$ and $\mathrm{Im}\{\zeta(s)\} = \sum_{n=1}^{\infty} n^{-\sigma} \sin(t \, ln(n))$. As Eq. (1) is defined only for $\sigma > 1$ where zeros never occur, we will not carry out further treatment here.

Apply n^s (Euler) to Eq. (3). Then $\zeta(s) = \gamma \cdot \eta(s) = \gamma \cdot [\mathrm{Re}\{\eta(s)\} + i \cdot \mathrm{Im}\{\eta(s)\}]$ with
$\mathrm{Re}\{\eta(s)\} = \sum_{n=1}^{\infty}((2n-1)^{-\sigma} \cos(t \, ln(2n-1)) - (2n)^{-\sigma} \cos(t \, ln(2n)))$;
$\mathrm{Im}\{\eta(s)\} = \sum_{n=1}^{\infty}((2n)^{-\sigma} \sin(t \, ln(2n)) - (2n-1)^{-\sigma} \sin(t \, ln(2n)))$;

357

and proportionality factor $\gamma = \dfrac{1}{(1-2^{1-s})}$.

Complex number s in critical strip is designated by $s = \sigma + \imath t$ for $0 < t < +\infty$ and $s = \sigma - \imath t$ for $-\infty < t < 0$. Nontrivial zeros equating to $\zeta(s) = 0$ give rise to our desired $\eta(s) = 0$. Modulus of $\eta(s)$, $|\eta(s)|$, is defined as $\sqrt{(Re\{\eta(s)\})^2 + (Im\{\eta(s)\})^2)}$ with $|\eta(s)|^2 = (Re\{\eta(s)\})^2 + (Im\{\eta(s)\})^2$. Mathematically $|\eta(s)| = |\eta(s)|^2 = 0$ is an unique condition giving rise to $\eta(s) = 0$ occurring only when $Re\{\eta(s)\} = Im\{\eta(s)\} = 0$ as any non-zero values for $Re\{\eta(s)\}$ and/or $Im\{\eta(s)\}$ will always result in $|\eta(s)|$ and $|\eta(s)|^2$ having nonzero values. Important implication is that sum of $Re\{\eta(s)\}$ and $Im\{\eta(s)\}$ equating to zero [given by Eq. (4)] must always hold when $|\eta(s)| = |\eta(s)|^2 = 0$ and consequently $\eta(s) = 0$.

$$\sum ReIm\{\eta(s)\} = Re\{\eta(s)\} + Im\{\eta(s)\} = 0 \qquad (4)$$

In principle, advocating for existence of theoretical s values leading to non-zero values in $Re\{\eta(s)\}$ and $Im\{\eta(s)\}$ depicted as possibility $+Re\{\eta(s)\} = -Im\{\eta(s)\}$ or $-Re\{\eta(s)\} = +Im\{\eta(s)\}$ could satisfy Eq. (4). This reverse implication is not necessarily true as these s values will not result in $|\eta(s)| = |\eta(s)|^2 = 0$. In any event, we need not consider these two possibilities since solving Riemann hypothesis involves nontrivial zeros defined by $\eta(s) = 0$ with non-zero values in $Re\{\eta(s)\}$ and/or $Im\{\eta(s)\}$ being not compatible with $\eta(s) = 0$.

Riemann hypothesis proposed all nontrivial zeros to be located on critical line. This location is conjectured to be uniquely associated with presence of exact DA homogeneity in derived equation and inequation of Dirichlet Sigma-Power Law with Eq. (4) intrinsically incorporated into this Law as the $\eta(s) = 0$ definition for nontrivial zeros equates to Eq. (4).

Apply trigonometry identity $\cos(x) - \sin(x) = \sqrt{2}\sin(x + \frac{3}{4}\pi)$ to $\mathrm{Re}\{\eta(s)\}+\mathrm{Im}\{\eta(s)\}$ to get Eq. (5) with terms in last line built by mixture of terms from $\mathrm{Re}\{\eta(s)\}$ and $\mathrm{Im}\{\eta(s)\}$. $\sum\mathrm{ReIm}\{\eta(s)\}$

$$= \sum_{n=1}^{\infty}[((2n-1)^{-\sigma}\cos(t\ln(2n-1)) - (2n-1)^{-\sigma}\sin(t\ln(2n-1))$$

$$-(2n)^{-\sigma}\cos(t\ln(2n)) + (2n)^{-\sigma}\sin(t\ln(2n))]$$

$$= \sum_{n=1}^{\infty}[((2n-1)^{-\sigma}\sqrt{2}\sin(t\ln(2n-1) + \frac{3}{4}\pi) -$$

$$(2n)^{-\sigma}\sqrt{2}\sin\left(t\ln(2n) + \frac{3}{4}\pi\right)] \qquad (5)$$

When depicted in terms of Eq. (4), Eq. (5) becomes

$$\sum_{n=1}^{\infty}(2n)^{-\sigma}\sqrt{2}\sin\left(t\ln(2n) + \frac{3}{4}\pi\right) = \sum_{n=1}^{\infty}(2n-1)^{-\upsilon}\sqrt{2}\sin(t\ln(2n-1) + \frac{3}{4}\pi)$$

$$\sum_{n=1}^{\infty}(2n)^{-\sigma}\sqrt{2}\sin\left(t\ln(2n) + \frac{3}{4}\pi\right) - \sum_{n=1}^{\infty}(2n-1)^{-\sigma}\sqrt{2}\sin\left(t\ln(2n-1) + \frac{3}{4}\pi\right) = 0 \qquad (6)$$

Eq. (6) in discrete (summation) format is a non-Hybrid integer sequence equation – see Appendix C. $\eta(s)$ calculations for all σ values result in infinitely many non-Hybrid integer sequence equations for $0 < \sigma < 1$ critical strip region of interest with n = 1, 2, 3, 4, 5,..., ∞ as discrete integer number values, or n = 1 to ∞ as continuous real numbers values with Riemann integral application. These equations will geometrically represent entire plane of critical strip, thus (at least) allowing our proposed proof to be of a complete nature.

Eq. (6) being the "simplified" Dirichlet eta function derived directly from $\eta(s)$ will intrinsically incorporate *actual location [but not actual positions]* of all nontrivial zeros. *The proof is now complete for Lemma 3.1*□.

359

Proposition 3.2. Dirichlet Sigma-Power Law in continuous (integral) format given as equation and inequation can both be derived directly from "simplified" Dirichlet eta function in discrete (summation) format with Riemann integral application.

Proof. In Calculus, integration is reverse process of differentiation viewed geometrically as area enclosed by curve of function and x-axis. Apply definite integral I between points a and b is to compute its value when $\Delta x \rightarrow 0$, i.e. $I = \lim_{\Delta x \rightarrow 0} \sum f(x_i)\Delta x_i = \int_a^b f(x)dx$. This is Riemann integral of function f(x) in interval [a, b] where a<b. Apply Riemann integral to "simplified" Dirichlet eta function in [$\Delta x \rightarrow 1$] discrete (summation) format which intrinsically incorporates *actual location [but not actual positions]* of all nontrivial zeros criterion to obtain Dirichlet Sigma-Power Law in [$\Delta x \rightarrow 0$] continuous (integral) format with the later validly representing the former. Then Dirichlet Sigma-Power Law will also fullfil this criterion. Due to resemblance to power law functions in σ from $s = \sigma + \imath t$ being exponent of a power function n^σ, logarithm scale use, and harmonic $\zeta(s)$ series connection in Zipf's law; we elect to call this Law by its given name. A characteristic and crucial step of this Law is its exact formula expression in usual mathematical language [$y = f(x_1, x_2)$ format description for a 2-variable function with (2n) and (2n−1) parameters] consist of $y = f(t, \sigma)$ with discrete n = 1, 2, 3, 4, 5,..., ∞ or continuous n = 1 to ∞; -∞ < t < +∞; and $0 < \sigma < 1$.

With steps of manual integration shown using indefinite integrals [for simplicity], solve definite integral below based on numerator portion of R1 with (2n) parameter in Eq. (6):

360

$$\int_1^\infty \frac{2^{\frac{1}{2}-\sigma}\sin(t\ln(2n)+\frac{3}{4}\pi)}{n^\sigma}\,dn \;=\; \int_1^\infty -\frac{\sin(t\ln(2n))-\cos\,(t\ln(2n))}{2^\sigma n^\sigma}\,dn.$$ We

deduce most other important integrals to be "variations" of this

particular integral containing (i) deletion of $(2n)^{-\sigma}$, $\sqrt{2}$ or $\frac{3}{4}\pi$ terms,

and/or (ii) interchange of sine and cosine function. We check all

derived antiderivatives to be correct using computer algebra system

Maxima.

Simplifying and applying linearity, we obtain $2^{\frac{1}{2}-\sigma}\int\frac{\sin(t\ln(2n)+\frac{3}{4}\pi)}{n^\sigma}dn.$

Now solving $\int\frac{\sin(t\ln(2n)+\frac{3}{4}\pi)}{n^\sigma}dn.$ Substitute $u=t\ln(2n)+\frac{3}{4}\pi\to$

$dn=\frac{n}{t}du$, use $n^{1-\sigma}=e^{\frac{(1-\sigma)(u-t\ln(2n)-\frac{3}{4}\pi)}{t}}=$

$\dfrac{e^{\frac{(\sigma-1)(4t\ln(2n)-3\pi)}{4t}}}{t}\displaystyle\int e^{\frac{(1-\sigma)u}{t}}\sin(u)\,du.$

Now solving $\int e^{\frac{(1-\sigma)u}{t}}\sin(u)\,du.$ We integrate by parts twice in

a row: $\int f\,g'\;=\;f\,g-\int f'\,g.$

First time: $f=\sin\,(u)$, $g'=e^{\frac{(1-\sigma)u}{t}}.$

Then $f'=$

$$\boxed{\cos\,(u)}$$

$g=$

$$\boxed{\dfrac{(1-\sigma)te^{\frac{(1-\sigma)u}{t}}}{\sigma^2-2\sigma+1}}$$

$=\dfrac{(1-\sigma)te^{\frac{(1-\sigma)u}{t}}\sin(u)}{\sigma^2-2\sigma+1}-\int\dfrac{(1-\sigma)te^{\frac{(1-\sigma)u}{t}}\cos(u)}{\sigma^2-2\sigma+1}\,du$

Second time: $f=$

361

$$\boxed{\cos(u)}$$

$$g' = \boxed{\dfrac{(1-\sigma)te^{\frac{(1-\sigma)u}{t}}}{\sigma^2-2\sigma+1}}$$

Then $f' = -\sin(u)$, $g = \dfrac{t^2 e^{\frac{(1-\sigma)u}{t}}}{\sigma^2-2\sigma+1}$:

$$= \frac{(1-\sigma)te^{\frac{(1-\sigma)u}{t}}\sin(u)}{\sigma^2-2\sigma+1} - \left(\frac{t^2 e^{\frac{(1-\sigma)u}{t}}\cos(u)}{\sigma^2-2\sigma+1} - \int -\frac{t^2 e^{\frac{(1-\sigma)u}{t}}\sin(u)}{\sigma^2-2\sigma+1}\,du\right)$$

Apply linearity:

$$= \frac{(1-\sigma)te^{\frac{(1-\sigma)u}{t}}\sin(u)}{\sigma^2-2\sigma+1} - \left(\frac{t^2 e^{\frac{(1-\sigma)u}{t}}\cos(u)}{\sigma^2-2\sigma+1} + \right.$$

$$\left. \frac{t^2}{\sigma^2-2\sigma+1}\int e^{\frac{(1-\sigma)u}{t}}\sin(u)\,du\right)$$

As integral $\int e^{\frac{(1-\sigma)u}{t}}\sin(u)\,du$ appears again on Right Hand Side, we can solve for it:

$$= \frac{\dfrac{(1-\sigma)e^{\frac{(1-\sigma)u}{t}}\sin(u)}{t} - e^{\frac{(1-\sigma)u}{t}}\cos(u)}{\dfrac{\sigma^2-2\sigma+1}{t^2}+1}$$

Plug in solved integrals: $\dfrac{e^{\frac{(\sigma-1)4t\ln(2)+3\pi}{4t}}}{4t}\int e^{\frac{(1-\sigma)u}{t}}\sin(u)\,du$

$$= \frac{e^{\frac{(\sigma-1)4t\ln(2)+3\pi}{4t}}\left(\dfrac{(1-\sigma)e^{\frac{(1-\sigma)u}{t}}\sin(u)}{t} - e^{\frac{(1-\sigma)u}{t}}\cos(u)\right)}{\left(\dfrac{\sigma^2-2\sigma+1}{t^2}+1\right)t}$$

Undo substitution $u = t\,\ln(2n) - \dfrac{3}{4}\pi$ and simplifying:

$$= \frac{e^{\frac{(\sigma-1)4t\ln(2)+3\pi}{4t}}\left(\dfrac{(1-\sigma)e^{\frac{(\sigma-1)(t\ln(2)+\frac{3}{4}\pi)}{t}}\sin(t\,\ln(2)+\frac{3}{4}\pi)}{t} - e^{\frac{(\sigma-1)(t\ln(2)+\frac{3}{4}\pi)}{t}}\cos(t\ln(2)+\frac{3}{4}\pi)\right)}{\left(\dfrac{\sigma^2-2\sigma+1}{t^2}+1\right)t}$$

Plug in solved integrals:

362

$$2^{\frac{1}{2}-\sigma}\int \frac{\sin(t\ln(2n)+\frac{3}{4}\pi)}{n^{\sigma}}dn$$

$$=\frac{2^{\frac{1}{2}-\sigma}e^{\frac{(\sigma-1)4t\ln(2)+3\pi}{4t}}\left((1-\sigma)e^{\frac{(\sigma-1)(t\ln(2)+\frac{3}{4}\pi)}{t}}\frac{\sin(t\ln(2)+\frac{3}{4}\pi)}{t}-e^{\frac{(\sigma-1)(t\ln(2)+\frac{3}{4}\pi)}{t}}\cos\left(t\ln(2)+\frac{3}{4}\pi\right)\right)}{\left(\frac{\sigma^2-2\sigma+1}{t^2}+1\right)t}$$

By rewriting and simplifying, $\int_1^{\infty} \frac{2^{\frac{1}{2}-\sigma}\sin(t\ln(2n)+\frac{3}{4}\pi)}{n^{\sigma}}dn$ is finally

solved as

$$\left[\frac{(2n)^{1-\sigma}((t+\sigma-1)sin(t\,ln(2n))+(t-\sigma+1)cos(t\,ln(2n)))}{2(t^2+(\sigma-1)^2)}+C\right]_1^{\infty} \quad (7)$$

For denominator portion of R1 with (2n-1) parameter in Eq. (6), Eq.

(7) equates to

$$\left[\frac{(2n-1)^{1-\sigma}((t+\sigma-1)sin(t\,ln(2n-1))+(t-\sigma+1)cos(t\,ln(2n-1)))}{2(t^2+(\sigma-1)^2)}+C\right]_1^{\infty} \quad (8)$$

Dirichlet Sigma-Power Law as equation derived from Eq. (6) is

given by:

$$\frac{1}{2\left(t^2+(\sigma-1)^2\right)}\cdot\left[(2n)^{1-\sigma}\left((t+\sigma-1)\sin\left(t\ln(2n)\right)+(t-\sigma+1)\cos\left(t\ln(2n)\right)\right)-\right.$$

$$\left.(2n-1)^{1-\sigma}\left((t+\sigma-1)\sin\left(t\ln(2n-1)\right)+(t-\sigma+1)\cos\left(t\ln(2n-1)\right)\right)\right]_1^{\infty}=0 \quad (9)$$

Apply Ratio Study to Eq. (6) – see Segment A3, Appendix A. This

involves [intentional] incorrect but "balanced" rearrangement of terms

in Eq. (6) giving rise to Eq. (10) which is a non-Hybrid integer sequence

inequation. Left-hand side contains 'cyclical' sine function in first term

(Ratio R1) and 'non-cyclical' power function in second term (Ratio R2).

$$\frac{\sqrt{2}\sin\left(t\ln(2n)+\frac{3}{4}\pi\right)}{\sqrt{2}\sin\left(t\ln(2n-1)+\frac{3}{4}\pi\right)}-\frac{(2n)^{-\sigma}}{(2n-1)^{-\sigma}}\neq 0 \quad (10)$$

Apply Riemann integral to selected parts of Eq. (10) without depicting

steps of calculation:

$$\int_1^\infty \sqrt{2}\sin\left(t\ln(2n) + \tfrac{3}{4}\pi\right)dn =$$

$$\left[\frac{(2n)((t-1)\sin(t\,\ln(2n)) + (t+1)\cos(t\,\ln(2n)))}{2(t^2+1)} + C\right]_1^\infty \text{ and}$$

$$\int_1^\infty \sqrt{2}\sin\left(t\ln(2n-1) + \tfrac{3}{4}\pi\right)dn =$$

$$\left[\frac{(2n-1)((t-1)\sin(t\,\ln(2n-1)) + (t+1)\cos(t\,\ln(2n-1)))}{2(t^2+1)} + C\right]_1^\infty.$$

$$\int_1^\infty (2n)^\sigma dn = \left[\frac{(2n)^{\sigma+1}}{2(\sigma+1)} + C\right]_1^\infty \text{ and}$$

$$\int_1^\infty (2n-1)^\sigma dn = \left[\frac{(2n-1)^{\sigma+1}}{2(\sigma+1)} + C\right]_1^\infty.$$

Dirichlet Sigma-Power Law as inequation derived from Eq. (10) is given by:

$$\left[\frac{(2n)((t-1)\sin(t\,\ln(2n)) + (t+1)\cos(t\,\ln(2n)))}{(2n-1)((t-1)\sin(t\,\ln(2n-1)) + (t+1)\cos(t\,\ln(2n-1)))} - \frac{(2n)^{\sigma+1}}{(2n-1)^{\sigma+1}}\right]_1^\infty$$

$$\neq 0 \hspace{5cm} (11)$$

Intended derivation of Dirichlet Sigma-Power Law as equation and inequation have been successful. *The proof is now complete for Proposition 3.2*□.

Proposition 3.3. Exact Dimensional analysis homogeneity at $\sigma = \frac{1}{2}$ in Dirichlet SigmaPower Law as equation and inequation is (respectively) indicated by $\sum$(all fractional exponents) = whole number '1' and '3'.

Proof. Dirichlet Sigma-Power Law as equation for $\sigma = \frac{1}{2}$ value is given by:

$$\frac{1}{2t^2 + \frac{1}{2})} \cdot \left[(2n)^{\frac{1}{2}}\left((t - \tfrac{1}{2})\sin(t\,\ln(2n)) + (t + \tfrac{1}{2})\cos(t\,\ln(2n))\right)\right.$$

364

$-(2n-1)^{\frac{1}{2}}((t-\frac{1}{2})sin(t\ ln(2n-1)) + (t+\frac{1}{2})cos(t\ ln(2n-1)))\}^{\infty}_{1} = 0$ (12)

Respectively evaluation of definite integrals Eq. (12), Eq. (24) and Eq. (26) using limit as n $\rightarrow$ +∞ for $0 < t < +\infty$ enable countless computations resulting in t values for CIS of nontrivial zeros, Gram[y=0] points and Gram[x=0] points. We illustrate this for Eq. (12) as expanded antiderivative [depicted as linear combination of sine and cosine waves: a sin x + b cos x = c sin (x + φ) with c = $\sqrt{(a^2 + b^2)}$ and φ = atan2 (b, a) = $tan^{-1}(\frac{b}{a})$ for a > 0].

$(2\infty)^{\frac{1}{2}} sin((t\ ln2\infty) + tan^{-1}(\frac{t+\frac{1}{2}}{t-\frac{1}{2}}) - (2\infty - 1)^{\frac{1}{2}} sin((t\ ln2\infty-$

$1) + tan^{-1}(\frac{t+\frac{1}{2}}{t-\frac{1}{2}}) - 2^{\frac{1}{2}} sin((t\ ln2) + tan^{-1}(\frac{t+\frac{1}{2}}{t-\frac{1}{2}}) + \frac{t+\frac{1}{2}}{2t^2+\frac{1}{2}} = 0$

(2∞) and $(2\infty - 1)$ involve exponent $\frac{1}{2}$, sin and ln functions. At relevant t values for all nontrivial zeros, (first term - second term) = (- third term + fourth term).

Dirichlet Sigma-Power Law as inequation for $\sigma = \frac{1}{2}$ value is given by:

$$\left[\frac{(2n)((t-1)sin(t\ ln(2n))+(t+1)cos(t\ ln(2n)))}{(2n-1)((t-1)sin(t\ ln(2n-1))+(t+1)cos(t\ ln(2n-1)))} - \frac{(2n)^{\frac{1}{2}}}{(2n-1)^{\frac{1}{2}}}\right]^{\infty}_{1} \neq 0$$

(13)

$\sum$(all fractional exponents) as $2(1 - \sigma) =$ whole number '1' for Eq. (12) and $2(\sigma + 1) =$ whole number '3' for Eq. (13). These findings signify presence of complete set nontrivial zeros for Eq. (12) and Eq. (13). *The proof is now complete for Proposition 3.3□.*

Corollary 3.4. Inexact Dimensional analysis homogeneity at $\sigma \neq \frac{1}{2}$ [illustrated using $\sigma = \frac{2}{5}$] in Dirichlet Sigma-Power Law as equation and inequation is (respectively) indicated by $\sum$(all fractional exponents) = fractional number '$\neq$ 1' and '$\neq$ 3'.

Proof. Dirichlet Sigma-Power Law as equation for $\sigma = \frac{2}{5}$ value is given by:

$$\frac{1}{2t^2+\frac{18}{25}} \cdot \left[(2n)^{\frac{3}{5}}\left((t-\frac{3}{5})sin(t\ ln(2n)) + (t+\frac{3}{5})cos(t\ ln(2n))\right)\right.$$

$$-(2n-1)^{\frac{3}{5}}((t-\frac{3}{5})sin(t\ ln(2n-1)) + (t+\frac{3}{5})cos(t\ ln(2n-$$

$$\left.1)))\right\}_1^\infty = 0 \qquad (14)$$

Dirichlet Sigma-Power Law as inequation for $\sigma = \frac{2}{5}$ value is given by:

$$\left[\frac{(2n)((t-1)sin(t\ ln(2n))+(t+1)cos(t\ ln(2n)))}{(2n-1)((t-1)sin(t\ ln(2n-1))+(t+1)cos(t\ ln(2n-1)))} - \frac{(2n)^{\frac{7}{5}}}{(2n-1)^{\frac{7}{5}}}\right]_1^\infty \neq 0$$

$$(15)$$

$\sum$(all fractional exponents) as 2(1 - σ) = fractional number '$\neq$ 1' for Eq. (14) and 2(σ + 1) = fractional number '$\neq$ 3' for Eq. (15). These findings signify absence of complete set nontrivial zeros for Eq. (14) and Eq. (15). The proof is now complete for Corollary 3.4□.

4 Rigorous proof for Riemann hypothesis summarized as Theorem Riemann I – IV

$\zeta(s) = \frac{1}{s-1} + \frac{1}{2} + 2\int_0^\infty \frac{sin(s\ arctan\ t)}{(1+t^2)^{\frac{1}{2}}(e^{2\pi t}-1)}dt$ is integral relation (cf. Abel Plana summation formula[3], [4]) for all s $\in$ C and s $\neq$ 1. This integral is insufficient for our purpose as it involves integration w.r.t. t [instead of n] for

366

$\zeta(s)$ [instead of $\eta(s)$]. Rigorous proof for Riemann hypothesis is summarized by Theorem Riemann I – IV. One could obtain this proof with only using Dirichlet Sigma-Power Law [solely] as equation. For completeness and clarification of this proof, we supply following important mathematical arguments.

For $0 < \sigma < 1$, then $0 < 2(1-\sigma) < 2$. The only whole number between 0 and 2 is '1' which coincide with $\sigma = \frac{1}{2}$. When $0 < \sigma < \frac{1}{2}$ and $\frac{1}{2} < \sigma < 1$, then $0 < 2(1-\sigma) < 1$ and $1 < 2(1-\sigma) < 2$.

For $0 < \sigma < 1$, $2 < 2(\sigma + 1) < 4$. The only whole number between 2 and 4 is '3' which coincide with $\sigma = \frac{1}{2}$. When $0 < \sigma < \frac{1}{2}$ and $\frac{1}{2} < \sigma < 1$, then $2 < 2(\sigma +1) < 3$ and $3 < 2(\sigma +1) < 4$.

Legend: R = all real numbers. For $0 < \sigma < 1$, σ consist of $0 < R < 1$. For $0 < 2(1-\sigma) < 2$ and $2 < 2(\sigma +1) < 4$, $2(1-\sigma)$ and $2(\sigma +1)$ must (respectively) consist of $0 < R < 2$ and $2 < R < 4$. An important caveat is that previously used phrases such as "fractional exponent σ" and "$\sum$(all fractional exponents) = whole number '1' [or '3'] and fractional number '$\neq$ 1' [or '$\neq$ 3']", although not incorrect *per se*, should respectively be replaced by "real number exponent σ" and "$\sum$(all real number exponents) = whole number '1' [or '3'] and real number '$\neq$ 1' [or '$\neq$ 3']^[5]" for complete accuracy. We apply this caveat to Theorem Riemann I – IV.

Footnote 5: As whole numbers $\subset$ real numbers, one could also depict this phrase as "$\sum$(all real number exponents) = real number '1' [or '3'] and real number '$\neq$ 1' [or '$\neq$ 3']".

Theorem Riemann I. Derived from *proxy* Dirichlet eta function, "simplified" Dirichlet eta function will exclusively contain *de novo* property for actual location [but not actual positions] of all nontrivial zeros.

Proof. The phrase "actual location [but not actual positions] of all nontrivial zeros" can be validly shortened to "actual location of all nontrivial zeros" as used in Theorem Riemann II, III and IV. *The proof for Theorem Riemann I is now complete as it successfully incorporates proof for Lemma 3.1□.*

Theorem Riemann II. Dirichlet Sigma-Power Law [in continuous (integral) format] as equation and inequation which are both derived from "simplified" Dirichlet eta function [in discrete (summation) format] will exclusively manifest exact DA homogeneity in equation and inequation only when real number exponent $\sigma = \frac{1}{2}$.

Proof. *The proof for Theorem Riemann II is now complete as it successfully incorporates proofs from Proposition 3.2 on derivation for equation and inequation of Dirichlet Sigma-Power Law [with both containing de novo property for "actual location of all nontrivial zeros"] and Proposition 3.3 on manifestation of exact DA homogeneity in Dirichlet Sigma-Power Law as equation and inequation when real number exponent $\sigma = \frac{1}{2}$□.*

Theorem Riemann III. Real number exponent $\sigma = \frac{1}{2}$ in Dirichlet Sigma-Power Law as equation and inequation satisfying exact DA

368

homogeneity is identical to σ variable in Riemann hypothesis which propose σ to also have exclusive value of $\frac{1}{2}$ (representing critical line) for "actual location of all nontrivial zeros", thus fully supporting Riemann hypothesis to be true with further clarification by Theorem Riemann IV.

Proof. Since s = σ ± ιt, complete set of nontrivial zeros which is defined by η(s) = 0 is exclusively associated with one (and only one) particular η(σ ± ιt) = 0 value solution, and by default one (and only one) particular σ [conjecturally] = $\frac{1}{2}$ value solution. When performing exact DA homogeneity on Dirichlet Sigma-Power Law as equation and inequation [with both containing *de novo* property for "actual location of all nontrivial zeros"], the phrase "If real number exponent σ has exclusively $\frac{1}{2}$ value, only then will exact DA homogeneity be satisfied" implies one (and only one) possible mathematical solution. Theorem Riemann III reflect Theorem Riemann II on presence of exact DA homogeneity for σ = $\frac{1}{2}$ in Dirichlet Sigma-Power Law as equation and inequation. This Law has identical σ variable as that referred to by Riemann hypothesis [whereby σ here uniquely refer to critical line]. *The proof for Theorem Riemann III is now complete as it independently refers to simultaneous association of confirmed (i) solitary σ = $\frac{1}{2}$ value in Dirichlet Sigma-Power Law as equation and inequation satisfying exact DA homogeneity and (ii) critical line defined by solitary σ = $\frac{1}{2}$ value being the "actual location [but with no request to determine actual positions]" of all nontrivial zeros as proposed in original Riemann hypothesis□.*

Theorem Riemann IV. Condition 1. All $\sigma \neq \frac{1}{2}$ values (non-critical lines), viz. $0 < \sigma < \frac{1}{2}$ and $\frac{1}{2} < \sigma < 1$ values, exclusively does not contain "actual location of all nontrivial zeros" [manifesting *de novo* inexact DA homogeneity in equation and inequation], together with Condition 2. One (and only one) $\sigma = \frac{1}{2}$ value (critical line) exclusively contains "actual location of all nontrivial zeros" [manifesting *de novo* exact DA homogeneity in equation and inequation], fully support Riemann hypothesis to be true when these two mutually inclusive conditions are met.

Proof. Condition 2 Theorem Riemann IV simply reflect proof from Theorem Riemann III [incorporating Proposition 3.3] for "actual location of all nontrivial zeros" exclusively on critical line manifesting *de novo* exact DA homogeneity $\sum$(all real number exponents) = whole number '1' for equation [or '3' for inequation]. *The proof for Condition 2 Theorem Riemann IV is now complete*□. Corollary 3.4 confirms *de novo* inexact DA homogeneity manifested as $\sum$(all real number exponents) = real number '$\neq 1$' for equation [or '$\neq 3$' for inequation] by all $\sigma \neq \frac{1}{2}$ values (non-critical lines) that are exclusively not associated with "actual location of all nontrivial zeros". Applying inclusion-exclusion principle: Exclusive presence of nontrivial zeros on critical line for Condition 2 Theorem Riemann IV implies exclusive absence of nontrivial zeros on non-critical lines for Condition 1 Theorem Riemann IV. *The proof for Condition 1 Theorem Riemann IV is now complete*□.

We logically deduce that explicit mathematical explanation why presence and absence of nontrivial zeros[6] should (respectively) coincide precisely with $\sigma = \frac{1}{2}$ and $\sigma \neq \frac{1}{2}$ [literally the Completely Predictable meta-properties ('overall' *complex properties*)] will require "complex" mathematical arguments. Attempting to provide explicit mathematical explanation with "simple" mathematical arguments would intuitively mean nontrivial zeros have to be (incorrectly and impossibly) treated as Completely Predictable entities.

Footnote 6: Completely Predictable meta-properties for Gram and virtual Gram points equating to "Presence of Gram[y=0] and Gram[x=0] points, and virtual Gram[y=0] and virtual Gram[x=0] points (respectively) coincide precisely with $\sigma = \frac{1}{2}$, and $\sigma \neq \frac{1}{2}$".

5 Conclusions

In our Hybrid method of Integer Sequence classification, a formula is either non-Hybrid or Hybrid integer sequence. Inequation with two 'necessary' Ratio (R) or equation with one 'unnecessary' R contains non-Hybrid integer sequence. Equation with one 'necessary' R contains Hybrid integer sequence. "In the limit" Hybrid integer sequence approach unique Position X, it becomes non-Hybrid integer sequence for all Positions $\geq$ Position X.

Consider kinetic energy (KE) in MJ with m_o = rest mass in kg and v = velocity in ms^{-1}. In classical mechanics concerning low velocity with v $<<$ c, Newtonian KE $= \frac{1}{2}m_o v^2$. In relativistic mechanics concerning high velocity with v $\geq$ 0.01c, Relativistic KE $= \frac{m_o c^2}{\sqrt{(1-(v^2/c^2))}} - m_o c^2$. Obtained from the later by binomial approximation or by taking first two terms of

371

Taylor expansion for reciprocal square root, the former approximates the later well at low speed.

We arbitrarily divide DA homogeneity into inexact DA homogeneity for ["<100% accuracy"] Newtonian KE and exact DA homogeneity for ["100% accuracy"] Relativistic KE. "In the limit" ['<100% accuracy'] Newtonian KE at low speed approach ['100% accuracy'] Relativistic KE at high speed, we achieve *perfection*.

Analogy: "In the limit" all three version of Dirichlet Sigma-Power Laws for Gram[y=0] points, Gram[x=0] points and nontrivial zeros as '*<100% accuracy*' inequations approach *perfection* as '*100% accuracy*' equations, compliance with inexact DA homogeneity becomes compliance with exact DA homogeneity. We note R1 terms in all inequations contain (2n) and (2n-1) 'base quantities' but these are not endowed with fractional exponent ($\sigma+1$) as relevant 'unit of measurement'. As Incompletely Predictable problems, we gave relatively elementary proof of Riemann hypothesis and explain closely related Gram points whereby various "meta-properties" such as exact and inexact DA homogeneity occur in (respectively) equations and inequations of relevant Dirichlet Sigma-Power Laws. Harnessed key benefit from successful proof for Riemann hypothesis is often stated as "With this one solution, we have proven five hundred theorems or more at once". This apply to important theorems in number theory that rely on properties of Riemann zeta function or Dirichlet eta function such as location of trivial and nontrivial zeros. E.g., we delineate prime number theorem by prime counting function $\pi(x)$ [which is defined as number of primes $\leq$ x].

Appendix A: Definitions and Supplementary materials

Exposition on definitions and related commentaries is crucial to help solve Riemann hypothesis and explain closely related Gram points as Incompletely Predictable problems.

Segment A1. *Completely Predictable and Incompletely Predictable numbers*

Completely Unpredictable numbers arising from totally chaotic physical processes give rise to countable infinite set (CIS) of measured true random numbers. In this sense, computational pseudorandom number generators using deterministic logic are not regarded as sources for true random numbers. Two types of Predictable numbers: CIS of Completely, and CIS of Incompletely Predictable numbers with former "contained" in *simple* equations or algorithms obeying '*Simple* Elementary Fundamental Laws', and later "contained" in *complex* equations or algorithms obeying '*Complex* Elementary Fundamental Laws'.

A Completely, and Incompletely Predictable number is locationally defined as a number whose position is *independently* determined by simple calculations using simple equation or algorithm without, and *dependently* by complex calculations using complex equation or algorithm with needing to know related positions of all preceding numbers in neighborhood. Both types of Predictable number exist as either rational [integers or fractions of two integers] numbers (Q) or irrational [algebraic or transcendental] numbers (R – Q). A well-defined set of R – Q will twice obey relevant location definition in CIS R – Q themselves and in CIS numerical digits after decimal point of each R – Q.

97 is an Incompletely Predictable number whose precise position is determined by computing positions of all preceding 24 prime numbers (P) using complex algorithm Sieve of Eratosthenes to conclude that 97 is 25th P. Calculated using simple algorithm, 97 is also [i = (97+1)/2] 49th odd number (O) which is a Completely Predictable number. 98 & 99 are respectively [i = 98/2] 49th even number (E) & [i = (99+1)/2] 50th O which are Completely Predictable numbers calculated using simple algorithm. Determined indirectly using complex algorithm Sieve of Eratosthenes, 98 & 99 are respectively also 72nd & 73rd composite numbers (C) which are Incompletely Predictable numbers.

Computing Riemann zeta function (or specifically its *proxy* Dirichlet eta function) and Sieve of Eratosthenes will, respectively, supply Incompletely Predictable nontrivial zeros, Gram[y=0] & Gram[x=0] points and P & C. CIS of nontrivial zeros (denoted by imaginary part parameter t) = CIS of transcendental numbers = 14.134725, 21.022040, 25.010858, 30.424876, 32.935062, 37.586178,... [rounded off to six decimal places]. CIS of all P = Countable Finite Set (CFS) of all even P + CIS of all odd P = 2, 3, 5, 7, 11, 13,... whereby P '2' when treated as E is also regarded as a Completely Predictable number.

The three sets of nontrivial zeros, Gram[y=0] points and Gram[x=0] points, respectively, will *dependently* constitute three sets of Origin intercepts (or simultaneous x- & y-axes intercepts), x-axis

374

intercepts and y-axis intercepts. Traditional 'Gram points' [see Segment A2 below] are x-axis intercepts with choice of index 'n' for 'Gram points' historically chosen such that first 'Gram point' corresponds to t value which is larger than (first) nontrivial zero located at t = 14.134725. By convention, first six Gram[y=0] points will occur with following values [rounded off to six decimal places]: at n = -3, t = 0; at n = -2, t = 3.436218; at n = -1, t = 9.666908; at n = 0, t = 17.845599; at n = 1, t = 23.170282; at n = 2, t = 27.670182.

The two sets of P 2, 3, 5, 7, 11, 13,... and C 4, 6, 8, 9, 10, 12,... will *dependently* constitute set of natural numbers (N) 1, 2, 3, 4, 5, 6,... minus first N '1'. Whole numbers (W) = N plus '0'. '0' & '1' are special numbers being neither P nor C as they represent nothingness (zero) and wholeness (one), and the idea of having factors for '0' & '1' is meaningless. Treating '0' & '1' here as Completely or Incompletely Predictable numbers is also meaningless.

CIS of numbers derived from well-defined simple/complex algorithms or equations are "dual numbers" displayed as Completely & Incompletely Predictable number. Examples of Q '2' as P (& E), '97' as P (& O), '98' as C (& E) and '99' as C (& O) are described above. Examples of R – Q are described next. First & only negative Gram[y=0] point (by convention at n = -3) with Completely Predictable y = 0 value is obtained by substituting Completely Predictable t = 0 resulting in $\zeta(\frac{1}{2} + \imath t) = \zeta(\frac{1}{2}) = $ -1.4603545, an Incompletely Predictable transcendental number [rounded off to seven decimal places] calculated as a limit similar to limit for Euler-Mascheroni constant or Euler gamma – its precise (1st) position can only be determined by computing positions of all preceding (nil)

Gram[y=0] points in this case. With exception of this first Gram[y=0] point, all t values from Gram[y=0] points, Gram[x=0] points, and nontrivial zeros (Gram[x=0,y=0] points) are Incompletely Predictable transcendental numbers – these are respectively associated with Completely Predictable x = 0, y = 0, and simultaneous x = 0 & y = 0 values. First 'Gram point' (by convention at n = 0 & is associated with Completely Predictable x = 0 value from Incompletely Predictable t = 17.845599 substitution) is actually the 4th Gram[y=0] point whose precise (4th) position can only be determined by computing positions of all preceding (three) Gram[y=0] points in this case. First nontrivial zero associated with simultaneous x = 0 & y = 0 value [equating to ζ(s) = 0 whereby s = σ + ıt = $\frac{1}{2}$ + it] is Completely Predictable occurring with Incompletely Predictable t = 14.134725 value substitution – its precise (1st) position can only be determined by computing positions of all preceding (nil) nontrivial zeros in this case.

Remark A.1. Countable finite set (CFS) of Completely Predictable *simple properties* intrinsically present in simple equations or algorithms help us solve Completely Predictable problems containing countable infinite set (CIS) of Completely Predictable numbers; whereas CFS of Completely Predictable *complex properties* intrinsically present in complex equations or algorithms help us solve Incompletely Predictable problems containing CIS of Incompletely Predictable numbers.

Simple properties are inferred from a phrase like: "...the simple equation or algorithm by itself will intrinsically incorporate actual location [and actual positions] of all Completely Predictable numbers". Solving Completely Predictable problems endowed with simple properties which are amendable to *simple* treatments using *usual* mathematical tools such as Calculus will result in their 'Simple Elementary Fundamental Laws'-based solutions. Complex properties are inferred from a phrase like: "...the complex equation or algorithm by itself will intrinsically incorporate actual location [but not actual positions] of all Incompletely Predictable numbers". Solving Incompletely Predictable problems endowed with complex properties which are amendable to *complex* treatments using *unusual* mathematical tools such as deriving complex equation Dirichlet Sigma-Power Law as well as using *usual* mathematical tools such as Calculus will result in their 'Complex Elementary Fundamental Laws'-based solutions.

Consider x for real number (R) values ≥ 1. Let y be Set R such that (simple equation) y = 2x or y = 2x - 1. This simple equation will "contain" the complete uncountable infinite set (UIS) of R [straight line of infinite length] commencing from Cartesian point (x=1, y=2) or (x=1, y=1). Computing y = 2x or y = 2x - 1 an infinite number of times – a *mathematical impasse* – will not *per se* result in its 'Simple Elementary Fundamental Laws'-based solution for gradient or slope = 2. This gradient (simple property) is obtained by trigonometrically calculating tangent of y = 2x or y = 2x - 1 straight line which = 2 or analyzing y = 2x or y = 2x - 1 equation using Differential Calculus viz. dy/dx = d(2x)/dx or d(2x-1)/dx = 2. Note: applying Integral Calculus from Fundamental Theorem of Calculus to y = 2x or y = 2x - 1 for interval [1, +∞], viz. $\int_1^\infty (2x)dx$ or $\int_1^\infty (2x-1)dx = [x^2 + C]_1^\infty$ or $[x^2 + x + C]_1^\infty = (\infty^2 + C) - (1^2 + C)$ or $(\infty^2 - \infty + C) - (1^2 - 1 + C)$ = ∞ result in 'Simple Elementary Fundamental Laws'-based solution

377

for area (simple property) of infinite size enclosed by the straight line and x-axis.

Consider x≥1 integer number (Z) values for (simple algorithm) y = 2x or y = 2x - 1. We obtain "contained" complete Set E or Set O. Computing E or O infinitely often – a *mathematical impasse* – will not *per se* result in 'Simple Elementary Fundamental Laws'-based solution for gap between any two consecutive E (E gap) or O (O gap) will both = 2. This gradient-equivalent E gaps or O gaps (simple property) is obtained by transforming those algorithms from their "discrete" into "continuous" format [viz. "discrete" $\Delta x \rightarrow 1$ into "continuous" $\Delta x \rightarrow 0$] resulting in their gradients using either tangent method or Differential Calculus method. Then E or O gaps, both = 2, is numerically identical and mathematically equivalent to relevant gradients, both also = 2. Similar method of transforming from "discrete" into "continuous" format to help solve Riemann hypothesis involves applying Riemann integral to discrete-like equation of "simplified" Dirichlet eta function (in summation format) to obtain Dirichlet Sigma-Power Law [which is the continuous-like equation of "simplified" Dirichlet eta function (in integral format)].

Similar to Incompletely Predictable 'varying gaps' [equivalent to 'varying gradients'] between consecutive P (P gaps) & consecutive C (C gaps) [relevant to research on Polignac's and Twin prime conjectures], we have Incompletely Predictable 'varying gaps' [equivalent to 'varying gradients'] between consecutive nontrivial zeros (nontrivial zero gaps), consecutive Gram[y=0] points (Gram[y=0] points gaps) & consecutive Gram[x=0] points (Gram[x=0] points gaps). These 'varying gaps' or 'varying gradients' (complex properties) are geometrically related to different shapes/sizes of spirals depicted in Figure 2.

Segment A2. *Gram's Law and traditional 'Gram points'*

Named after Danish mathematician Jørgen Pedersen Gram (June 27, 1850 – April 29, 1916), traditional 'Gram points' (or Gram[y=0] points which are **x-axis intercepts** shown in figure above) are other conjugate pairs values on critical line defined by $Im\{\zeta(\frac{1}{2} \pm \imath t)\} = 0$. They obey Gram's Rule and Rosser's Rule with interesting characteristic properties as outlined by our brief exposition below.

Z function is used to study Riemann zeta function on critical line. Defined in terms of Riemann-Siegel theta function & Riemann zeta function by $Z(t) = e^{\imath\theta(t)}\zeta(\frac{1}{2} + \imath t)$ whereby $\theta(t) = arg(\Gamma \ (\frac{(2\imath t+1)}{4})) - \frac{ln\pi}{2}t$; it is also called Riemann-Siegel Z function, Riemann-Siegel zeta function, Hardy function, Hardy Z function, & Hardy zeta function.

The algorithm to compute $Z(t)$ is called Riemann-Siegel formula. Riemann zeta function on critical line, $\zeta(\frac{1}{2} + \imath t)$, will be real when $sin(\theta(t)) = 0$. Positive real values of t where this occurs are called 'Gram points' and can also be described as points where $\frac{\theta(t)}{\pi}$ is an integer. Real part of this function on critical line tends to be positive, while imaginary part alternates more regularly between positive & negative values. That means sign of $Z(t)$ must be opposite to that of sine function most of the time, so one would expect nontrivial zeros of $Z(t)$ to alternate with zeros of sine term, i.e. when θ takes on integer multiples of π. This turns out to hold most of the time and is known as Gram's Rule (Law) – a law which is violated infinitely often though. Thus Gram's Law is statement that nontrivial zeros of $Z(t)$ alternate with 'Gram

379

points'. 'Gram points' which satisfy Gram's Law are called 'good', while those that do not are called 'bad'. A Gram block is an interval such that its very first & last points are good 'Gram points' and all 'Gram points' inside this interval are bad. Counting nontrivial zeros then reduces to counting all 'Gram points' where Gram's Law is satisfied and adding the count of nontrivial zeros inside each Gram block. With this process we do not have to locate nontrivial zeros, and we just have to accurately compute $Z(t)$ to show that it changes sign.

Ratio Study and Inequations

A mathematical equation, containing one or more variables, is a statement that values of two ['left-hand side' (LHS) and 'right-hand side' (RHS)] mathematical expressions is related as equality: LHS = RHS; or as inequalities: LHS < RHS, LHS > RHS, LHS ≤ RHS, or LHS ≥ RHS. A ratio is one mathematical expression divided by another. The term 'unnecessary' Ratio (R) for any given equation is explained by two examples: (1) LHS = RHS and with rearrangement, 'unnecessary' R is given by $\frac{LHS}{RHS} = 1$ or $\frac{RHS}{LHS} = 1$; and (2) LHS > RHS and with rearrangement, 'unnecessary' R is given by $\frac{LHS}{RHS} > 1$ or $\frac{RHS}{LHS} < 1$.

Consider exponent $y \in$ all **R** values & base $x \in$ **R**≥0 values for mathematical expression $\frac{x}{y}$. Equations such as $x^1 = x$, $x^0 = 1$ & $0^y = 0$ are all valid. Simultaneously letting both x & y = 0 is an incorrect mathematical action because xy as function of two-variables is not continuous & is thus undefined at Origin. But if we elect to intentionally carry out this "balanced" action [equally] on x & y, we obtain (simple) inequation $0^0 \neq 1$ with associated perpetual obeyance of '=' equality symbol in x^y for all applicable **R** values

380

except when both x & y = 0. The Number '1' value in this inequation is justified by two arguments: I. Limit of x^y value as both x & y tend to zero (from right) is 1 [thus fully satisfying criterion "x^y is right continuous at the Origin"]; and II. Expression x^y is product of x with itself y times [and thus x^0, the "empty product", should be 1 (no matter what value is given to x)]. Mathematical operator 'summation' must obey the law: We can break up a summation across a sum or difference but not across a product or quotient viz, factoring a sum of quotients into a corresponding quotient of sums is an incorrect mathematical action. But if we elect to carry out this action equally on LHS & RHS products or quotients in a suitable equation, we obtain two (unique) 'necessary' R denoted by R1 for LHS and R2 for RHS whereby R1 ≠ R2 relationship will always hold. We define 'Ratio Study' as intentionally performing this incorrect [but "balanced"] mathematical action on suitable equation [equivalent to one (non-unique) 'unnecessary' R] to obtain its inequation [equivalent to two (unique) 'necessary' R]. Let C denote complex numbers. Set C is a field (but not an ordered field). Thus it is not possible to define a relation between two given (z1 & z2) C as z1 < z2 since inequality operation here is not compatible with addition and multiplication. But performing Ratio Study to obtain inequations involving C does not involve defining a relation between two C.

Appendix B: Prerequisite lemma, corollary and propositions for Gram[y=0] and Gram[x=0] points

For Gram[y=0] and Gram[x=0] points (and corresponding virtual Gram[y=0] and virtual Gram[x=0] points with totally different values), we apply a parallel procedure carried out on nontrivial zeros but only depict abbreviated treatments and discussions.

Lemma B.1. "Simplified" Gram[y=0] and Gram[x=0] points-Dirichlet eta functions are derived directly from Dirichlet eta function with Euler formula application and (respectively) they will intrinsically incorporate actual location [but not actual positions] of all Gram[y=0] and Gram[x=0] points.

Proof. For Gram[y=0] points, the equivalent of Eq. (4) and Eq. (6) are respectively given by Eq. (16) and Eq. (17) below.

$$\sum \text{ReIm}\{\eta(s)\} = \text{Re}\{\eta(s)\}+0, \text{ or simply } \text{Im}\{\eta(s)\} = 0 \quad (16)$$

$$\sum_{n=1}^{\infty} (2n)^{-\sigma}\sin(t\ln(2n)) = \sum_{n=1}^{\infty} (2n-1)^{-\sigma}\sin(t\ln(2n-1))$$

$$\sum_{n=1}^{\infty}(2n)^{-\sigma}\sin(t\ln(2n)) - \sum_{n=1}^{\infty}(2n-1)^{-\sigma}\sin(t\ln(2n-1)) = 0 \quad (17)$$

For Gram[x=0] points, the equivalent of Eq. (4) and Eq. (6) are respectively given by Eq. (18) and Eq. (19) below.

$$\sum \text{ReIm}\{\eta(s)\} = 0+\text{Im}\{\eta(s)\}, \text{ or simply } \text{Re}\{\eta(s)\} = 0 \quad (18)$$

$$\sum_{n=1}^{\infty} (2n)^{-\sigma}\cos(t\ln(2n)) = \sum_{n=1}^{\infty} (2n-1)^{-\sigma}\cos(t\ln(2n-1))$$

$$\sum_{n=1}^{\infty}(2n)^{-\sigma}\cos(t\ln(2n)) - \sum_{n=1}^{\infty}(2n-1)^{-\sigma}\cos(t\ln(2n-1)) = 0 \quad (19)$$

Eq. (17) and Eq. (19) being the "simplified" Gram[y=0] and Gram[x=0] points-Dirichlet eta functions derived directly from $\eta(s)$ will intrinsically incorporate *actual location [but not actual positions]* of (respectively) all Gram[y=0] and Gram[x=0] points. *The proof is now complete for Lemma B.1*□.

Proposition B.2. Gram[y=0] and Gram[x=0] points-Dirichlet Sigma-Power Laws in continuous (integral) format given as equations and inequations can both be (respectively) derived directly from "simplified" Gram[y=0] and Gram[x=0] points-Dirichlet eta functions in discrete (summation) format with Riemann integral application.

Proof. Antiderivatives below using (2n) parameter help obtain all subsequent equations: first two for Gram[y=0] points and second two for Gram[x=0] points.

$$\int_1^\infty (2n)^{-\sigma} \sin(t \ln(2n)) dn = \left[-\frac{(2n)^{1-\sigma}((\sigma-1)\sin(t \ln(2n)) + t \cos(t \ln(2n)))}{2\left(t^2 + (\sigma-1)^2\right)} + C \right]_1^\infty$$

$$\int_1^\infty \sin(t \ln(2n)) dn = \left[\frac{(2n)(\sin(t \ln(2n)) - t \cos(t \ln(2n)))}{2(t^2+1)} + C \right]_1^\infty$$

$$\int_1^\infty (2n)^{-\sigma} \cos(t \ln(2n)) dn = \left[\frac{(2n)^{1-\sigma}(t \sin(t \ln(2n)) - (\sigma-1) \cos(t \ln(2n)))}{2\left(t^2 + (\sigma-1)^2\right)} + C \right]_1^\infty$$

$$\int_1^\infty \cos(t \ln(2n)) dn = \left[\frac{(2n)(t \sin(t \ln(2n)) + \cos(t \ln(2n)))}{2(t^2+1)} + C \right]_1^\infty$$

For Gram[y=0] points-Dirichlet Sigma-Power Law, the equivalent of Eq. (9) and Eq. (11) are respectively given by Eq. (20) as equation and Eq. (21) as inequation.

$$-\frac{1}{2\left(t^2 + (\sigma-1)^2\right)} \cdot [(2n)^{1-\sigma}((\sigma-1)\sin(t \ln(2n)) + t \cos(t \ln(2n))) -$$

$$(2n-1)^{1-\sigma}((\sigma-1)\sin(t \ln(2n-1)) + t \cos(t \ln(2n-1)))]\big|_1^\infty = 0 \quad (20)$$

$$\left[\frac{(2n)(\sin(t \ln(2n)) - t \cos(t \ln(2n)))}{(2n-1)(\sin(t \ln(2n-1)) - t \cos(t \ln(2n-1)))} - \frac{(2n)^{\sigma+1}}{(2n-1)^{\sigma+1}} \right]_1^\infty \neq$$

$$0 \qquad\qquad\qquad (21)$$

For Gram[x=0] points-Dirichlet Sigma-Power Law, the equivalent of Eq. (9) and Eq. (11) are respectively given by Eq. (22) as equation and Eq. (23) as inequation.

$$\frac{1}{2\left(t^2+(\sigma-1)^2\right)} \cdot [(2n)^{1-\sigma}\left(t\sin\left(t\ln\left(2n\right)\right)-(\sigma-1)\cos\left(t\ln\left(2n\right)\right)\right)-$$

$$(2n-1)^{1-\sigma}\left(t\sin\left(t\ln\left(2n-1\right)\right)-(\sigma-1)\cos\left(t\ln\left(2n-1\right)\right)\right)]_1^{\infty}=0$$

$$(22)$$

$$\left[\frac{(2n)(t\sin\left(t\ln\left(2n\right)\right)+\cos\left(t\ln\left(2n\right)\right))}{(2n-1)(t\sin\left(t\ln\left(2n-1\right)\right)+\cos\left(t\ln\left(2n-1\right)\right))}-\frac{(2n)^{\sigma+1}}{(2n-1)^{\sigma+1}}\right]_1^{\infty}\neq$$

$$0 \qquad\qquad\qquad\qquad (23)$$

Intended derivation of Gram[y=0] and Gram[x=0] points-Dirichlet Sigma-Power Laws as equations and inequations is successful. *The proof is now complete for Lemma B.2*□.

Proposition B.3. Exact Dimensional analysis homogeneity at $\sigma=\frac{1}{2}$ in Gram[y=0] and Gram[x=0] points-Dirichlet Sigma-Power Laws as equations and inequations are (respectively) indicated by $\sum$(all fractional exponents) = whole number '1' and '3'.

Proof. Gram[y=0] points-Dirichlet Sigma-Power Law as equation for $\sigma=\frac{1}{2}$ value is given by:

$$-\frac{1}{2t^2+\frac{1}{2}} \cdot [(2n)^{\frac{1}{2}}\left(t\cos(t\ln(2n))-\frac{1}{2}\sin(t\ln(2n))\right)-$$

$$(2n-1)^{\frac{1}{2}}\left(t\cos(t\ln(2n-1))-\frac{1}{2}\sin(t\ln(2n-1))\right)]_1^{\infty}=$$

$$0 \qquad\qquad\qquad\qquad (24)$$

Gram[y=0] points-Dirichlet Sigma-Power Law as inequation for $\sigma=\frac{1}{2}$ value is given by:

384

$$\left[\frac{(2n)\left(\sin\left(t\ln\left(2n\right)\right)-t\cos\left(t\ln\left(2n\right)\right)\right)}{(2n-1)\left(\sin\left(t\ln\left(2n-1\right)\right)-t\cos\left(t\ln\left(2n-1\right)\right)\right)}-\frac{(2n)^{\frac{3}{2}}}{(2n-1)^{\frac{3}{2}}}\right]_{1}^{\infty}\neq$$

$$0 \qquad\qquad\qquad\qquad\qquad\qquad (25)$$

Gram[x=0] points-Dirichlet Sigma-Power Law as equation for $\sigma=\frac{1}{2}$

value is given by: $\frac{1}{2t^2+\frac{1}{2}}\cdot\left[(2n)^{\frac{1}{2}}\left(t\,sin\left(t\,ln(2n)\right)+\frac{1}{2}cos\left(t\,ln(2n)\right)\right)\right.$

$$\left.-(2n-1)^{\frac{1}{2}}(t\,sin(t\,ln(2n-1))+\frac{1}{2}cos(t\,ln(2n-1)))\right\}_{1}^{\infty}=0 \quad (26)$$

Gram[x=0] points-Dirichlet Sigma-Power Law as inequation for σ

$=\frac{1}{2}$ value is given by:

$$\left[\frac{(2n)\left(t\sin\left(t\ln\left(2n\right)\right)+\cos\left(t\ln\left(2n\right)\right)\right)}{(2n-1)\left(t\sin\left(t\ln\left(2n-1\right)\right)+\cos\left(t\ln\left(2n-1\right)\right)\right)}-\frac{(2n)^{\frac{3}{2}}}{(2n-1)^{\frac{3}{2}}}\right]_{1}^{\infty}\neq$$

$$0 \qquad\qquad\qquad\qquad\qquad\qquad (27)$$

$\sum$(all fractional exponents) as $2(1-\sigma)$ = whole number '1' for Eqs. (24) and (26), and $2(\sigma+1)$ = whole number '3' for Eqs. (25) and (27). These findings signify presence of complete sets Gram[y=0] points for Eqs. (24) and (25) and Gram[x=0] points for Eqs. (26) and (27). *The proof is now complete for Proposition B.3□.*

Corollary B.4. Inexact Dimensional analysis homogeneity at $\sigma=\frac{1}{2}$

[illustrated using $\sigma=\frac{2}{5}$] in Gram[y=0] and Gram[x=0] points-Dirichlet Sigma-Power Laws as equations and inequations are (respectively) indicated by $\sum$(all fractional exponents) = fractional number '$\neq$ 1' and '$\neq$ 3'.

Proof. Gram[y=0] points-Dirichlet Sigma-Power Law as equation

for $\sigma=\frac{2}{5}$ value is given by:

$$-\frac{1}{2t^2+\frac{18}{25}}\cdot\left[(2n)^{\frac{3}{5}}\left(t\cos(t\ln(2n))-\frac{3}{5}\sin(t\ln(2n))\right)-\right.$$

$$\left.(2n-1)^{\frac{3}{5}}\left(t\cos(t\ln(2n-1))-\frac{3}{5}\sin(t\ln(2n-1))\right)\right]_1^\infty=$$

$$0 \qquad\qquad\qquad\qquad (28)$$

Gram[y=0] points-Dirichlet Sigma-Power Law as inequation for $\sigma=\frac{2}{5}$ value is given by:

$$\left[\frac{(2n)(\sin(t\ln(2n))-t\cos(t\ln(2n)))}{(2n-1)(\sin(t\ln(2n-1))-t\cos(t\ln(2n-1)))}-\frac{(2n)^{\frac{7}{5}}}{(2n-1)^{\frac{7}{5}}}\right]_1^\infty\neq$$

$$0 \qquad\qquad\qquad\qquad (29)$$

Gram[x=0] points-Dirichlet Sigma-Power Law as equation for $\sigma=\frac{2}{5}$ value is given by:

$$\frac{1}{2t^2+\frac{18}{25})}\cdot\left[(2n)^{\frac{3}{5}}\left(t\sin(t\,ln(2n))+\frac{3}{5}\cos(t\,ln(2n))\right)\right.$$

$$\left.-(2n-1)^{\frac{3}{5}}\left(t\sin(t\,ln(2n-1))+\frac{3}{5}\cos(t\,ln(2n-1))\right)\right]_1^\infty=0 \quad(30)$$

Gram[x=0] points-Dirichlet Sigma-Power Law as inequation for σ

$=\frac{2}{5}$ value is given by:

$$\left[\frac{(2n)(t\sin(t\ln(2n))+\cos(t\ln(2n)))}{(2n-1)(t\sin(t\ln(2n-1))+\cos(t\ln(2n-1)))}-\frac{(2n)^{\frac{7}{5}}}{(2n-1)^{\frac{7}{5}}}\right]_1^\infty\neq$$

$$0 \qquad\qquad\qquad\qquad (31)$$

$\sum$(all fractional exponents) as $2(1-\sigma)$ = fractional number '$\neq 1$' for Eqs. (28) and (30), and $2(\sigma+1)$ = fractional number '$\neq 3$' for Eqs. (29) and (31). These findings signify presence of complete sets virtual Gram[y=0] points for Eqs. (28) and (29) and virtual Gram[x=0] points for Eqs. (30) and (31). *The proof is now complete for Corollary B.4□.*

Appendix C: Hybrid method of Integer Sequence classification

The Hybrid method of Integer Sequence classification enables meaningful division of all integer sequences into either Hybrid or non-Hybrid integer sequences. My exotic A228186 [5] integer sequence was published on The On-line Encyclopedia of Integer Sequences website in 2013. It is the first ever [infinite length] Hybrid integer sequence synthesized from Combinatorics Ratio. In 'Position i' notation, let i = 0, 1, 2, 3, 4, 5,..., ∞ be complete set of natural numbers. A228186 "Greatest k > n such that ratio R < 2 is a maximum rational number with R = $\frac{\textit{Combinations With Repetition}}{\textit{Combinations Without Repetition}}$," is equal to [infinite length] non-Hybrid (usual garden-variety) integer sequence A100967 except for finite 21 'exceptional' terms at Positions 0, 11, 13, 19, 21, 28, 30, 37, 39, 45, 50, 51, 52, 55, 57, 62, 66, 70, 73, 77, and 81 with their values given by relevant A100967 terms plus 1. The first 49 terms [from Position 0 to Position 48] of A100967 [6] "Least k such that binomial(2k+1, k-n) ≥ binomial(2k, k)" are listed below: 3, 9, 18, 29, 44, 61, 81, 104, 130, 159, 191, 225, 263, 303, 347, 393, 442, 494, 549, 606, 667, 730, 797, 866, 938, 1013, 1091, 1172, 1255, 1342, 1431, 1524, 1619, 1717, 1818, 1922, 2029, 2138, 2251, 2366, 2485, 2606, 2730, 2857, 2987, 3119, 3255, 3394, and 3535. For those 21 'exceptional' terms: at Position 0, A228186 (= 4) is given by A100967 (= 3) + 1; at Position 11, A228186 (= 226) is given by A100967 (= 225) + 1; at Position 13, A228186 (= 304) is given by A100967 (= 303) + 1; at Position 19, A228186 (= 607) is given by A100967 (= 606) + 1; etc. Here is a useful concept: Commencing from Position 0 onwards "in the limit" that this Position approaches 82, A228186 Hybrid integer sequence becomes (and is identical to) A100967 non-Hybrid integer sequence for all Positions ≥ 82.

Acknowledgements I am indebted to Mr. Rodney Williams and Mr. Tony O'Hagan for reviewing this paper (dedicated to my daughter Jelena born 13 weeks early on May 14, 2012).

References

1. Hardy, G. H. (1914), Sur les Zeros de la Fonction $\zeta(s)$ de Riemann, *C. R. Acad. Sci. Paris, 158:* 1012 – 1014, JFM 45.0716.04 Reprinted in (Borwein et al. 2008)

2. Hardy, G. H.; Littlewood, J. E. (1921), The zeros of Riemann's zeta-function on the critical line, *Math. Z., 10* (34): 283 – 317, http://dx.doi:10.1007/BF01211614

3. Abel, N.H. (1823), Solution de quelques problmes laide dintgrales dfinies, *Magazin Naturvidensk*, 1: 55 – 68

4. Plana, G.A.A. (1820), 'Sur une nouvelle expression analytique des nombres Bernoulliens, propre exprimer en termes finis la formule gnrale pour la sommation des suites', *Mem. Accad. Sci. Torino*, 25: 403 – 418

5. Helpful, B (August 15, 2013), A228186, OEIS Foundation Inc. (2011), The On-Line Encyclopedia of Integer Sequences.

6. Noe, T (November 23, 2004), A100967, OEIS Foundation Inc. (2011), The On-Line Encyclopedia of Integer Sequences, https://oeis.org/A100967

Solving Incompletely Predictable problems Polignac's and Twin
Prime conjectures using Information-Complexity conservation

Dr. Bernhard Helpful

Published in viXra in 2019

Abstract Prime numbers are Incompletely Predictable numbers
calculated using complex algorithm Sieve of Eratosthenes. Involving
proposals that prime gaps and associated sets of prime numbers are infinite
in magnitude, Twin prime conjecture deals with even prime gap 2 and is a
subset of Polignac's conjecture which deals with all even prime gaps 2, 4, 6,
8, 10,.... Treated as Incompletely Predictable problems, we solve these
conjectures with research method Information-Complexity conservation to
get Plus Gap 2 Composite Number Continuous Law and Plus-Minus Gap 2
Composite Number Alternating Law.

Keywords Dimensional analysis, Incompletely Predictable problems,
Information-Complexity conservation, Plus Gap 2 Composite Number
Continuous Law, Plus-Minus Gap 2 Composite Number Alternating Law,
Polignac's and Twin prime conjectures

2010 Mathematics Subject Classification. 11A41, 11M26

1 Introduction

Uncountable complex numbers (C) include uncountable real numbers (R). R = countable rational numbers (Q) + uncountable irrational numbers (R − Q). R − Q = countable algebraic numbers + uncountable transcendental numbers. Q include countable integers (Z) which include countable whole numbers (W) which in turn include countable natural numbers (N). N is constituted by either countable even numbers (E) and countable odd numbers (O) or countable prime numbers (P), countable composite numbers (C) and Number '1'. Then (i) Set N = Set E + Set O, (ii) Set N = Set P + Set C + Number '1', and (iii) Set N ⊂ Set W ⊂ Set Z ⊂ Set Q ⊂ Set R ⊂ Set C.

With increasing magnitude, arbitrary Set X belongs to countable finite set (CFS), countable infinite set (CIS) or uncountable infinite set (UIS). Cardinality of Set X, $|X|$, measures the "number of elements" in Set X. E.g. Set even P has CFS of even P with $|\text{even P}| = 1$, Set N has CIS of N with $|N| = \aleph_0$, and Set R has UIS of R with $|R| = c$ (cardinality of the continuum). Respectively, CIS of P and C are *Incompletely Predictable numbers* dependently calculated directly and indirectly from *complex algorithm* Sieve of Eratosthenes. Involving proposals that prime gaps and associated sets of prime numbers are infinite in magnitude, Twin prime conjecture deals with even prime gap 2 and is a subset of Polignac's conjecture dealing with all even prime gaps 2, 4, 6, 8, 10,.... Activities to prove these open problems in number theory equate to solving *Incompletely Predictable problems*.

All claims arising from these activities are made meaningful by providing definitions on above mentioned terms. Respectively, an Incompletely (Completely) Predictable number is locationally defined as a number whose position is *dependently* (*independently*) determined by complex (simple) calculations using complex (simple) equation or algorithm with (without)

390

needing to know related positions of all preceding numbers in neighborhood. Simple properties are inferred from a phrase such as: "...simple equation or algorithm by itself will intrinsically incorporate actual location [and actual positions] of all Completely Predictable numbers". Solving Completely Predictable problems with simple properties amendable to *simple* treatments using *usual* mathematical tools such as Calculus result in 'Simple Elementary Fundamental Laws'-based solutions. Complex properties are inferred from a phrase such as: "...complex equation or algorithm by itself will intrinsically incorporate actual location [but not actual positions] of all Incompletely Predictable numbers". Solving Incompletely Predictable problems with complex properties amendable to *complex* treatments using *unusual* mathematical tools such as our novel research method Information-Complexity conservation as well as using *usual* mathematical tools such as Calculus result in 'Complex Elementary Fundamental Laws'-based solutions.

1.1 Dimensional analysis on Cardinality and "Dimensions"

For 'base quantities' such as *length*, *mass* and *time*; their fundamental SI 'units of measurement' are [respectively] given by meter (m), kilogram (kg) and second (s). The word 'dimension' is commonly used to denote 'units of measurement' in well-defined equations. Dimensional analysis (DA) is an analytic tool with resulting DA homogeneity and nonhomogeneity (respectively) denoting valid and invalid equation when 'units of measurements' are "balanced" and "unbalanced" across both sides of the equation. E.g. 2 m + 3 m = 5 m is a valid equation but 2 m + 3 kg = 5 mkg is an invalid equation.

We use "Dimensions" to denote well-defined Incompletely Predictable entities obtained from Information-Complexity conservation. Relevant

391

"Dimensions" *dependently* represent Number '1', P and C. Then *by default* any (sub)sets of P and C in well-defined equations can also be represented by their corresponding "Dimensions".

Remark 1.1. We can apply Dimensional analysis to "Dimensions" from Information-Complexity conservation and cardinality of relevant sets in certain well-defined equations.

Let X denote E, O, N [which are classified as Completely Predictable numbers], P and C [which are classified as Incompletely Predictable numbers]. For x = 1, 2, 3, 4, 5,..., ∞; consider all X ≤ x. Then this "all X ≤ x" is definition for X-$\pi(x)$ [denoting "X counting function"] resulting in following two types of equations coined as (I) 'Exact' equation N$\pi(x)$ = E-$\pi(x)$ + O-$\pi(x)$ with "non-varying" relationships E-$\pi(x)$ = O-$\pi(x)$ for all x = E and E-$\pi(x)$ = O-$\pi(x)$ - 1 for all x = O, and (II) 'Inexact' equation N-$\pi(x)$ = 1 + P-$\pi(x)$ + C-$\pi(x)$ with "varying" relationships P-$\pi(x)$ > C-$\pi(x)$ for all x ≤ 8; P-$\pi(x)$ = C-$\pi(x)$ for x = 9, 11, and 13; and P-$\pi(x)$ < C-$\pi(x)$ for x = 10, 12, and all x ≥ 14.

Let "Dimensions" and different (sub)sets of E, O, N, P and C be 'base quantities'. Then exponent '1' of "Dimensions" and cardinality of these (sub)sets in well-defined equations are corresponding 'units of measurement'. Performing DA on "Dimensions" for PC pairing are depicted later on. Performing DA on cardinality are depicted next.

For Set N = Set E + Set O, then |N| = |E| + |O| =⇒ $\aleph_0 = \aleph_0 + \aleph_0$ thus conforming with DA homogeneity.

For Set N = Set P + Set C + Number '1', then Set N - Number '1' = Set P + Set C and |N - Number '1'| = |P| + |C| =⇒ $\aleph_0 = \aleph_0 + \aleph_0$ thus conforming with DA homogeneity.

For Set N - Set even P - Number '1'= Set odd P + Set even C + Set odd C, then $|$N - even P - Number '1'$|$ = $|$odd P$|$ + $|$even C$|$ + $|$odd C$|$ =⇒ $\aleph_0 = \aleph_0 + \aleph_0 + \aleph_0$ thus conforming with DA homogeneity. Symbolically represented by all available O prime gap = 1 and E prime gaps = 2, 4, 6, 8, 10,...; O composite gap = 1 and E composite gap = 2; and O natural gap = 1; then $|$Gap 1 N - Gap 1 P - Number '1'$|$ = $|$Gap 2 P$|$ + $|$Gap 4 P$|$ + $|$Gap 6 P$|$ + $|$Gap 8 P$|$ + $|$Gap 10 P$|$ + ... + $|$Gap 1 C$|$ + $|$Gap 2 C$|$ =⇒ $\aleph_0 = \aleph_0 + \aleph_0 + \aleph_0 + \aleph_0 + \aleph_0 + ... \aleph_0 + \aleph_0$ thus conforming with DA homogeneity. It is known that $|$Gap 1 P$|$ = $|$Number '1'$|$ = 1 and $|$Gap 1 N$|$ = $|$Gap 1 C$|$ = $|$Gap 2 C$|$ = $\aleph_0$. Then solving Polignac's and Twin prime conjectures translate to successfully proving $|$Gap 2 P$|$ = $|$Gap 4 P$|$ = $|$Gap 6 P$|$ = $|$Gap 8 P$|$ = $|$Gap 10 P$|$ = ... = $\aleph_0$ with $|$E prime gaps$|$ = $\aleph_0$.

Outline of proof for Polignac's and Twin prime conjectures. Requires simultaneously satisfying two mutually inclusive conditions:

I. With rigid manifestation of DA homogeneity, quantitive[1] fulfillment by considering i ∈ E for each Subset odd Pi generated by E prime gap = i from Set E prime gaps occurs only if solitary cardinality value is present in equation Set odd P = $\sum_{i=2}^{\infty} Subset\ odd\ Pi$ with $|$odd P$|$ = $|$odd Pi$|$ = $|$E prime gaps$|$ = $\aleph_0$.

II. With rigid manifestation of DA non-homogeneity, quantitive[1] fulfillment by considering i ∈ E for each Subset odd Pi generated by E prime gap = i from Set E prime gaps does not occur if more than one cardinality values are present in equation Set odd P > $\sum_{i=2}^{\infty} Subset\ odd\ Pi$ with $|$E prime gaps$|$ = $\aleph$o having incorrect $|$Subset(s) odd P$|$ = N (finite value) and/or Set odd P > $\sum_{i=2}^{N} Subset\ odd\ Pi$ with $|$odd Pi$|$ = $\aleph_0$ having incorrect $|$E prime gaps| = N (finite value).

Footnote 1: Qualitative fulfilment of |odd P| = |odd P_i| = |all E prime gaps| = $\aleph_0$ equates to Plus-Minus Gap 2 Composite Number Alternating Law being precisely obeyed by all E prime gaps apart from first E prime gap precisely obeying Plus Gap 2 Composite Number Continuous Law. Derived using Information-Complexity conservation, these Laws symbolize "end-result" proof on Polignac's and Twin prime conjectures. *Law of Continuity* is a heuristic principle *whatever succeed for the finite, also succeed for the infinite*. Then these Laws which inherently manifest 'Gap 2 Composite Number' on finite and infinite time scale should in principle "succeed for the finite, also succeed for the infinite".

Polignac's and Twin prime conjectures mathematical foot-prints. Six identifiable steps to prove these conjectures: *Step 1* Considering x $\in$ N, obtain Dimensions $(2x - 2)^1$, $(2x - 4)^1$, $(2x - 5)^1$, $(2x - 7)^1$, $(2x - 8)^1$, $(2x - 9)^1$, ..., $(2x - \infty)^1$ with specific groupings to constitute all elements of Set P [culminating in obtaining all prime gaps (= E prime gaps + Solitary O prime gap) with |all prime gaps| = $\aleph_0$]. Note Dimension $(2x - 2)^1$ represents x = 1 (Number '1') which is neither P nor C. *Step 2* Considering i $\in$ E, confirm perpetual recurrences of individual E prime gap = i (associated with its unique odd P_i) occur only when depicted as specific groupings of these Dimensions endowed with exponent '1' for all ranges of x. *Step 3* Perform DA on exponent '1' in these Dimensions. *Step 4* Perform DA on equation

Set odd P $= \sum^{\infty}$ Subset odd P_i to obtain |**odd P**| = |**odd** $\mathbf{P}_i$| = $\aleph_0$ whereby Subset odd P_i is derived from its associated unique E prime gap = i with |**E prime gaps**| = $\aleph_0$. *Step 5* Confirm 'Prime number' variable and 'Prime gap' variable complex algorithm "containing" all P with knowing their overall

394

actual location [but not actual positions][2]. *Step 6* Derive Plus-Minus Gap 2 Composite Number Alternating Law and Plus Gap 2 Composite Number Continuous Law using Information-Complexity conservation.

Footnote 2: This phrase implies all P (and C) are treated as Incompletely Predictable numbers. Actual positions will require using complex algorithm Sieve of Eratosthenes to *dependently* calculate positions of all preceding P (and C) in the neighborhood.

'Complex Elementary Fundamental Laws'-based solutions of Plus-Minus Gap 2 Composite Number Alternating Law and Plus Gap 2 Composite Number Continuous Law are obtained by undertaking the non-negotiable mathematical steps outlined above. These Laws are literally Completely Predictable meta-properties ('overall' *complex properties*) arising from "interactions" between P and C producing relevant patterns of Gap 2 Composite Number perpetual appearances [albeit with Incompletely Predictable timing]. We logically deduce explicit mathematical explanation for these meta-properties requires "complex" mathematical arguments. Attempts to give explicit mathematical explanation with "simple" mathematical arguments would intuitively mean Incompletely Predictable numbers P and C be (incorrectly and impossibly) treated as Completely Predictable numbers.

1.2 Brief overview of Polignac's and Twin prime conjectures

Occurring over 2000 years ago (c. 300 BC), ancient Euclid's proof on infinitude of P in totality [viz. $|P| = \aleph_0$ for Set P] predominantly by *reductio ad absurdum* (proof by contradiction) is earliest known but not the only proof for this simple problem in number theory. Since then dozens of proofs have been devised such as three chronologically listed: Goldbach's Proof using

Fermat numbers (written in a letter to Swiss mathematician Leonhard Euler, July 1730), Furstenberg's Topological Proof in 1955[1], and Filip Saidak's Proof in 2006[2]. The strangest candidate is likely to be Furstenberg's Topological Proof.

In 2013, Yitang Zhang proved a landmark result showing some unknown even number 'N' < 70 million such that there are infinitely many pairs of P that differ by 'N'[3]. By optimizing Zhang's bound, subsequent Polymath Project collaborative efforts using a new refinement of GPY sieve in 2013 lowered 'N' to 246; and assuming Elliott-Halberstam conjecture and its generalized form have further lower 'N' to 12 and 6, respectively. Then 'N' has intuitively more than one valid values such that there are infinitely many pairs of P that differ by each of those 'N' values [thus proving existence of more than one Subset **odd P**$_i$ with $|$ **odd P**$_i| = \aleph_0$]. We can only theoretically lower 'N' to 2 (in regards to P with 'small gaps') but there are still an infinite number of E prime gaps (in regards to P with 'large gaps') that require "the proof that each will generate its unique set of infinite P".

Remark 1.2. Existence of maximal and non-maximal prime gaps supply crucial indirect evidence to intuitively support but does not prove proposition "Each even prime gap will generate an infinite magnitude of odd prime numbers on its own accord". Comments relevant to Remark 1.2 are given in Section 2 below.

2 Supportive role of maximal and non-maximal prime gaps

We analyze data of all P obtained when extrapolated out over a wide range of x ≥ 2 integer values. As sequence of P carries on, P with ever larger prime gaps will appear. For given range of x integer values, prime gap = n2 is a 'maximal prime gap' if prime gap = n1 < prime gap = n2 for all n1 < n2. In other words, the largest such prime gaps in this range are called maximal prime gaps. The term 'first occurrence prime gaps' refers to first occurrences of maximal prime gaps whereby maximal prime gaps are prime gaps of "at least of this length".

Table 1 First 17 prime gaps depicted in the format utilizing maximal prime gaps [depicted with asterisk symbol (*)] and non-maximal prime gaps [depicted without this asterisk symbol].

Prime gap	Following the prime number	Prime gap	Following the prime number
1*	2	18*	523
2*	3	20*	887
4*	7	22*	1129
6*	23	24	1669
8*	89	26	2477
10	139	28	2971
12	199	30	4297
14*	113	32	5591
16	1831		

We use maximal prime gaps to denote 'first occurrence prime gaps'. CIS non-maximal prime gaps (endorsed with nickname 'slow jumpers') will always lag behind CIS maximal prime gaps for onset appearances in P sequence. These are shown for first 17 prime gaps in Table 1. Apart from O prime gap = 1 representing solitary even P '2', remaining P depicted in Table 1 consist of representative single odd P for each E prime gap. These odd P will individually make one-off appearance in P sequence in a *perpetual albeit Incompletely Predictable manner*. Initial seven of [majority] "missing" odd P are 5, 11, 13, 17, 19, 29, 31,... belonging to Subset P with 'residual' prime gaps are potential source of odd P in relation to proposal that each E prime gap from Set E prime gaps will generate its specific Subset odd P. Set all P from all prime gaps = Subset P from maximal prime gaps + Subset P from non-maximal prime gaps + Subset P from 'residual' prime gaps. Subset P from 'residual' prime gaps with representation from all E prime gaps must include all correctly selected "missing" odd P. These observations support but does not prove proposition that each E prime gap will generate its own Subset odd P with $|$ odd P $| = \aleph_0$.

For i $\in$ N; primordial $P_i\#$ is analog of usual factorial for P = 2, 3, 5, 7, 11, 13,.... Then $P_1\# = 2$, $P_2\# = 2 \times 3 = 6$, $P_3\# = 2 \times 3 \times 5 = 30$, $P_4\# = 2 \times 3 \times 5 \times 7 = 210$, $P_5\# = 2 \times 3 \times 5 \times 7 \times 11 = 2310$, $P_6\# = 2 \times 3 \times 5 \times 7 \times 11 \times 13 = 30030$, etc. English mathematician John Horton Conway coined the term 'jumping champion' in 1993. An integer n is a 'jumping champion' if n is the most frequently occurring difference (prime gap) between consecutive P<x for some x integer values. Example: for any x with 7<x<131, n = 2 (indicating twin P) is the 'jumping champion'. It has been conjectured that (i) the only 'jumping champions' are 1, 4 and primorials 2, 6, 30, 210, 2310, 30030,... and (ii) 'jumping champions' tend to infinity. Their required proofs will likely need proof of k-tuple conjecture. P from 'jumping champion' prime gaps have their onset appearances in P sequence in a *perpetual albeit Incompletely Predictable manner* [as another example to that outlined in previous paragraph].

398

3 Information-Complexity conservation

A formula, as equation or algorithm, is simply a Black Box generating necessary Output (with qualitative structural 'Complexity') when supplied with given Input (with quantitative data 'Information'). This 'Information' and 'Complexity' are what is referred to in the term 'Information-Complexity conservation'.

N (CIS): 1, 2, 3,..., $+\infty$. Let x be from Set X such that $x \in N$. Consider x for the upper boundary of interest in Set X whereby X is chosen from N, E, O, P or C.

Lemma 3.1. Natural counting function $N\text{-}\pi(x)$, defined as $|N \leq x|$, is Completely Predictable by independently using simple algorithm to be equal to x.

Proof Formula to generate N with 100% certainty is $N_i = i$ whereby N_i is the ith N and i = 1, 2, 3,..., ∞. For a given N_i, its ith position is simply i. Natural gap $(G_{Ni}) = N_{i+1} - N_i$, with G_{Ni} always = 1. There are x N $\leq$ x. Thus $N\text{-}\pi(x) = |N \leq x| = x$. *The proof is now complete for Lemma 3.1*□.

Lemma 3.2. Even counting function $E\text{-}\pi(x)$, defined as $|E \leq x|$, is Completely Predictable by independently using simple algorithm to be equal to floor(x/2).

Proof. Formula to generate E with 100% certainty is $E_i = i \times 2$ whereby E_i is the ith E and i = 1, 2, 3,..., ∞ abiding to mathematical label "All N always ending with a digit 0, 2, 4, 6 or 8". For a given E_i, its ith position is calculated as $i = E_i/2$. Even gap $(G_{Ei}) = E_{i+1} - E_i$, with G_{Ei} always = 2. There are $\lfloor \frac{x}{2} \rfloor$ E $\leq$ x. Thus $E\text{-}\pi(x) = |E \leq x| = $ floor(x/2). *The proof is now complete for Lemma 3.2*□.

Lemma 3.3. Odd counting function O-$\pi(x)$, defined as $|O \leq x|$, is Completely Predictable by independently using simple algorithm to be equal to ceiling(x/2).

Proof. Formula to generate O with 100% certainty is $O_i = (i \times 2) - 1$ whereby O_i is the ith odd number and i = 1, 2, 3,..., ∞ abiding to mathematical label "All N always ending with a digit 1, 3, 5, 7, or 9". For a given O_i number, its ith position is calculated as $i = (O_i + 1)/2$. Odd gap $(G_{Oi}) = O_{i+1} - O_i$, with G_{Oi} always = 2. There are $\lceil \frac{x}{2} \rceil$ O $\leq$ x. Thus O-$\pi(x)$ = $|O \leq x|$ = ceiling(x/2). *The proof is now complete for Lemma 3.3*□.

Lemma 3.4. Prime counting function P-$\pi(x)$, defined as $|P \leq x|$, is Incompletely Predictable with Set P dependently obtained using complex algorithm Sieve of Eratosthenes.

Proof. Algorithm to generate P_i whereby P_1 (= 2), P_2 (= 3), P_3 (= 5), P_4 (= 7),..., ∞ with 100% certainty is based on Sieve of Eratosthenes abiding to mathematical label "All N apart from 1 that are evenly divisible by itself and by 1". Although we can check primality of a given O by trial division, we can never determine its position without knowing positions of preceding P. Prime gap $(G_{Pi}) = P_{i+1} - P_i$, with G_{Pi} constituted by all E except $1^{st} G_{P1}$ = 3 - 2 = 1. P-$\pi(x)$ = $|P \leq x|$. This is Incompletely Predictable and is calculated via mentioned algorithm. Using definition of prime gap, every P [represented here with aid of 'n' notation instead of usual 'i' notation] is written as P_{n+1} = $2 + \sum_{i=1}^{n} GPi$ with '2' denoting P_1. Here i & n = 1, 2, 3, 4, 5, ..., ∞. *The proof is now complete for Lemma 3.4*□.

Lemma 3.5. Composite counting function C-$\pi(x)$, defined as $|C \leq x|$, is Incompletely Predictable with Set C derived as Set N - Set P [dependently obtained using complex algorithm Sieve of Eratosthenes] - Number '1'.

Proof. Composite numbers abide to mathematical label "All N apart from 1 that are evenly divisible by numbers other than itself and 1". Algorithm to generate C_i whereby C_1 (= 4), C_2 (= 6), C_3 (= 8), C_4 (= 9),..., ∞ with 100% certainty is based [indirectly] on Sieve of Eratosthenes via selecting non-prime N to be C. We define Composite gap G_{Ci} as C_{i+1} - C_i with G_{Ci} constituted by 1 & 2. C-$\pi(x)$ = C ≤ x. This is Incompletely Predictable and always need to be calculated indirectly via mentioned algorithm. Using definition of composite gap, every C [represented here with aid of 'n' notation instead usual 'i' notation] is written as $C_{n+1} = 4 + \sum_{i=1}^{n} GCi$ with '4' denoting C_1. Here i & n = 1, 2, 3, 4, 5, ..., ∞. *The proof is now complete for Lemma 3.5*□.

Denote X to be N, E, O, P or C. X-$\pi(x)$ = | X ≤ x | with x ∈ N. We define and compute entity 'Grand-Total Gaps for X at x' (Grand-Total ΣX_x-Gaps).

Proposition 3.6. For any given x ≥ 1 values in Set N, designated Complexity is represented by ΣN_x-Gaps = x - N with N = 1 being maximal.

Proof. Set N (for x = 1 to 12): 1, 2, 3, 4, 5, 6, 7, 8, 9, 10, 11, 12. N-$\pi(x)$ = 12. There are x - 1 = 11 N-Gaps each of '1' magnitude: 1, 1, 1, 1, 1, 1, 1, 1, 1, 1, 1. ΣN_x-Gaps = 11 X 1 = 11. This equates to "x - 1" – regarded as Complexity for N. *The proof is now complete for Proposition 3.6*□.

Proposition 3.7. For any given x ≥ 1 values in constituent Set E and Set O, designated Complexity is represented by ΣEO_x-Gaps = 2x - N with N = 4 being maximal.

Proof. Set E and Set O (for x = 1 to 12): 2, 4, 6, 8, 10, 12 and 1, 3, 5, 7, 9, 11. E-$\pi(x)$ = 6 and O-$\pi(x)$ = 6. There are $\lfloor \frac{x}{2} \rfloor$ - 1 = 5 E-Gaps each of '2' magnitude: 2, 2, 2, 2, 2. ΣE_x-Gaps = 5 X 2 = 10, and $\lceil \frac{x}{2} \rceil$ - 1 = 5)-Gaps each of '2' magnitude: 2, 2, 2, 2, 2. ΣO_x-Gaps = 5 X 2 = 10. Grand-Total

401

ΣEO_x-Gaps = 10 + 10 = 20. Depicted by Table 3 and Figure 2 in Appendix I, 2x - N = "2x - 4" [perpetual constant appearances of "N = 4 being maximal"] is Complexity for E and O. *The proof is now complete for Proposition 3.7□.*

Proposition 3.8. For selected $x \geq 2$ values in constituent Set P and Set C, designated Complexity is cyclically represented by ΣPC_x-Gaps = 2x - N with N = 7 being minimal.

Proof. Set P and Set C (for x = 2 to 12): 2, 3, 5, 7, 11 and 4, 6, 8, 9, 10, 12. P-$\pi(x)$ = 5 and C-$\pi(x)$ = 6. There are four P-Gaps of 1, 2, 2, 4 magnitude and five C-Gaps of 2, 2, 1, 1, 2 magnitude. ΣP_x-Gaps = 1 + 2 + 2 + 4 = 9. ΣC_x-Gaps = 2 + 2 + 1 + 1 + 2 = 8. Grand-Total ΣPC_x-Gaps = 9 + 8 = 17. Depicted by Table 2 and Figure 1, 2x - N = 2x - 7 [perpetual intermittent and cyclical appearances of "N = 7 being minimal"] is Complexity for P and C. *The proof is now complete for Proposition 3.8□.*

Designated Complexity is (i) x - N with N = 1 (maximal) for Completely Predictable N, (ii) 2x - N with N = 7 (minimal) for Incompletely Predictable P & C, and (iii) 2x - N with N = 4 (maximal) for Completely Predictable E & O. Interpretations: N has minimal Complexity, E & O have intermediate Complexity, and P & C have maximal [varying] Complexity. Defacto baseline "2x - 4" Grand-Total Gaps [minus 4 value] in E-O pairing > Defacto baseline "2x - ≥7" Grand-Total Gaps [minus ≥7 values] in P-C pairing.

Let both x & N $\in$ N. We tabulate in Table 2 and graph in Figure 1 [Incompletely Predictable] P-C mathematical landscape for a relatively larger x = 2 to 64 here (and ditto for [Completely Predictable] E-O mathematical landscape for relatively larger x = 1 to 64 in Appendix I). The term "mathematical landscape" denotes specific mathematical patterns in tabulated and graphed data. "Dimension" contextually denotes Dimension

402

2x - N whereby (i) allocated [infinite] N values result in Dimensions 2x - 7, 2x - 8, 2x - 9, ..., 2x - ∞ for P-C finite scale mathematical landscape and (ii) allocated [finite] N values for E-O finite scale mathematical landscape result in Dimension 2x - 4. For P-C pairing, initial one-off Dimensions 2x - 2, 2x - 4 and 2x - 5 (in consecutive order) are exceptions [with Dimension 2x - 2 validly representing Number '1' which is neither P nor C]. For E-O pairing, initial one-off Dimension 2x - 2 is an exception. P-C mathematical landscape consisting of Dimensions will intrinsically incorporate P and C in an integrated manner and there are infinite times whereby relevant Dimensions deviate away from 'baseline' Dimension 2x - 7 simply because P [and, by default, C] in totality are rigorously proven to be infinite in magnitude. In contrast, there is complete lack of deviation away from 'baseline' Dimension 2x - 4 apart from one-off deviation by initial Dimension 2x - 2 in Appendix I.

Table 2 Prime-Composite finite scale mathematical (tabulated) landscape using data obtained for x = 2 to 64. The Number '1' is neither prime nor composite. Legend: C = composite, P = prime, Y = Dimension 2x 7 (for visual clarity), N/A = Not Applicable.

x	P_i or C_i, Gaps	ΣPC_x -Gaps	Dimension
1	N/A	0	2x-2
2	P1, 1	0	2x-4
3	P2, 2	1	2x-5
4	C1, 2	1	Y
5	P3, 2	3	Y
6	C2, 2	5	Y

7	P4, 4	7	Y
8	C3, 1	9	Y
9	C4, 1	10	2x-8
10	C5, 2	11	2x-9
11	P5, 2	15	Y
12	C6, 2	17	Y
13	P6, 4	19	Y
14	C7, 1	21	Y
15	C8, 1	22	2x-8
16	C9, 1	23	2x-9
17	P7, 2	27	Y
18	C10, 2	29	Y
19	P8, 4	31	Y
20	C11, 1	33	Y
21	C12, 1	34	2x-8
22	C13, 2	35	2x-9
23	P9, 6	39	Y
24	C14, 1	41	Y
25	C15, 1	42	2x-8
26	C16, 1	43	2x-9
27	C17, 1	44	2x-10
28	C18, 2	45	2x-11
29	P10, 2	51	Y
30	C19, 2	53	Y
31	P11, 6	55	Y

32	C20, 1	57	Y
33	C21, 1	58	2x-8
34	C22, 1	59	2x-9
35	C23, 1	60	2x-10
36	C24, 2	61	2x-11
37	P12, 4	67	Y
38	C25, 1	69	Y
39	C26, 1	70	2x-8
40	C27, 1	71	2x-9
41	P13, 2	75	Y
42	C28, 2	77	Y
43	P14, 4	79	Y
44	C29, 1	81	Y
45	C30, 1	82	2x-8
46	C31, 2	83	2x-9
47	P15, 6	87	Y
48	C32, 1	89	Y
49	C33, 1	90	2x-8
50	C34, 1	91	2x-9
51	C35, 1	92	2x-10
52	C36, 1	93	2x-11
53	P16, 6	99	Y
54	C37, 1	101	Y
55	C38, 1	102	2x-8
56	C39, 1	103	2x-9
57	C40, 1	104	2x-10

405

58	C41, 1	105	2x-11
59	P17, 2	111	Y
60	C42, 2	113	Y
61	P18, 6	115	Y
62	C43, 1	117	Y
63	C44, 1	118	2x-8
64	C45, 1	119	2x-9

In Figure 1, Dimensions 2x - 7, 2x - 8, 2x - 9, ..., 2x - ∞ are symbolically represented by -7, -8, -9, ..., ∞ with 2x - 7 displayed as 'baseline' Dimension whereby Dimension trend (Cumulative Sum Gaps) must repeatedly reset itself onto this 'baseline' Dimension on a perpetual basis. Dimensions symbolically represented by ever larger negative integers will correspond to P associated with ever larger prime gaps and this phenomenon will generally happen at ever larger x values (with complete presence of Chaos and Fractals being manifested in our graph). At ever larger x values, P-$\pi(x)$ will overall become larger but with a *decelerating* trend whereas C-$\pi(x)$ will overall become larger but with an *accelerating* trend. This support ever larger prime gaps appearing at ever larger x values.

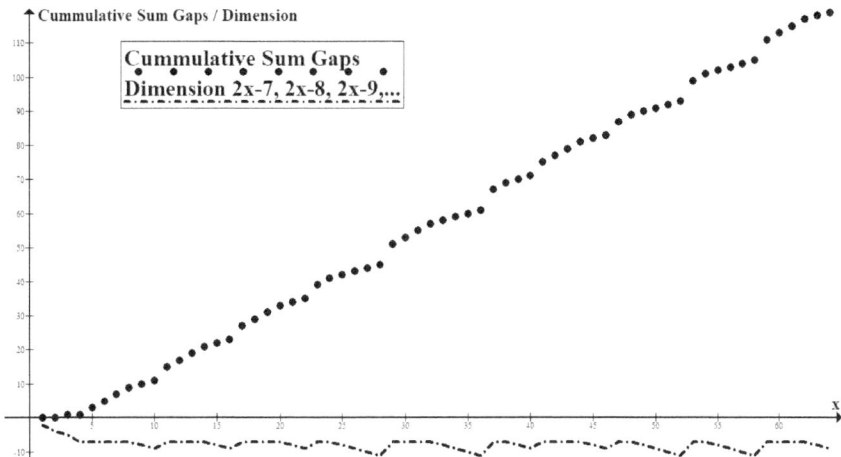

Fig. 1 Prime-Composite finite scale mathematical (graphed) landscape using data obtained for x = 2 to 64. Bottom graph symbolically represent "Dimensions" using ever larger negative integers.

Definitive derivation of data in Table 2 is illustrated by two examples for position x = 31 & 32. For i & x $\in$ N; ΣPC_x-Gap = ΣPC_{x-1}-Gap + Gap value at P_{i-1} or Gap value at C_{i-1} whereby (i) P_i or C_i at position x is determined by whether relevant x value belongs to a P or C, and (ii) both ΣPC_1-Gap and ΣPC_2-Gap = 0. Example, for position x = 31: 31 is P (P11). Desired Gap value at P10 = 2. Thus ΣPC_{31}-Gap (55) = ΣPC_{30}-Gap (53) + Gap value at P10 (2). Example, for position x = 32: 32 is C (C20). Desired Gap value at C19 = 2. Thus ΣPC_{32}-Gap (57) = ΣPC_{31}-Gap (55) + Gap value at C20 (2). 'Overall magnitude of C will always be greater than that of P' will hold true from x = 14 onwards. For instance, position x = 61 corresponds to P 61 which is 18th P, whereas [the one lower] position x = 60 corresponding to C 60 is the [much higher] 42nd C.

407

4 Polignac's and Twin prime conjectures

Previous section alludes to P-C finite scale mathematical landscape. This section alludes to P-C infinite scale mathematical landscape. Let 'Y' symbolizes (baseline) Dimension $2x - 7$. Let prime gap at $P_i = P_{i+1} - P_i$ with P_i & P_{i+1} respectively symbolizes consecutive "first" & "second" P in any P_i-P_{i+1} pairings. We denote (i) Dimensions YY grouping [depicted by $2x - 7$ initially appearing twice in (iii)] to represent signal for appearances of P pairings other than twin P such as cousin P, sexy P, etc; (ii) Dimension YYYY grouping to represent signal for appearances of P pairings as twin P; and (iii) Dimension $(2x - \geq 7)$-Progressive Grouping allocated to $2x - 7$, $2x - 7$, $2x - 8$, $2x - 9$, $2x - 10$, $2x - 11$,..., $2x - \infty$ as elements of *precise* and *proportionate* CFS Dimensions representation of an individual P_i with its associated prime gap namely, Dimensions $2x - 7$ & $2x - 7$ pairing = twin P (with both its prime gap & CFS cardinality = 2); $2x - 7$, $2x - 7$, $2x - 8$ & $2x - 9$ pairing = cousin P (with both its prime gap & CFS cardinality = 4); $2x - 7$, $2x - 7$, $2x - 8$, $2x - 9$, $2x - 10$ & $2x - 11$ pairing = sexy P (with both its prime gap & CFS cardinality = 6); and so on. The higher order [traditionally defined as closest possible] prime groupings of three P as P triplets, of four P numbers as prime quadruplets, of five P numbers as prime quintuplets, etc consist of relevant serendipitous groupings abiding to mathematical rule: With exception of three 'outlier' P 3, 5, & 7; groupings of any three P as P, P+2, P+4 combination (viz. manifesting two consecutive twin P) is a mathematical impossibility. The 'anomaly' one of every three consecutive O is a multiple of three, and hence this particular number cannot be P, explains this impossibility. Then closest possible P grouping [viz. for prime triplet] must be either P, P+2, P+6 format or P, P+4, P+6 format.

P groupings not respecting traditional closest-possible-prime groupings are also the norm occurring infinitely often, indicating continual presence of prime gaps ≥ 6. As P become sparser at larger range, perpetual presence of (i) prime gaps ≥ 6 [which we propose to arbitrarily represent 'large gaps'] and (ii) prime gaps 2 & 4 [which we propose to arbitrarily represent 'small gaps'] with progressive greater magnitude will cumulatively occur for each prime gap but always in a decelerating manner. With permanent requirement at larger range of intermittently resetting to baseline Dimension 2x - 7 occurring [either two or] four times in a row, nature seems to dictate, at the very least, perpetual twin P or one other non-twin P occurrences is inevitable.

We dissect Dimension YYYY unique signal for twin P appearances. Initial two CFS Dimensions YY components of YYYY represent "first" P component of twin P pairing. Last two Dimensions YY components of YYYY signifying appearance of "second" P component of twin P pairing is also the initial first-two-element component of full CFS Dimensions representation for "first" P component of following non-twin P pairing. Twin P are uniquely represented by repeating *single* type Dimension 2x - 7. In all other 'higher order' P pairings (with prime gaps ≥ 4), they require *multiple* types Dimension representation. There is qualitative aspect association of *single* type Dimension representation for twin P resulting in "less colorful" Plus Gap 2 Composite Number *Continuous Law* as opposed to *multiple* types Dimension representation for all other 'higher order' P pairings resulting in "more colorful" Plus-Minus Gap 2 Composite Number *Alternating Law*. 'Gap 2 Composite Number' occurrences in both Laws on finite scale are (directly) observed in Figure 1 & Table 2 for x = 2 to 64, and on infinite scale are (indirectly) deduced using logical arguments for all x values.

409

We endow all "Dimensions" with exponent of '1' for perusal in on-going mathematical arguments. $P_1 = 2$ is represented by CFS as Dimension $(2x - 4)^1$ (with both prime gap & CFS cardinality = 1); $P_2 = 3$ is represented by CFS as Dimensions $(2x - 5)^1$ & $(2x - 7)^1$ (with both prime gap & CFS cardinality = 2); $P_3 = 5$ is represented by CFS Dimension $(2x - 7)^1$ & $(2x - 7)^1$ (with both prime gap & CFS cardinality = 2), etc.

Proposition 4.1. Let Case 1 be Completely Predictable E & O pairing and Case 2 be Incompletely Predictable P & C pairing. Furthermore, let Case 1 and Case 2 be independent of each other. Then for any given x value, there exist grand total number of Dimensions [Complexity] such that it exactly equal to either two combined subtotal number of Dimensions [Complexity] to precisely represent E & O in Case 1, or combined subtotal number of Dimensions [Complexity] to precisely represent P & C & Number '1' in Case 2.

Proof. N is directly constituted from either combined E & O in Case 1 or combined P & C & Number '1' in Case 2 – Number '1' is neither P nor C. Correctly designated infinitely many CFS of Dimensions used to represent combined E & O in Case 1 and combined P & C & Number '1' in Case 2 must also directly and proportionately be representative of relevant N arising from combined subtotal of E & O in Case 1 and from combined subtotal of P & C & Number '1' in Case 2. *The proof is now complete for Proposition 4.1□.*

Proposition 4.2. Let Case 1 be Completely Predictable E & O pairing and Case 2 be Incompletely Predictable P & C pairing. Furthermore, let Case 1 and Case 2 be independent of each other. Part I: For any given x value apart from x = 1 value in Case 1 and x = 1, 2, and 3 values in Case 2; Dimension $(2x - N)^1$ [Complexity] representations of all Completely Predictable E & O in Case 1 and all Incompletely Predictable P & C & Number '1' in Case

410

2 are such that they are given by N = 4 in Case 1 and by N ≥ 7 in Case 2. Part II: Odd P obeys 'Plus-Minus Composite Gap 2 Number Alternating Law' for prime gaps ≥ 4 and 'Plus Composite Gap 2 Number Continuous Law' for prime gap = 2.

Proof. Apart from first Dimension $(2x - 2)^1$ representation in E & O pairing in Case 1 and first three Dimension $(2x - 2)^1$, Dimension $(2x - 4)^1$ and Dimension $(2x - 5)^1$ representations in P & C pairing in Case 2; possible N value in Dimension $(2x - N)^1$ representation has been shown to be (constantly) maximal 4 for Case 1 and (variably) minimal 7 for Case 2. For Case 2, we again note Dimension $(2x - 2)^1$ to (validly) represent Number '1' which is neither P nor C. These nominated Dimensions simply represent possible (constant) baseline "2x - 4" Grand-Total Gaps as per Proposition 3.7 for Case 1 & (variable) baseline "2x 7" Grand-Total Gaps as per Proposition 3.8 for Case 2. Note that all CFS of Dimensions that can be used to precisely represent combined E & O in Case 1 will persistently consist of same [solitary] Dimension $(2x - 4)^1$ after first Dimension $(2x - 2)^1$. Perpetual repeated deviation of N values away from N = 7 (minimum) in Case 2 is simply representing infinite magnitude of P & C. *The proof is now complete for Part I of Proposition 4.2*□.

Derived Dimensions will comply with Incompletely Predictable property as explained using P '61'. At Position x = 61 equating to $P_{18} = 61$, it is represented by CFS Dimensions $(2x - 7)^1$, $(2x - 7)^1$, $(2x - 8)^1$, $(2x - 9)^1$, $(2x - 10)^1$ & $(2x - 11)^1$ (with both prime gap & CFS cardinality = 6). This representation indicates an 'unknown but correct" P with prime gap = 6 when we intentionally conceal full information '61' = 31st O = 18th P with prime gap = 6. But to arrive at this representation requires calculations of all

411

preceding CFS Dimensions thus manifesting hallmark Incompletely Predictable property of CFS Dimensions.

Overall sum total of individual CFS Dimensions required to represent every P is infinite in magnitude as $|$ **all P** $| = \aleph_0$. Standalone Dimensions YY groupings [representing signals for "higher order" non-twin P appearances] &/or as front Dimensions YY (sub)groupings [which by itself is fully representative of twin P as Dimensions YYYY appearances] need to recur on an indefinite basis. Then twin P and "higher order" cousin P, sexy P, etc should aesthetically all be infinite in magnitude because (respectively) they regularly and universally arise as part of Dimension YYYY and Dimension YY appearances. An isolated P is defined as a P such that neither P - 2 nor P + 2 is P. In other words, isolated P is not part of a twin P pair. Example 23 is an isolated P since 21 and 25 are both C. Then repeated inevitable presence of Dimension YY grouping is nothing more than indicating repeated occurrences of isolated P. This constitutes another view on Dimension YY.

CIS of Gap 1 Composite Numbers are fully associated with non-twin P as they eternally occur in between any two consecutive non-twin P. CIS of Gap 2 Composite Numbers are (i) fully associated with twin P as they are eternally present in between any twin P pair, and (ii) partially associated with non-twin P as they are eternally present alternatingly or intermittently in between any two consecutive non-twin P. Then (i) Gap 1 Composite Numbers do not have valid representation by E prime gap = 2, and (ii) Gap 2 Composite Numbers have valid representations by all E prime gaps = ["consistently" only for] 2, ["inconsistently" for each of] 4, 6, 8, 10,.... This is an alternative view on P from perspective of CFS composite gaps [instead of CIS prime gaps] with intrinsic patterns having *alternating presence* and *absence* of Gap 2 Composite Numbers associated with every CFS Dimensions

412

representations of P with prime gaps ≥ 4, viz. 'Plus-Minus Gap 2 Composite Number Alternating Law'. CFS Dimensions representations of Twin P are always associated with Gap 2 Composite Numbers, viz. 'Plus Gap 2 Composite Number Continuous Law'.

Examples for both Laws: A twin P (prime gap = 2) in its unique CFS Dimensions format always has Gap 2 Composite Numbers in a [constant] pattern. A cousin P (prime gap = 4) in its unique CFS Dimensions format always has two Gap 1 Composite Numbers & then one Gap 2 Composite Number [combined] pattern *alternating* with three consecutive Gap 1 Composite Numbers [non-combined] pattern. From this simple observation alone, we deduce we can generate an infinite magnitude of C from each composite gaps 1 & 2. Gap 2 Composite Numbers *alternating* pattern behavior in cousin P will not hold true unless twin P & all other non-cousin P are infinite in magnitude and integratedly supplying essential "driving mechanism" to eternally sustain this Gap 2 Composite Numbers *alternating* pattern behavior in cousin P. Thus we establish twin P and cousin P in their CFS Dimensions formats are CIS intertwined together when depicted using C with composite gaps = 1 & 2 with each supplying their own peculiar (infinite) share of associated Gap 2 Composite Numbers [thus contributing to overall pool of Gap 2 Composite Numbers].

An inevitable statement in relation to "Gap 2 Composite Numbers pool contribution" based on above reasoning: At the bare minimum, *either* twin P *or* at least one of non-twin P must be infinite in magnitude. An inevitable impression: All generated subsets of P from 'small gaps' [of 2 & 4] and 'large gaps' [of ≥ 6] alike should each be CIS thus allowing true uniformity in P distribution. Again we see in Table 2 above depicting P-C data for x = 2 to 64 that, for instance, P with prime gap = 6 must also persistently have this

413

'last place' Gap 2 Composite Numbers intermittently appearing in certain rhythmic *alternating* patterns, thus complying with Plus-Minus Gap 2 Composite Number Alternating Law. This CFS Dimensions representation for P with prime gaps = 6 will again generate their infinite share of associated Gap 2 Composite Numbers to contribute to this pool. The presence of this last-place Gap 2 Composite Numbers in various alternating pattern in appearances & nonappearances must *self-generatingly* be similarly extended in a mathematically consistent fashion *ad infinitum* to all other remaining infinite number of prime gaps [which were not discussed in details above]. *The proof is now complete for Part II of Proposition 4.2*□.

5 Rigorous Proofs now named as Polignac's and Twin prime hypotheses

The proofs on lemmas and propositions from previous section supply all necessary evidences to fully support Theorem Polignac-Twin prime I to IV below thus depicting proofs for Polignac's and Twin prime conjectures in a rigorous manner. Gap 1 Composite Numbers do not have valid representation by E prime gap = 2, and Gap 2 Composite Numbers have valid representations by all E prime gaps = ["consistently" only for] 2, ["inconsistently" for each of] 4, 6, 8, 10,.... Plus-Minus Gap 2 Composite Number Alternating Law confirms that Gap 2 Composite Numbers present in each P with prime gaps ≥ 4 situation must appear as some sort of "rhythmic patterns of alternating presence and absence" for Gap 2 Composite Numbers. Twin P with prime gap = 2 obeying Plus Gap 2 Composite Number Continuous Law can be understood as special situation of "(non-)rhythmic patterns with continual presence" for relevant Gap 2 Composite Numbers.

414

In 1849 when French mathematician Alphonse de Polignac (1826 - 1863) was admitted to Polytechnique, he made what is known as Polignac's conjecture which relates complete set of odd P to all E prime gaps. Twin prime conjecture, which relates twin prime numbers to prime gap = 2, is nothing more than a subset of Polignac's conjecture.

Theorem Polignac-Twin prime I. Incompletely Predictable prime numbers $Pn = 2, 3, 5, 7, 11, ..., \infty$ or composite numbers $Cn = 4, 6, 8, 9, 10, ..., \infty$ are CIS with overall actual location [but not actual positions] of all prime or composite numbers accurately represented by complex algorithm involving prime gaps GPi viz. $P_{n+1} = 2 + \sum_{i=1}^{n} GPi$ or involving composite gaps GCi viz. $C_{n+1} = 4 + \sum_{i=1}^{n} GCi$ whereby prime & composite numbers are symbolically represented here with aid of 'n' notation instead of usual 'i' notation; and i & n = 1, 2, 3, 4, 5, ..., ∞. Number '2' in first algorithm represents P1, the very first (and only even) P. Number '4' in second algorithm represent C1, the very first (and even) C.

Proof. We treat above algorithms as unique mathematical objects looking for key intrinsic properties and behaviors. Each P or C is assigned a unique prime or composite gap. Absolute number of P or C and (thus) prime or composite gaps are infinite in magnitude. As original formulae containing all P or C by themselves (viz. without supplying prime or composite gaps as "input information" to generate P or C as "output complexity"), these algorithms intrinsically incorporate overall actual location [but not actual positions] of all P or C. *The proof is now complete for Theorem Polignac-Twin prime I*□.

Theorem Polignac-Twin prime II. Set of prime gaps $G_{Pi} = 2, 4, 6, 8, 10, ..., \infty$ is infinite in magnitude whereby these prime gaps accurately and completely represented by Dimensions $(2x - 7)^1$, $(2x - 8)^1$, $(2x - 9)^1$, ..., $(2x -$

415

∞)^1 must satisfy Information-Complexity conservation in a consistent manner.

Proof. Part I of Proposition 4.2 proved all P are represented by Dimension $(2x - N)^1$ with $N \geq 7$ for any given x value (except for x = 2 & 3 values). Note that although x = 1 is neither P nor C, it is validly represented by Dimension $(2x - 2)^1$. If each P is endowed with a specific prime gap value, then each such prime gap must [via logical mathematical deduction] be represented by Dimension $(2x - N)^1$. We advocate this nominated method of prime gap representation using Dimensions be [purportedly] the only way to achieve Information-Complexity conservation. The preceding mathematical statements are correct as there is a unique prime gap value associated with each P. Proposition 5.1 below based on principles from Set theory provides further supporting materials that prime gaps are infinite in magnitude. *The proof is now complete for Theorem Polignac-Twin prime II*□.

Theorem Polignac-Twin prime III. To maintain Dimensional analysis (DA) homogeneity, those Dimensions $(2x - N)^1$ from Theorem Polignac-Twin prime II must contain eternal repetitions of well-ordered sets constituted by Dimensions $(2x - 7)^1$, $(2x - 8)^1$, $(2x\ 9)^1$, $(2x - 10)^1$, $(2x - 11)^1$, ..., $(2x - \infty)^1$.

Proof. This Theorem is stated in greater details as "To maintain DA homogeneity, those aforementioned [endowed with exponent 1] Dimensions $(2x - N)^1$ from Theorem Polignac-Twin prime II must repeat themselves indefinitely in following specific combinations – (i) Dimension $(2x - 7)^1$ only appearing as twin [two-times-in-a-row] and quadruplet [fourtimes-in-a-row] sequences, and (ii) Dimensions $(2x - 8)^1$, $(2x - 9)^1$, $(2x - 10)^1$, $(2x - 11)^1$,..., $(2x - \infty)^1$ appearing as progressive groupings of E 2, 4, 6, 8, 10,..., ∞." To accommodate the only even P '2', exceptions to this DA homogeneity

416

compliance will expectedly occur right at beginning of P sequence – (i) one-off appearance of Dimensions $(2x - 2)^1$, $(2x - 4)^1$ and $(2x - 5)^1$ and (ii) one-off appearance of Dimension $(2x - 7)^1$ as a quintuplet [five-times-in-a-row] sequence which is equivalent to (eternal) non-appearance of Dimension $(2x - 6)^1$ at $x = 4$. [We again note Dimension $(2x - 2)^1$ validly represent Number '1' which is neither P nor C.] These sequentially arranged sets are CFS whereby from $x = 11$ onwards, each set always commence initially as 'baseline' Dimension $(2x - 7)^1$ at $x = O$ values and always end with its last Dimension at $x = E$ values. Each set also have varying cardinality with values derived from all E; and correctly combined sets always intrinsically generate two infinite sets of P and, by default, C in an integrated manner. Our Theorem Polignac-Twin prime III simply represent a mathematical summary derived from Section 3 & 4 of all expressed characteristics of Dimension $(2x - N)^1$ when used to represent P with intrinsic display of DA homogeneity. See Proposition 5.2 below for further details on DA aspect. *The proof is now complete for Theorem Polignac-Twin prime III*□.

Theorem Polignac-Twin prime IV. Aspect 1. The "quantitive" aspect to existence of both prime gaps and their associated prime numbers as sets of infinite magnitude will be shown to be correct by utilizing principles from Set theory. Aspect 2. The "qualitative" aspect to existence of both prime gaps and their associated prime numbers as sets of infinite magnitude will be shown to be correct by 'Plus-Minus Gap 2 Composite Number Alternating Law' and 'Plus Gap 2 Composite Number Continuous Law'.

Proof. Required concepts from Set theory involve cardinality of a set with its 'well-ordering principle' application. Supporting materials for these concepts based on 'pigeonhole principle' in relation to Aspect 1 are outlined in Proposition 5.1 below. 'Plus-Minus Gap 2 Composite Number Alternating

Law' is applicable to all E prime gaps [apart from first E prime gap = 2 for twin primes]. The prime gap = 2 situation will obey 'Plus Gap 2 Composite Number Continuous Law'. These Laws are essentially Laws of Continuity inferring underlying intrinsic driving mechanisms that enables infinity magnitude association for both prime gaps & prime numbers to co-exist. By the same token, these Laws have the important implication that they must be applicable to those relevant prime gaps on a perpetual time scale. Supporting materials in relation to Aspect 2 are found in Proposition 4.2 above. *The proof is now complete for Theorem Polignac-Twin prime IV*□.

We note two mutually inclusive conditions: Condition 1. Presence of all Dimensions that repeat themselves on an indefinite basis and with exponent of '1' will give rise to complete sets of P & C ["DA-wise one & only one mathematical possibility argument" associated with inevitable *de novo* DA homogeneity], and Condition 2. Presence of any Dimension(s) that do not repeat itself (themselves) on an indefinite basis or with exponent other than '1' will give rise to incomplete set of P & C or incorrect set of non-P & non-C ["DA-wise mathematical impossibility argument" associated with inevitable *de novo* DA non-homogeneity]. When met, these two conditions will fully support the point that CFS Dimensions representations of P & C [with respective prime & composite gaps] are totally accurate. Condition 1 reflect proof from Theorem Polignac-Twin prime III above as all P & C are associated with DA homogeneity when their Dimensions are endowed with exponent of '1'. Condition 2 invoke corollary on inevitable appearance of incomplete P or C or non-P or non-C [associated with DA non-homogeneity] being tightly incorporated into this mathematical framework. See Propositions 5.1 and 5.2, and Corollary 5.3 below for supporting materials on DA homogeneity & non-homogeneity.

418

We analyze P (& C) in terms of (i) measurements based on cardinality of CIS and (ii) pigeonhole principle which states that if n items are put into m containers, with n>m, then at least one container must contain more than one item. We note that ordinality of all infinite P (& C) is "fixed" implying that each one of the infinite well-ordered Dimension sets conforming to CFS type as constituted by Dimensions $(2x - 7)^1$, $(2x - 8)^1$, $(2x - 9)^1$, $(2x - 10)^1$, $(2x - 11)^1$, ..., $(2x - \infty)^1$ on respective gaps for P (& C) must also be "fixed".

Proposition 5.1. "Even number prime gaps are infinite in magnitude with each even number prime gap generating odd prime numbers which are again infinite in magnitude" is supported by principles from Set theory and two Laws based on Gap 2 Composite Number.

Proof. We validly exclude even P '2' here. Let (i) cardinality $T = \aleph_0$ for Set all odd P derived from E prime gaps 2, 4, 6,..., ∞, (ii) cardinality $T_2 = \aleph_0$ for Subset odd P derived from E prime gap 2, cardinality $T_4 = \aleph_0$ for Subset odd P derived from E prime gap 4, cardinality $T_6 = \aleph_0$ for Subset odd P derived from E prime gap 6, etc. Paradoxically $T = T_2 + T_4 + T_6 +... + T_\infty$ equation is valid despite $T = T_2 = T_4 = T_6 =... = T_\infty$ [with well-ordering principle "stating that every non-empty set of positive integers contains a least element" fulfilled by each (sub)set]. But if Subset odd P derived from one or more E prime gap(s) are finite in magnitude, this will breach $\aleph_0$ 'uniformity' resulting in (i) DA non-homogeneity and (ii) inequality $T > T_2 + T_4 + T_6 +... + T_\infty$. In language of pigeonhole principle "stating that if n items are put into m containers with n>m, then at least one container must contain more than one item", residual odd P (still CIS in magnitude) not accounted for by CFS type E prime gap(s) will have to be [incorrectly] contained in one (or more) of composite gap(s). These arguments using cardinality constitute proof that

419

(i) E prime gaps and (ii) odd P generated from each E prime gap, must all be CIS. *The proof [on "quantitative" aspect] is now complete for Proposition 5.1□.*

Complete set of P is represented by Dimensions $(2x - N)^1$. Table 2 & Figure 1 on PC finite scale mathematical landscape depict perpetual repeating features used in "qualitative" statements supporting (i) Plus-Minus Gap 2 Composite Number Alternating Law (stated as C with composite gaps = 2 present in each of P with prime gaps $\geq$ 4 situation must be observed to appear as some sort of rhythmic patterns of alternating presence and absence of this type of C), and (ii) Plus Gap 2 Composite Number Continuous Law (stated as C with composite gaps = 2 continual appearances in each of (twin) P with prime gap = 2 situation). Plus-Minus Gap 2 Composite Number Alternating Law has intrinsic mechanism to automatically generate all prime gaps $\geq$ 4 in a mathematically consistent *ad infinitum* manner. Plus Gap 2 Composite Number Continuous Law has built-in intrinsic mechanism to further generate prime gap = 2 appearances in a mathematically consistent *ad infinitum* manner. *The proof [on "qualitative" aspect] is now complete for Proposition 5.1□.*

Proposition 5.2. The presence of Dimensional analysis homogeneity will always result in correct and complete set of prime (and composite) numbers.

Proof. DA homogeneity is completely dependent on all Dimensions being consistently endowed with exponent '1'. As all P (& C) are "fixed", we deduce from Figure 1 & Table 2 that there is one (& only one) way to represent Information-Complexity conservation using our defined Dimensions. Thus, there is one (& only one) way to depict all P (& C) using these Dimensions in a self-consistent manner and this is achieved with the one (& only one) DA homogeneity possibility. *The proof is now complete for Proposition 5.2□.*

Corollary 5.3. The presence of Dimensional analysis non-homogeneity will always result in incorrect and/or incomplete set of prime (and composite) numbers.

Proof. For optimal clarity, we endow all Dimensions with exponent '1' depicted as $(2x - 7)^1$, $(2x - 8)^1$, $(2x - 9)^1$, $(2x - 10)^1$, $(2x - 11)^1$,..., $(2x - \infty)^1$. Proposition 5.2 equates DA homogeneity with correct & complete set of P (& C). There are "more than one" DA possibilities when, for instance, a particular [first] term from $(2x - 7)^0$, $(2x - 8)^1$, $(2x - 9)^1$,..., $(2x - \infty)^1$ "terminates" prematurely and does not perpetually repeat [with loss of continuity]. There are intuitively two 'broad' DA possibilities here; namely, (one) DA homogeneity possibility and (one) DA non-homogeneity possibility – Dimension $(2x - 7)^0$ [= 1] with its exponent arbitrarily set as '0' against-all-trend in this case. Thus Dimension $(2x - 7)^1$ that stop recurring at some point in P (or C) sequence may cause well-ordered CFS sets from progressive groupings of [E] 2, 4, 6, 8, 10,..., ∞ for Dimensions $(2x - 8)^1$, $(2x - 9)^1$, $(2x - 10)^1$, $(2x - 11)^1$,..., $(2x - \infty)^1$ to stop existing (and ultimately for sequential P (or C) to stop appearing) at that point with ensuing outcome that P (or C) may overall be incorrectly finite or incomplete in magnitude. Finally also manifesting DA non-homogeneity, a Dimension endowed with fractional exponent values other than '1' such as '$\frac{2}{5}$' or '$\frac{3}{5}$' will result in non-P (or non-C) [fractional] numbers. *The proof is now complete for Corollary 5.3*□.

Each [fixed] finite scale mathematical landscape "page" as part of [fixed] infinite scale mathematical landscape "pages" for P & C display Chaos [sensitivity to initial conditions viz. positions of subsequent P & C are "sensitive" to positions of initial P & C] and Fractals [manifesting fractal dimensions with self-similarity viz. those aforementioned Dimensions for P & C are always present, albeit in non-identical manner, for all ranges of x $\geq$

421

2]. Advocated in another manner, Chaos and Fractals phenomena of those Dimensions for P & C are always present signifying accurate composition of P & C in different [predetermined] finite scale mathematical landscape "(snapshot) pages" for P & C that are self-similar but never identical – and there are an infinite number of these finite scale mathematical landscape "(snapshot) pages". The crucial mathematical step in representing all P (& C) and prime (& composite) gaps with "Dimensions" based on Information-Complexity conservation allows us to obtain the two Laws based on Gap 2 Composite Numbers and perform DA on these entities. The 'strong' principle argument is DA homogeneity equates to complete set of P (& C) whereas DA non-homogeneity does not equate to complete set of P (& C). We could also advocate for a 'weak' principle argument supporting DA homogeneity for P (& C) in that nature should not "favor" any particular Dimension(s) to terminate and therefore DA non-homogeneity does not, and cannot, exist for P (& C). Abiding to our advocated convention that 'conjecture' be termed 'hypothesis' once proven; we now call Polignac's & Twin prime conjectures as Polignac's & Twin prime hypotheses.

6 Conclusions

Harnassed property: CIS of [Completely Predictable] natural numbers 1, 2, 3, 4, 5, 6, 7,... having CIS of [Completely Predictable] natural gaps 1, 1, 1, 1, 1, 1,... are constituted by three dependent sets of numbers: (i) CIS of [Incompletely Predictable] odd prime numbers 3, 5, 7, 11, 13, 17,... having CIS of [Incompletely Predictable] prime gaps 2, 2, 4, 2, 4,... plus CFS of solitary [Incompletely Predictable] even prime number 2 having CFS of [Incompletely Predictable] prime gap 1 (ii) CIS of [Incompletely Predictable] even and odd composite numbers 4, 6, 8, 9, 10, 12,... having CIS of

[Incompletely Predictable] composite gaps 2, 2, 1, 1, 2, 2,.... and (iii) CFS of solitary odd number '1' [neither prime nor composite]. Treated as Incompletely Predictable problems endowed with "meta-properties", we gave relatively elementary proofs on Polignac's and Twin prime conjectures based on this harnessed property by performing Dimensional analysis on (sub)sets and "Dimensions" of prime and composite numbers, and obtaining 'Plus-Minus Gap 2 Composite Number Alternating Law' and 'Plus Gap 2 Composite Number Continuous Law'.

Prime number theorem describes asymptotic distribution of prime numbers among positive integers by formalizing intuitive idea that prime numbers become less common as they become larger through precisely quantifying rate at which this occurs using probability. Nontrivial zeros [from 'Axes intercept relationship interface' relevant to Riemann hypothesis] and prime numbers [from 'Numerical relationship interface' relevant to prime number theorem] are Incompletely Predictable entities and numbers. Deep-seated connections exist between Riemann hypothesis and prime number theorem (which is fully delineated by prime counting function [denoted here with $\pi(x)$]). Solving Incompletely Predictable problem Riemann hypothesis is instrumental in proving efficacy of techniques that estimate $\pi(x)$ efficiently. This should now confirm "best possible" bound for error ("smallest possible" error) of prime number theorem.

In mathematics, logarithmic integral function or integral logarithm li(x) is a special function. Relevant to problems of physics and with number theoretic significance, it occurs in prime number theorem as an estimate of $\pi(x)$ whereby the form of this special function is defined so that li(2) = 0; viz. li(x) $= \int_2^x \frac{du}{\ln u} =$ li(x) - li(2). There are less accurate ways of estimating $\pi(x)$ such as conjectured by Gauss and Legendre at end of 18th century. This $\pi(x)$ is approximately x/ln x in the sense sense $\lim_{x\to\infty} \left(\dfrac{\pi(x)}{nx/\ln x}\right) = 1$. Skewes' number is any of several extremely large numbers used by South African mathematician Stanley Skewes as upper bounds for smallest natural number x for which li(x)$<\pi(x)$. These bounds have since been improved by others: there is a crossing near $e^{727.95133}$ but it is not known whether this is the smallest. John Edensor Littlewood, who was Skewes' research supervisor, proved in 1914[4] that there is such a [first] number; and found that sign of difference $\pi(x)$ - li(x) changes infinitely often. This refute all prior numerical evidence that seem to suggest li(x) was always more than $\pi(x)$. The key point is [100% accurate] $\pi(x)$ mathematical tool being "wrapped around" by [less-than-100% accurate] approximate mathematical tool li(x) infinitely often via this 'sign of difference' changes meant that li(x) is the most efficient approximate mathematical tool. Contrast this with "crude" x/ln$x$ approximate mathematical tool where values obtained diverge away from $\pi(x)$ at increasingly greater rate when larger range of prime numbers are studied.

Table 3 Even-Odd mathematical (tabulated) landscape using data obtained for x = 1 to 64. Legend: E=even, O=odd, Y=Dimension 2x-4.

x	E_i or O_i, Gaps	ΣEO_x - Gaps	Dimension
1	O1, 2	0	2x-2
2	E1, 2	0	Y
3	O2, 2	2	Y
4	E2, 2	4	Y

5	O3, 2	6	Y
6	E3, 2	8	Y
7	O4, 2	10	Y
8	E4, 2	12	Y
9	O5, 2	14	Y
10	E5, 2	16	Y
11	O6, 2	18	Y
12	E6, 2	20	Y
13	O7, 2	22	Y
14	E7, 2	24	Y
15	O8, 2	26	Y
16	E8, 2	28	Y
17	O9, 2	30	Y
18	E9, 2	32	Y
19	O10, 2	34	Y
20	E10, 2	36	Y
21	O11, 2	38	Y
22	E11, 2	40	Y
23	O12, 2	42	Y
24	E12, 2	44	Y
25	O13, 2	46	Y
26	E13, 2	48	Y
27	O14, 2	50	Y
28	E14, 2	52	Y
29	O15, 2	54	Y

30	E15, 2	56	Y
31	O16, 2	58	Y
32	E16, 2	60	Y
33	O17, 2	62	Y
34	O17, 2	64	Y
35	O17, 2	66	Y
36	O17, 2	68	Y
37	O17, 2	70	Y
38	O17, 2	72	Y
39	O17, 2	74	Y
40	O17, 2	76	Y
41	O17, 2	78	Y
42	O17, 2	80	Y
43	O17, 2	82	Y
44	O17, 2	84	Y
45	O17, 2	86	Y
46	O17, 2	88	Y
47	O17, 2	90	Y
48	O17, 2	92	Y
49	O17, 2	94	Y
50	O17, 2	96	Y
51	O17, 2	98	Y
52	O17, 2	100	Y
53	O17, 2	102	Y
54	O17, 2	104	Y

55	O17, 2	106	Y
56	O17, 2	108	Y
57	O17, 2	110	Y
58	O17, 2	112	Y
59	O17, 2	114	Y
60	O17, 2	116	Y
61	O17, 2	118	Y
62	O17, 2	120	Y
63	O17, 2	122	Y
64	O17, 2	124	Y

Appendix I: Tabulated and graphical data on Even-Odd mathematical landscape

We tabulate (in Table 3) and graph (in Figure 2) [Completely Predictable] E-O mathematical landscape for x = 1 to 64. Involved Dimensions are 2x - 2 & 2x - 4 with Y denoting Dimension 2x - 4 for visual clarity. This mathematical landscape of Dimension 2x - 4 (except for first and only Dimension 2x - 2) will intrinsically incorporate E & O in an integrated manner. Except for first O, all Completely Predictable E & O and all their associated gaps are represented by countable finite set of [single] Dimension 2x - 4. Dimensions 2x - 2 & 2x - 4 are symbolically represented by -2 & -4 with 2x - 4 displayed as 'baseline' Dimension whereby Dimension trend (Cumulative Sum Gaps) must reset itself onto this (Grand-Total Gaps) 'baseline' Dimension after initial Dimension 2x - 2 on a permanent basis. Graphical appearances of Dimensions symbolically represented by two negative integers are Completely Predictable with both Even-$\pi(x)$ and Odd-

427

$\pi(x)$ becoming larger at a constant rate. There is a complete absence of Chaos and Fractals phenomena.

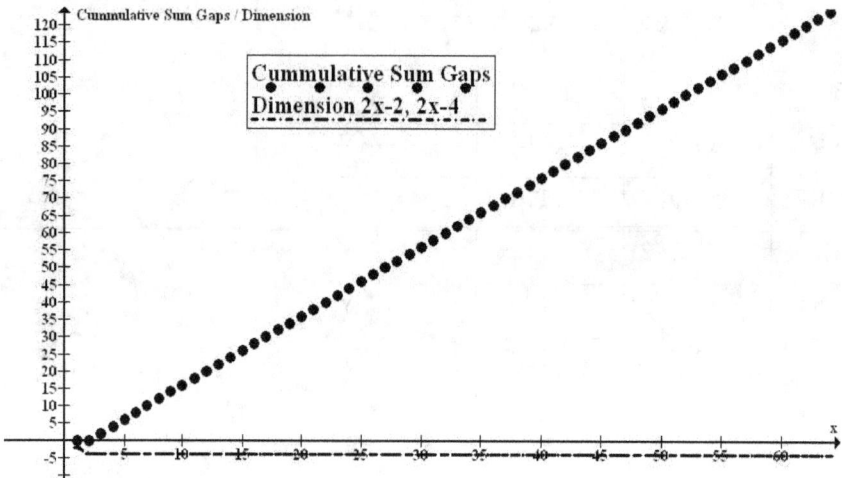

Fig. 2 Even-Odd mathematical (graphed) landscape using data obtained for x = 1 to 64.

Definitive derivation of data in Table 3 is illustrated by two examples for position x = 31 & 32. For i & x ∈ 1, 2, 3, ..., ∞; ΣEO_x-Gap = ΣEO_{x-1}-Gap + Gap value at E_{i-1} or Gap value at O_{i-1} whereby (i) E_i or O_i at position x is determined by whether relevant x value belongs to E or O, and (ii) both ΣEO_1-Gap and ΣEO_2-Gap = 0. Example, for position x = 31: 31 is O (O16). Our desired Gap value at O15 = 2. Thus ΣEO_{31}-Gap (58) = ΣEO_{30}-Gap (56) + Gap value at O15 (2). Example, for position x = 32: 32 is E (E16). Our desired Gap value at E15 = 2. Thus ΣEO_{32}-Gap (60) = ΣEO_{31}-Gap (58) + Gap value at E15 (2).

Acknowledgements

I am indebted to Mr. Rodney Williams and Mr. Tony O'Hagan for reviewing this paper (dedicated to my daughter Jelena born 13 weeks early on May 14, 2012).

References

1. Furstenberg, H. (1955). On the infinitude of primes. *Amer. Math. Monthly, 62*, (5) 353, http://dx.doi.org/10.2307/2307043

2. Saidak, F. (2006), A New Proof of Euclid's theorem, *Amer. Math. Monthly, 113*, (10) 937, http://dx.doi.org/10.2307/27642094

3. Zhang, Y. (2014), Bounded gaps between primes, *Ann. Math. 179*(3) (2014) 1121 – 1174, http://dx.doi.org/10.4007/annals.2014.179.3.7

4. Littlewood, J. E. (1914), Sur la distribution des nombres premiers. Comptes Rendus de l'Acad. Sci. Paris, 158, 1869 – 1872

ABOUT THE AUTHOR

Image of Author portrayed as a Doctor-Scientist (obtained from Pixabay licensed as free for commercial use).

Dr. Bernhard Helpful is a Researcher on Fundamental Laws of Nature. His novel Hybrid integer sequence A228186 was published in The On-Line Encyclopedia of Integer Sequences in 2013. From 2016 to 2019, he carries out extensive mathematical research with published papers in Number theory on Riemann Hypothesis, Polignac's and Twin prime conjectures. He lives in Australia with his wife and five children. He possesses Medical degree, General Practice qualification, Primary Anesthesia Fellowship Examination and Opioid Replacement license. His work experiences involve the specialty area of Anesthesia, Intensive Care, Pain Medicine, Medicinal Cannabis and Addiction Medicine. His medical publication in 2012 as primary author with the Professor of Nephrology as secondary author include "Supramaximal elevation in B-type natriuretic peptide and its N-terminal fragment levels in anephric patients with heart failure: a case series".

www.ingramcontent.com/pod-product-compliance
Lightning Source LLC
Chambersburg PA
CBHW081506220326
45467CB00010B/2811